Improving Disaster Resilience and Mitigation - IT Means and Tools

NATO Science for Peace and Security Series

This Series presents the results of scientific meetings supported under the NATO Programme: Science for Peace and Security (SPS).

The NATO SPS Programme supports meetings in the following Key Priority areas: (1) Defence Against Terrorism; (2) Countering other Threats to Security and (3) NATO, Partner and Mediterranean Dialogue Country Priorities. The types of meeting supported are generally "Advanced Study Institutes" and "Advanced Research Workshops". The NATO SPS Series collects together the results of these meetings. The meetings are co-organized by scientists from NATO countries and scientists from NATO's "Partner" or "Mediterranean Dialogue" countries. The observations and recommendations made at the meetings, as well as the contents of the volumes in the Series, reflect those of participants and contributors only; they should not necessarily be regarded as reflecting NATO views or policy.

Advanced Study Institutes (ASI) are high-level tutorial courses to convey the latest developments in a subject to an advanced-level audience

Advanced Research Workshops (ARW) are expert meetings where an intense but informal exchange of views at the frontiers of a subject aims at identifying directions for future action

Following a transformation of the programme in 2006 the Series has been re-named and re-organised. Recent volumes on topics not related to security, which result from meetings supported under the programme earlier, may be found in the NATO Science Series.

The Series is published by IOS Press, Amsterdam, and Springer, Dordrecht, in conjunction with the NATO Emerging Security Challenges Division.

Sub-Series

A.	Chemistry and Biology	Springer
B.	Physics and Biophysics	Springer
C.	Environmental Security	Springer
D.	Information and Communication Security	IOS Press
E.	Human and Societal Dynamics	IOS Press

http://www.nato.int/science
http://www.springer.com
http://www.iospress.nl

Series C: Environmental Security

Improving Disaster Resilience and Mitigation - IT Means and Tools

edited by

Horia-Nicolai Teodorescu
Romanian Academy, Iasi Branch
Institute of Computer Science and
'Gheorghe Asachi' Technical University of Iasi
Romania

Alan Kirschenbaum
Faculty of Industrial Engineering and Management
Samuel Neaman Institute for Advanced Studies
Technion – Israel Institute of Technology
Haifa, Israel

Svetlana Cojocaru
Institute of Mathematics and Computer Science
of the Academy of Sciences of Moldova
Chisinau, Republic of Moldova

and

Claude Bruderlein
Department of Global Health and Population
Harvard University
Cambridge, MA, USA

Published in Cooperation with NATO Emerging Security Challenges Division

Proceedings of the NATO Advanced Research Workshop on
Improving Disaster Resilience and Mitigation – IT Means and Tools
Iasi, Romania
6–8 November 2013

Library of Congress Control Number: 2014946610

ISBN 978-94-017-9138-0 (PB)
ISBN 978-94-017-9135-9 (HB)
ISBN 978-94-017-9136-6 (e-book)
DOI 10.1007/978-94-017-9136-6

Published by Springer,
P.O. Box 17, 3300 AA Dordrecht, The Netherlands.

www.springer.com

Printed on acid-free paper

Preface

The main topics of this volume comprise fundamentals, models, and information technologies (IT) means and tools for disaster prediction and mitigation. A more detailed list of topics includes mathematical and computational modeling of processes leading to or producing disasters, modeling of disaster effects, IT means for disaster mitigation, including data mining tools, knowledge-based and expert systems for use in disaster circumstances, GIS-based systems for disaster prevention and mitigation, and equipment for disaster-prone areas. The volume is not focusing on a specific type or class of disasters (natural or human-made). It was conceived to offer a comprehensive, integrative view on disasters, seeking to determine what various disasters have in common. Because disaster resilience and mitigation involve humans, societies, and cultures, not only technologies and economic models, special attention was paid in this volume to gain a comprehensive view on these issues, as a foundation of the IT tool design.

The concept for this volume is that of a collection of "state-of-the-art" expositions combined with prospective, "out-of-the-box" presentations of new directions. The book explores the recent conclusions of various bodies on the use of IT in emergencies and disasters, the predictable developments, and the required advances that would make IT and communications (IT&C) more useful in disaster mitigation. Some chapters also make implicit or explicit recommendations.

It is Editors' view that a relevant purpose of the volumes in this series is to advance knowledge and science, as well as to help improving prevention and mitigation of disasters. This book was conceived taking into account that purpose. We hope that the audience of the book will find useful ideas and answers to several current questions related to IT means in disaster mitigation.

The structure of the volume reflects its purposes and scope. The first part of the book is devoted to fundamentals and modeling and includes chapters that do not bear directly to IT subjects, nevertheless relating intimately to the principles of modeling and use of IT in disasters. The second section addresses topics related to IT tools in disaster mitigations. The two sections are well balanced and they cover a wide range of issues.

In many respects, this volume is a 'first' as, with the notable exception of the volume edited by R. R. Rao, J. Eisenberg, and T. Schmitt on a similar but not identical topic (*Improving Disaster Management: The Role of IT in Mitigation, Preparedness, Response, and Recovery*. National Academy of Sciences. 2007), this is the first volume (and, relatedly, we had the first ARW) specifically dedicated to almost all aspects of IT use in the various phases of increasing the resilience of societies facing disasters.

This volume is ***not*** and should not be listed as a "conference proceedings". Every lecture in the related ARW was carefully drafted as a chapter, with an extended number of pages and deep coverage. The chapters where written after drafted versions of the general part and of the concept of the ARW and volume were circulated to the contributors by the first editor. All authors were asked to take into account the need for in-depth conclusions and recommendations. Moreover, the contributors, who are front-running scientists representing 14 countries, were selected based not only on top scientific merit, but on complementarity and variety of points of view too.

Iasi, Romania — Horia-Nicolai Teodorescu
Haifa, Israel — Alan (Avi) Kirschenbaum
Chisinau, Moldova — Svetlana Cojocaru
Cambridge, MA, USA — Claude Bruderlein
The Editors

Acknowledgements

The Editors and the contributors acknowledge the support of NATO Science Committee, which made possible this volume and the related Advanced Research Workshop with the same title.

The ARW organizing editors (Horia-Nicolai Teodorescu and Svetlana Cojocaru) express their thanks not only for the financial support, but in the first place for the continuous attention and courteous helpfulness during the whole process of organizing the related ARW. The SPS officers in Brussels have been very friendly and supportive in the preliminary phase of the organization of the ARW and answered in a timely manner all questions asked by the local organizing committee of the ARW. Special thanks to Ms. Lynne Campbell-Nolan, who was most kind and caring, as well to the SPS leadership, who responded obligingly to several of our proposals during the organization process. The advice and backing offered by Dr. Deniz Yüksel-Beten, Senior Science for Peace and Security and Partnership Cooperation Advisor, and by Dr. Eyup Turmus, Science Advisor, during a meeting in Bucharest, was instrumental. Also, the ARW benefited of the participation of Mr. Jahier Khan, Staff Officer Civil Protection Group, Civil-Military Planning and Support Section (CMPS), Operations Division – NATO HQ, who gave an introductory lecture and helped with stimulating discussion topics.

The ARW organization benefited of the excellent support of the Romanian Academy – Iasi branch, especially of the chief of Financial Department, Meda Gâlea, and of the officer Giorgiana Donceag. The local organizing committee, composed of researchers from the Institute of Computer Science of the Romanian Academy (the first partner in the ARW), made a bright job. The ARW organization benefited of the graceful help of Iasi County Inspectorate for Emergencies, specifically of the Commander of the Inspectorate, Chief Inspector col. Dan Paul Iamandi and Adjunct Inspector Lt. col Dragos Rosu. We would explicitly like to thank the local personalities, especially the President of Iasi Branch of the Romanian Academy, the Rector of Al.I. Cuza University of Iasi and the Vice-Rector for International Relations of Gheorghe Asachi Technical University of Iasi, for participating in the opening round-table and for responding to questions.

The volume editors heartily thank all the contributors and the numerous referees and colleagues from the Institute of Computer Science of the Romanian Academy – Iasi Branch, and the Institute for Mathematics and Computer Science of the Science Academy of Republic of Moldova, who helped in the ARW organization process and partly in the editorial work, namely colleagues Vasile Apopei, Monica Feraru, Monica Fira, Adrian Ciobanu, Cecilia Bolea, Tatiana Verlan, Inga Țițchiev, Ramona Luca, Silviu Bejinariu, and many others.

Finally, we thank all contributors for their hard work and dedication to elucidating issues that contribute to the improvement of disaster mitigation. We also thank the book editor Annelies Kersbergen, from Springer, for the staunch support given to the editors in the last phases of editing the volume, as well as to Project Manager Marudhu Prakash for carefully and patiently correcting the whole volume.

The Editors

Contents

Part I Fundamentals and Modeling

Contributors

Martin Alfranseder Laboratory for Safe and Secure Systems LaS³ (www.las3.de), Faculty Electro- and Information-Technology, Ostbayerische Technische Hochschule Regensburg, Regensburg, Germany

Livia Andra Amarandei Technical University "Gheorghe Asachi", Iasi, Romania

Vakhtang Arabidze Agricultural University of Georgia, Tbilisi, Georgia

Peter I. Bidyuk Institute for Applied System Analysis NTUU "KPI" 37, Kyiv, Ukraine

Mykola M. Biliaiev Dnipropetrovsk National University of Railway Engineering, Dnipropetrovsk, Ukraine

Babiga Birregah CNRS Joint Research Unit in Sciences and Technologies for Risk Management, University of Technology of Troyes, Troyes, France

Frederic Bouder Department of Technology and Society Studies, Maastricht University, Maastricht, SZ, The Netherlands

Claude Bruderlein Department of Global Health and Population, Harvard University, Cambridge, MA, USA

Cristian Butincu Gheorghe Asachi Technical University of Iasi, Iasi, Romania

Josep Casanovas Universitat Politècnica de Catalunya, Barcelona, Spain

Svetlana Cojocaru Institute of Mathematics and Computer Science of the Academy of Sciences of Moldova, Chisinau, Republic of Moldova

Victor P. Cojocaru D. Ghitu Institute of the Electronic Engineering and Nanotechnologies of the Academy of Sciences of Moldova, Chisinau, Republic of Moldova

Gabriela Covatariu Faculty of Civil Engineering and Building Services, "Gheorghe Asachi" Technical University of Iasi, Iasi, Romania

Mitica Craus Gheorghe Asachi Technical University of Iasi, Iasi, Romania

Ernesto Damiani Computer Science Department, Università degli Studi di Milano, Crema (CR), Italy

Jaume Figueras Jové Universitat Politècnica de Catalunya, Barcelona, Spain

Pau Fonseca i Casas Universitat Politècnica de Catalunya, Barcelona, Spain

Fulvio Frati Computer Science Department, Università degli Studi di Milano, Crema (CR), Italy

Constantin Gaindric Institute of Mathematics and Computer Science of the Academy of Sciences of Moldova, Chisinau, Republic of Moldova

Antoni Guasch Petit Universitat Politècnica de Catalunya, Barcelona, Spain

Tengiz Gugeshashvili Seismic Hazard and Disaster Risks, M. Nodia Institute of Geophysics of I. Javakishvili Tbilisi State University, Tbilisi, Georgia

Alexander Gventcadze Seismic Hazard and Disaster Risks, M. Nodia Institute of Geophysics of I. Javakishvili Tbilisi State University, Tbilisi, Georgia

Marius Gheorghe Hăgan Institute of Computer Science of the Romanian Academy – Iasi Branch, Iasi, Romania

Keisuke Himoto Building Research Institute, Tsukuba, Ibaraki, Japan

Peter Hlousek Department of Applied Electronics and Telecommunications, Technical University of Pilsen, University of West Bohemia, Pilsen, Czech Republic

Abraham Kandel School of Computer Science, Florida International University Miami, Miami, FL, USA

Mustafa Asim Kazancigil Computer Science Department, Università degli Studi di Milano, Crema (CR), Italy

M. Kharytonov Dnipropetrovsk National University of Railway Engineering, Dnipropetrovsk, Ukraine

Jumpei Kimata Construction Division, Aichi Prefectural Government, Naka-ku, Nagoya, Aichi, Japan

Alan Kirschenbaum Faculty of Industrial Engineering and Management, Samuel Neaman Institute for Advanced Studies, Technion – Israel Institute of Technology, Haifa, Israel

Rainer Knauf Faculty of Computer Science and Automation, Technische Universität Ilmenau, Ilmenau, Germany

Ivan Konecny Department of Applied Electronics and Telecommunications, Technical University of Pilsen, University of West Bohemia, Pilsen, Czech Republic

Irina Lungu Faculty of Civil Engineering and Building Services, "Gheorghe Asachi" Technical University of Iasi, Iasi, Romania

Morgan Magnin École Centrale de Nantes, Nantes, France

Mykhailo P. Makukha Institute for Applied System Analysis NTUU "KPI" 37, Kyiv, Ukraine

Lyudmila Y. Malafeeva Institute for Applied System Analysis NTUU "KPI" 37, Kyiv, Ukraine

Guillaume Moreau Computer Science and Mathematics Department, École Centrale de Nantes, Nantes, France

Jürgen Mottok Laboratory for Safe and Secure Systems LaS³ (www.las3.de), Faculty Electro- and Information-Technology, Ostbayerische Technische Hochschule Regensburg, Regensburg, Germany

Matthias Mucha Laboratory for Safe and Secure Systems LaS³ (www.las3.de), Faculty Electro- and Information-Technology, Ostbayerische Technische Hochschule Regensburg, Regensburg, Germany

Teimuraz Mukhadze Seismic Hazard and Disaster Risks, M. Nodia Institute of Geophysics of I. Javakishvili Tbilisi State University, Tbilisi, Georgia

Jörg Rainer Noennig Department of Architecture Junior Professorship for Knowledge Architecture, Wissensarchitektur, Technische Universität Dresden, Dresden, Germany

Jean-Marie Normand École Centrale de Nantes, Nantes, France

Natalya D. Pankratova Institute for Applied System Analysis NTUU "KPI" 37, Kyiv, Ukraine

Olga Popcova Institute of Mathematics and Computer Science of the Academy of Sciences of Moldova, Chisinau, Republic of Moldova

Serghei Puiu Moldova State University of Medicine and Pharmacy "N. Testemitsanu", Chisinau, Republic of Moldova

Carmit Rapaport Faculty of Industrial Engineering and Management, Samuel Neaman Institute for Advanced Studies, Technion – Israel Institute of Technology, Haifa, Israel

Wolfgang P. Reinhardt AGIS/Institut für Angewandte Informatik, University of the Bundeswehr Munich, Neubiberg, Germany

Naphtali D. Rishe School of Computer Science, Florida International University Miami, Miami, FL, USA

N. Rostochilo Dnipropetrovsk National University of Railway Engineering, Dnipropetrovsk, Ukraine

Leon J.M. Rothkrantz Intelligent Interaction, Delft University of Technology, Delft, The Netherlands

SEWACO, The Netherlands Defence Academy, Den Helder, The Netherlands

Andreas Sailer Laboratory for Safe and Secure Systems LaS³ (www.las3.de), Faculty Electro- and Information-Technology, Ostbayerische Technische Hochschule Regensburg, Regensburg, Germany

Volodymyr V. Savastiyanov Institute for Applied System Analysis NTUU "KPI" 37, Kyiv, Ukraine

Illia O. Savchenko Institute for Applied System Analysis NTUU "KPI" 37, Kyiv, Ukraine

Stefan Schmidhuber Laboratory for Safe and Secure Systems LaS³ (www.las3.de), Faculty Electro- and Information-Technology, Ostbayerische Technische Hochschule Regensburg, Regensburg, Germany

Peter Schmiedgen Department of Architecture Junior Professorship for Knowledge Architecture, Wissensarchitektur, Technische Universität Dresden, Dresden, Germany

Iulian Secrieru Institute of Mathematics and Computer Science of the Academy of Sciences of Moldova, Chisinau, Republic of Moldova

Yurij M. Selin Institute for Applied System Analysis NTUU "KPI" 37, Kyiv, Ukraine

Myriam Servières École Centrale de Nantes, Nantes, France

Anghel Stanciu Faculty of Civil Engineering and Building Services, "Gheorghe Asachi" Technical University of Iasi, Iasi, Romania

Dan Tamir Department of Computer Science, Texas State University, San Marcos, TX, USA

Horia-Nicolai L. Teodorescu Institute of Computer Science, Romanian Academy – Iasi Branch, Iasi, Romania

Gheorghe Asachi Technical University of Iasi, Iasi, Romania

Iancu-Bogdan Teodoru Faculty of Civil Engineering and Building Services, "Gheorghe Asachi" Technical University of Iasi, Iasi, Romania

Nino Tsereteli Seismic Hazard and Disaster Risks, M. Nodia Institute of Geophysics of I. Javakishvili Tbilisi State University, Tbilisi, Georgia

Setsuo Tsuruta School of Information Environment, Tokyo Denki University, Tokyo, Japan

Otar Varazanashvili Seismic Hazard and Disaster Risks, M. Nodia Institute of Geophysics of I. Javakishvili Tbilisi State University, Tbilisi, Georgia

Kenji Watanabe Graduate School of Social Engineering, Nagoya Institute of Technology, Nagoya, Japan

Part I
Fundamentals and Modeling

Chapter 1
Survey of IC&T in Disaster Mitigation and Disaster Situation Management

Horia-Nicolai L. Teodorescu

Abstract This chapter serves as an introduction to the topic of the volume and sets the framework of the discussion. It is largely a review and synthesis of the literature, also including some personal suggestions and conclusions. It is a working paper in many aspects, where the term working paper is used in the sense of draft of the whole volume as well as an article intended to elicit feedback and discussions in the other chapters. This chapter also serves as a partial analysis of the literature. In the first section of the chapter, we list the roles of IT&C in disaster situations, review some of the international and national regulations on the subject, and plead for specific regulations regarding IT for disaster mitigation. The second part emphasizes the various aspects of disasters and the association with the design of specific IT means. The third part addresses resilience and robustness of the society and of the IT systems in disaster circumstances and the roles of IT means in improving disaster resilience.

1.1 The Role of IT&C in Disaster Areas and the International Regulatory Framework

In this section, we argue that the individuality of the information technology is poorly understood in the context of disaster mitigation, is misrepresented as a part of communication technology or of the equipment, and lacks appropriate regulations and development efforts. Therefore, we maintain that the potential of IT in disaster prevention, management, and mitigation remains largely untapped today.

H.-N.L. Teodorescu (✉)
Institute of Computer Science, Romanian Academy – Iasi Branch, Bd. Carol I nr. 8, Iasi, Romania

Gheorghe Asachi Technical University of Iasi, Iasi, Romania
e-mail: hteodor@etti.tuiasi.ro

H.-N. Teodorescu et al. (eds.), *Improving Disaster Resilience and Mitigation - IT Means and Tools*, NATO Science for Peace and Security Series C: Environmental Security, DOI 10.1007/978-94-017-9136-6_1,

Back in 1953, NATO recognized the threat of natural and manmade disasters for the stability of the partner countries and established a disaster assistance scheme. After more than 40 years, recognizing the cornerstone importance of communications in disaster area for disaster relief and mitigation, the United Nations forged in 1998 the "Tampere Convention on the Provision of Telecommunication Resources for Disaster Mitigation and Relief Operations" [1]. However, nowhere in Tampere Convention information technology is named, not a single time. Information in general is named several times, and the Web is referred once in the context "ReliefWeb as the global information system for the dissemination of reliable and timely information on emergencies and natural disaster" [1].

During the last few decades, IT became an essential tool in many fields of activity, while information technology and communications (IT&C) were recognized as essential means in disaster mitigation. Therefore, the time is ripe to ask if IT per se is required, as much as communications, for disaster mitigation. IT main roles are, among others, to manage the collection of data from the disaster area, to process the crude information gathered, to help develop models and predictions of the disaster situation dynamics, to propose prevention methods, and to support management and decision-making during the response to the disaster situations.

This volume aims to help answer a chain of questions: Is IT developed to the level where IT is essential in disaster relief and mitigation? What tools does IT offer, and what tools are still needed? How IT means (e.g., large networks of sensors, or large models run on large computers) should be made accessible to nations or regions in need for the mitigation of disasters? Should nations, UN, and local actors (states, organizations) start consider these questions and prepare, through international meetings and conferences, declarations, resolutions and conventions, the collaborations in the field of IT for disaster situations? How should the scientific community react to the needs? What new standards are required?

The introductory part to the Tampere Convention is a milestone in establishing the role of communication in disasters. It is decisive in understanding the strong conviction of the need of such a convention, in case of communications for disaster relief. That introduction – extending on not less than four pages – exposes the main reasons of the state parties for taking action according to the convention. Among the reasons are

> "that the magnitude, complexity, frequency and impact of disasters are increasing at a dramatic rate, with particularly severe consequences in developing countries", "... that rapid, efficient, accurate and truthful information flows are essential to reducing loss of life, human suffering and damage to property and the environment caused by disasters". [1]

The first part of the Convention emphasizes that several international conferences, declarations and resolutions of the United Nations have prepared for a decade the Tampere Convention, among others the Resolution 44/236, "designating 1990–2000 the International Decade for Natural Disaster Reduction," and Resolution 46/182 regarding the improvement of "international coordination of humanitarian emergency assistance." Also, United Nations General Assembly, in its Resolution 51/194 asked for active "development of a transparent and timely procedure for implementing effective disaster relief coordination arrangements."

Tampere Convention clearly states its objectives and defines the concepts in the objectives. Specifically, it defines "telecommunication", "telecommunication assistance" (meaning "the provision of telecommunication resources or other resources or support intended to facilitate the use of telecommunication resources" [1]), and "telecommunication resources", meaning "personnel, equipment, materials, information, training, radio-frequency spectrum, network or transmission capacity or other resources necessary to telecommunications."

In Tampere Convention, there are no references to computer technology, sensor networks, or technical issues related to IT (except "information flow"), such as datacenters, specific software to support situation analysis and modeling, datamining, data visualization and decision-making, which are considered today by many analysts essential assets in disaster relief and mitigation. Nor more recent documents, such as [2], make explicit reference to means based on IT. Whether or not IT is an essential asset already in increasing disaster resilience and in disaster mitigation may be debatable.

The lack of explicit international regulatory framework for IT tools and systems devoted to disaster mitigation mirrors the lack of or unclear regulations at national levels. In several national regulations, the words communication, mobile/cellular phone, and computer/laptop appear, but we found in no regulation the words software, software application, GIS, expert system, decision support system, computer program and the like. This is surprising, because computers without appropriate software are of little use and only a hassle for mobility, a limitation factor for response time in disaster situations. This shows the perceived or actual low level of use of the capabilities of the information technology in emergencies.

All regulations show a level of understanding of the importance of the communication, but most limit the communications to radios, television programs and cellular phones. Few put an accent on Internet, web pages, and very few show an understanding of the integrated communication systems. Yet, there are probably few experts, if any, that believe that IT will not be a strategic component in disaster mitigation and resilience, in the near future.

Recommendation. The concerned organizations and the nations may consider, where deemed appropriate, to agree on bilateral or multilateral arrangements relating to the subject matter of IT in disaster resilience, relief and mitigation.

Subsequently, we address the uses of IT in increasing resilience to disasters and disaster mitigation. IT should address all phases of disaster fight, from the preparation for disasters, including disaster modeling and prediction, to disaster prevention, to removing the long-term effects of disasters, taking into account all levels of effects, from individual to society at large, see Fig. 1.1. Details of the roles of IT & C are further provided in Fig. 1.2. IT is able to become the backbone supporting the connection and action logic of various players that may act today with little relationship to the others, as planners, modelers, emergency services, relief supporters, disaster managers, political factors, and recovery managers.

A comprehensive, remarkable study and master plan for earthquakes mitigation realized in Turkey [3] emphasizes several aspects argued for in this chapter. Among others, it states that "Communication is the most important activity that should be

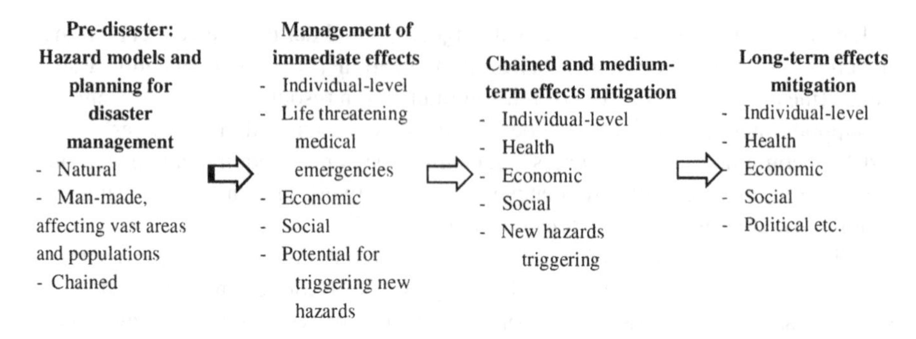

Fig. 1.1 Issues in planning for disaster preparedness and disaster mitigation

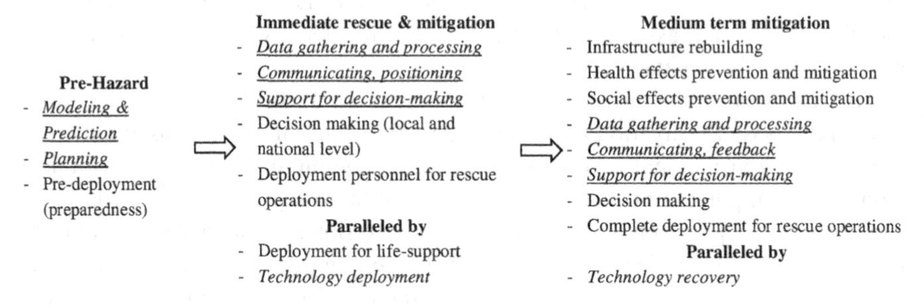

Fig. 1.2 Mitigation steps, with IT&C involved in the underlined phases

fulfilled" (p. 476) for insuring mitigation and management of the disasters. The cited study puts forward the idea of necessity of decision support systems in the forefront of the requirements:

> Establishing an extensive national information network and a decision support system which is fed by the correct and current information, should be Turkey's one of the first precedence objectives. Establishing a provincial information-decision support system in the short term should be followed by a national disaster prevention and mitigation information decision support system in the long run. [3], p. 519

The same study provides several examples of rewarding use of expert systems and decision support system. Among those examples are: transferring knowledge from one known disaster to the planning for another one, and for decision making and planning in an epidemic based on "*estimating the number of possible affected people [taking into account] atmospheric conditions, wind blow strength and direction, population [density] information*" [3].

While the range of IT tools envisaged by that study is ample and provides a good example of systematic planning and preventive measures, it remains limited in scope, especially at the levels of prevention, mitigation, and long-term recovery. Nevertheless, it is laudable by its thoroughness and amplitude and should be followed by other nations.

1.2 Disasters: Overview of Definitions and Classifications, and Their Impact on IT&C for Disaster Mitigation

Comprehensive definitions of disasters are needed because, beyond clarifying the concept, they allow determine the extent of modeling, prevention measures, and management capabilities, as well as the national and international legal limits of any action, including special measures enforced by specified agencies and institutions and during multi-national collaborations in rescue and relief operations. Because of the multi-purpose nature of these definitions, we may expect that various sources with different aims and roles on their agenda may produce different definitions. Having a clear definition allows political, organizational and management actors to declare or call off a disaster and thus to get or lose control of specific lines of action not allowed in non-disaster conditions.

Next, when designing decision support and expert systems for disaster mitigation, the scope, users, and purpose of use of these systems must guide their design. Accordingly, the understanding of the various dimensions of disasters and of their implications is essential in determining the comprehensive design of information support systems for disaster mitigation. Among others, decision support systems for disaster mitigation destined for the use of politicians and respectively those destined to operational disaster managers should answer different questions and support different decision-makings.

Moreover, a clear picture of the types of disasters allows trimming the actions according to specific national and international regulations and conventions. Very careful definitions also help avoiding legal issues, such as those triggered by bypassing under disaster situations the privacy laws. As a matter of example, under the recent (August–September 2013) fires in California, the use of drones was allowed over private properties as a special measure, but cautiously mentioned in press.

Definitions as precise as possible should clarify lawful actions in disaster regions. For example, disasters due to terrorist activities are recognized in US and in Europe. However, these disasters are due exclusively to human activities involving will to kill, harm, and destabilize societies. On the other hand, while almost the same pattern occurs in mass violence that have similar effects and scale of effects, mass violence situations have never been classified as disasters, at least in legal documents, except in US, at my best knowledge. This poses a legal problem: should conventions related to terrorism apply to mass violence events that are recurrent, for example in some African regions – and not only?

Not less, for academic purposes, a clear, comprehensive definition of disasters is required to allow disaster resilience building and disaster modeling. In a recent publication of NAP, an expression that may confuse appeared in the title, namely "Catastrophic Disaster" [4]. It is unclear if the authors wished to make a difference between non-catastrophic and catastrophic disasters, thus establishing a scale of disasters, or used the word disaster with the meaning of hazard. Anyway, that title may spread confusion and could create a dangerous precedent in the literature.

From the point of view of IT&C, the corresponding definitions must be embedded in the knowledge base of any decision support system for disasters and in every modeling tool, not to mention the effects on the use of specific communication systems, or the allowance to use the available spectrum in ways not permitted under normal conditions.

Disasters are major physical changes difficult for ecologies and societies to adjust to. Merriam-Webster Dictionary (M-WD) defines disaster as "a sudden calamitous event bringing great damage, loss, or destruction", and broadly as "a sudden or great misfortune or failure". The natural language is vaguer in the use of the word, and M-WD recognizes it by providing examples as "The program examined several bridge failures and other engineering disasters" and "The new regulations could be a disaster for smaller businesses," or "The dinner party was a complete disaster." The etymology sheds light on the human apprehension of disasters, and thus shows an important difference between natural events and disasters: disaster means, after all, in Latin, "bad luck" – disaster. Luck is a humanly perceived, personal or community related state of facts; natural (non-human, non-animated) processes have simply are, knowing neither good nor bad luck.

Disasters are defined as natural, technological, and human-initiated events that disrupt the normal functioning of the economy and society on a large scale [5]. In the detailed list provided by FEMA [6], the types of natural disasters are drought, earthquakes, extreme heat, floods, hurricanes, landslides & debris flow, severe weather, thunderstorms & lightning, tornadoes, tsunamis, volcanoes, wildfires, and winter storms & extreme cold, and solar flare in space (space weather). Pandemics are listed as a separate category. Technological, accidental, and terrorist hazards, complete the list on FEMA web page. In the category of technological and accidental hazards are included the blackouts, hazardous materials incidents, nuclear power plants [incidents]. "Explosives, flammable and combustible substances, poisons and radioactive materials" are all considered hazardous materials by FEMA [7]. Terrorist hazards include biological threats, chemical threats, cyber-attacks, explosions, nuclear blast, radiological dispersion device (RDD) [7].

According to [1], avalanches and insect infestations are also natural hazards that can produce disasters:

> "Natural hazard" means an event or process, such as an earthquake, fire, flood, wind, landslide, avalanche, cyclone, tsunami, insect infestation, drought or volcanic eruption, which has the potential for triggering a disaster. [1]

The sense of unpredictability and sudden strike are features of disasters. A winter flu season may wrack havoc from the economic point of view and may have a high death toll, yet it is not usually named a disaster, in common language, because it is more or less expected and because it has an insidious start. However, FEMA and UN include epidemics in the disaster category, possibly with the aim of allowing application of the international regulations of inter-nations help and deployment of personnel.

The conclusion of this sub-section is that communities, local, and national authorities, and responsible agencies need to have a large set of models, preventive measures, response and mitigation plans, and recovery plans for covering the diverse range of hazards and emergencies that the society may have to confront. The respective means and plans, many of them heavily based on IT, may take years to develop and put into operation.

1.3 Legal and Political Dimension of Disasters

It has been argued that disaster definition and declaration have strong political components and that, frequently, disaster declarations are politically motivated. In this subsection, these two distinct but entangled aspects are briefly presented and discussed.

Several international and national level documents define disasters, disaster mitigation, and emergencies, for use in political decision-making in the first place, thus allowing, among others, the public funds use for operations in the concerned area and for specific acquisitions, moreover for temporarily applying special regulations that restrict the individual's rights.

According to [1], the definition of disasters is not requiring the brisk onset,

> "Disaster" means "a serious disruption of the functioning of society, posing a significant, widespread threat to human life, health, property or the environment, whether caused by accident, nature or human activity, and whether developing suddenly or as the result of complex, long-term processes." [1]

while

> 7. "Disaster mitigation" means measures designed to prevent, predict, prepare for, respond to, monitor and/or mitigate the impact of, disasters. [1]

Province of Alberta [8], provides a very general definition of disasters, including in the definition a cover-all term, "welfare". In fact, this definition reduces the required amplitude of the impact on humans to "serious harm to the safety, health or welfare of people", but preserves a high amplitude for the economic impact, "widespread damage to property". The same document emphasizes on pre-emptive measures, in case of emergencies, where the notion of emergency is defined as "an event that requires prompt co-ordination of action or special regulation of persons or property to protect the safety, health or welfare of people or to limit damage to property."

Another state-level document, [9], further diminishes the requirements for the disaster definition to the conditions of "severe and prolonged", thus clarifying that the harm must have a longer duration,

> "Disaster" means a severe or prolonged, natural or human – caused, occurrence that threatens or negatively impacts life, health, property, infrastructure, the environment, the security of this state or a portion of this state, or critical systems, including computer, telecommunications, or agricultural systems. [9]

In yet another example, according to San Diego Municipal Code [10], Chap. 6, the legal component is essential in the definition of disasters, requiring the declaration of the disaster by a high authority,

> Disaster means a catastrophic, naturally occurring or man-made event, including earthquake, flood, fire, riot, or storm, for which a state of emergency has been declared by the President of the United States, the Governor of California, or the executive officer or legislative body of the City or County of San Diego. [10]

The above examples provide purely legal definitions, where the key identifier of the disasters is their declaration by political factors.

The political aspects in disaster situations should be cautiously addressed; see for example how carefully answered was a carelessly asked question ("Could you clarify how roles are distributed if NATO intervenes in a civil disaster?") answered by Maurits R. Jochems:

> I have some difficulty with the term "intervene" in your question. If a disaster occurs, NATO provides assistance following a request from a stricken nation. I underline that NATO does not intervene, it offers assistance on the basis of a request. [11]

A special mention deserves legal aspects of prioritization of the use of means at their disposal by planners, managers, and operative teams during a disaster. Frequently, prioritization issues are not covered by regulations, producing delay in operations, misinterpretations, confusion, and potentially risking lives. As a matter of example, Tampere Convention "is reaffirming the absolute priority accorded emergency life-saving communications in more than fifty international regulatory instruments, including the Constitution of the International Telecommunication Union." The legal duty to give absolute priority to life-saving communications must be taken into account in all models and plans for deploying communications in disaster areas. However, this priority is not easy to interpret or implement in the very dynamic scenes of some disasters. Difficult dilemma and questions may arise, as: Should a tool that, when employed, would save a few lives, be used in that purpose, taking into account that, if used in another way, at the same moment of time, it has a large chance to save many more lives in the immediate future? Answering such questions is beyond the scope of this article. However, the organizations in position to answer them should address the issue. Thus,

Recommendation. Legal bases for the prioritization of the use of tools available in limited amounts should be available to modelers, planners, software developers, and teams on the disaster scene.

1.3.1 Emotional Dimension of Disasters

There is no doubt that, whatever spread and with whatever death toll it is, a hazard is not a disaster if it is not emotionally elicited as such. A disaster is a reflection of an unexpected, fast onset hazard at the level of the societal collective feeling in the society at large; an event not arousing collective compassion is not a disaster, whatever strong is the hardship of those suffering the consequences of the hazard. The effects of the disasters are perceived as highly dramatic or tragic. Images play an essential role in conveying the sense of disaster. The collective feelings provoked by disasters determine and make possible the general support to the political factors to intervene and may mobilize significant parts of the society to help in various ways the recovery.

On the other hand, emotions on the disaster scene defining the disaster are fear, extreme anxiety, desperation, possibly panic, and horror. Emotions may worsen the effects of disasters, by leading to disorganization of the social tissue and by reducing responsiveness. Panic is especially disruptive. These emotions are expected to manifest themselves at the level of the used vocabulary as well at the level of pronunciation (voice).

The emotional character is not interesting only per se or for sociologists and psychologists. It offers huge amounts of specific data for mining with IT & C systems [12]. Possibly, the best determination of the disaster area in earthquakes or floods is not by spatial or aerial photography, but by the space where the human voices and speech reflect the hardship, suffering, despair and the panic. This vast reservoir of data has never been used, as far as I know, for determining the disaster-affected area, needs, and for rescue control. Locating disasters by speech-mining, specifically by emotional analysis should become a mean for early warning of the authorities and of reliably assessing the extent and profoundness of human misery brought by disasters.

Finally, emotional charges in the society may announce imminent social disasters and terrorism acts; therefore, text and speech emotion mining may be a tool in predicting and preventing dramatic social effects and increase the society robustness.

1.4 Human Life Loss Dimension

The dimension of human loss in typical disasters is not necessarily a criterion for disaster definition. Several diseases spreading over large regions and longer time intervals, as the hibernal flu pandemics are not categorized as disasters, although they produce a high death toll, in the range of tens of thousands for US alone. About some diseases causing tens of thousands of deaths per year, as diarrheal infection, the public is even little aware. Definitely, such long lasting processes also have a high human-life loss dimension, but may have small emotional dimension, making them appear to be non-disasters. In many respects, health events as those discussed above, as well as other events that may escape public attention and may be not labeled as disasters, as environment or ecological major events, should be treated with the same methods and tools as declared disasters.

1.4.1 Social Dimension

According to Greg Bankoff [13],

> "There are no such things as 'natural disasters'. Hazards are natural events, occurring more or less frequently and of greater or lesser magnitude, but disasters are not. What makes a hazard into a disaster depends primarily on the way a society is ordered. Human systems place some people more at risk than others . . . " [13]

While there is a confusion between what is meant by the hazard that represent the disaster primary cause (natural or not), Bankoff forgets that economic and social causes aggravating or even augmenting the effects of the primary cause are only part of the story. The effects of a natural hazard should not be confused with the whole social, biological (health) and economic process that a disaster represents. The resilience to disasters is an effect of the culture (as Bankoff himself points out in another paper, [14]), economic capacity, social coherence, and other factors. Bankoff proposes an answer to the question why some disasters are named natural disasters, when they are a complex human, social, and economical process, in the first place. His answer is brief, captivating, but probably simplistic, unfortunately, "The answer is simple: it suits some people to explain them that way." [13]

However, it is the merit of Bankoff of marking the importance of social vulnerability under disaster circumstances. For the efficiency of support and salvage under disaster situations, databases on the extant vulnerable population in the region must be prepared and updated in advance, including chronically sick, disabled, and aged people in the region. Maps highlighting positions of vulnerable people and groups should be a part of any support system for the management of disasters.

1.4.2 Economic Dimension

While mentioning this crucial facet of disaster mitigation, it is not discussed here, as it is already the object of two chapters (by Kirschenbaum & Rapaport, and by Watanabe) in this volume, moreover the subject it too vast to deal with in a few pages.

1.4.3 Conclusions on the Definition

Strikingly, agencies, state and national authorities in various countries use significantly different definitions for disaster situations, as well as for emergencies. That hampers the international collaboration in disaster mitigation and may considerably delay it, due to the need of ad hoc agreements after the disaster strikes.

With so many definitions, the design, deployment, and use of IT&C tools for disaster mitigation remains a moving target. The influence of that lack of harmonized definitions and related regulations on the IT role in disasters is more difficult to pinpoint. One impact on IT is that the IT industry cannot produce tools, like models, specific databases, decision-support systems, expert systems, resource allocation and management systems, and planning and task allocation tools that are acceptable under all legislations and national authorities [5]. Therefore, the corresponding market is highly fragmented and ineffective. This multiplies research and development, and depletes the funding. Worse, it makes the various national-level systems non-inter-operable, thus ineffective in collaborations on regional disaster mitigation. There is little doubt that harmonization of definitions, both by means of standards and by adoption by international bodies would tremendously help IT in the progress of providing better, more effective and powerful tools for disaster mitigation.

Synthesizing the various definitions and discussion in this section, we propose the following working definition as a basis and guide for the design of IT tools for disaster mitigation:

Disasters are complex processes

- originated by natural or man-made hazards, including physical, chemical, biological, ecological, of cosmic origin, or hazards of origins as yet unidentified,
- typically in an abrupt way,
- involving at a large scale life-threats, health-threats, and emotional hardship for individuals,
- producing vast economical destructions and crippling, societal destabilization and de-structuring, crossing national, cultural, organizational and personal boundaries, and
- requiring urgent local and political actions, including emergency or disaster situation declaration,
- where the declaration of the disaster by the authorities sets a specific, predetermined legal framework for the mitigation actions.

Summarizing this Section, virtually all disasters have a set of characteristics that tend to form a specific pattern, asking, in turn, for a pattern of responses, see Fig. 1.3.

1.5 Resilience and Robustness at Large (Society)

The topic of this volume regards the *resilience* and *robustness* of the IT & C systems in disaster areas as well as the role of IT & C in increasing the resilience of the overall society, population and economic system in the disaster areas. In this section, we clarify these terms and provide illustrative examples. Next, we address the resilience of the IT & C systems themselves.

The appeal of the terms resilience and robustness may partly relate to their fashionable use in recent years, but it is also due to the need of representing new concepts in dealing with disasters and hard-to-predict events that may have large

Disaster pattern	Resulting pattern of actions and requirements
• Are started by a hazard or at unexpected moment • Hazards are unexpected and cannot be accurately predicted	• Unavoidable • Need to educate to cope with them • Need to plan in advance • Have a surprise effect leading to unpreparedness (standby attitude) even when there are prepared plan to cope with the hazard
• Hazards are events of high impact	• Cannot be dealt with as minor hazards, as are the "accidents".
• Typically have a fast onset	• Leave no time for short-term measures • *Require vast amount of data collected in real-time or almost real-time* • *Require processing of large amounts of data in real time* • *Require urgent decision making and fast response, including personnel deployment* • *Decision-makers and operatives are overwhelmed during the first hours and days* • *As a consequence, IT means are needed to help data acquisition, data processing, and decision processes*
• Develop into a process of long duration	• Need for long-term planning and vast resources
• Cover from significant to very large geographical areas	• *Need for huge quantities of information* • *Information has a spatial (geographical) character* • *Need for sensor networks to gather data*
• Tend to produce chains of secondary hazards	• Urgent need to take measures to prevent the amplification effects by secondary and tertiary hazards
• Have a high emotional impact on individuals and can produce significant psychological stress consequences	• Need of education and training on large populations
• Primary and chained hazards are, and are perceived to be life threatening for / by large populations	• Need for rescue operations planned in advanced and rapidly deployed, to prevent life loss
• Produce physical and/or biological or both widespread harm	• There are different patterns of effects • *Need to build scenarios, models, decision support systems and databases specific for each scenario* • Multiplied difficulty to prepare for
• Pattern A: both physical and life destruction (earthquakes, floods, high winds, large fires, large explosions)	• *May affect all structures, including transportation, communication, health, security, control, economic and political structures* • *Need for increasing the reliability and survivability of these technical and social structures* • *The vulnerability to various hazards of transportation and IT & C has to be reduced*
• Pattern B: small or no physical destruction produced by the primary hazard, but high death toll (chemical, biological / epidemic, non-explosive nuclear hazard)	• May contaminate structures, buildings, supplies, and equipment, rending them unusable for a long time; food supplies may become unusable forever • May require massive or complete evacuation of the population • May impose isolation of the region and quarantine
• Tend to destabilize, cripple and destroy economic processes and facilities at a large scale	• Effects go beyond the affected area, due to the supply chains disrupted in the disaster area • Pose significant problems for sheltering and feeding the population in the area • May affect food supplies of the regions that are dependent on the affected area
• Have a significant impact on communities and local texture of the society, disrupting them	• Security personnel and various categories of personnel must be brought in the area from outside, further increasing the pressure on the already crippled food and shelter reserves
• Effects have a slow decline (long temporal tails)	• Long-term planning is required • Human misery may be of long duration for a large population
• Have a lasting human, societal, and economic impact a long time after the hazards disappear	• Need continuous support to the area for months or years

Fig. 1.3 Disaster patterns and related action patterns

impacts on the society. The most detailed definition of the terms we are aware of appears in the study [15] (see p. 43). In essence, a distinction is made with respect to the way the system (society, community etc.) operates after a powerful event such as a disaster, namely between a way of operation almost identical to the one before the event, or significantly different. When the system operates in the same way before and after the event, it is said that it was able to absorb the "transient", the disturbance presented by the event. In this case, the system is said to be robust to the respective type (and scale of) event. When the system cannot withstand the event without major disruptions in its operation, but is able to adapt and quickly adopt a new manner of operation, suitable to the conditions created by the event and its consequences, that is, to a partly new context, the system is named resilient. As the authors of the quoted study put it, "Resilience is the ability to bounce back." Therefore, resilience involves quick adaptation and changes in operation procedures, and fast recovery from a partial or total operation failure. These distinctions between resilience and robustness have been only recently clarified. As a consequence, many documents (regulations and recommendations of various national and international institutions) make no clear distinction between resilience and robustness, neither in the wording, nor in the concept.

In disaster events, the systems supporting the society may suffer significant partial operational failures, for instance utilities (water, power supply) in the disaster area may fail operating, communication and computer networks may stop operating due to hardware failing or power supply lack etc. When these systems are able to restart operations in a short period of time, with no apparent or with minimal degradation of the operation quality, they are resilient. As a matter of example, a highly redundant system made of ruggedized servers, workstations, handheld terminals, power supplies and cable may continue operation with no interruption or extremely short period of operation failure. In that case, they are robust. Both robustness and resilience have a cost, and the balance between the costs and the degree of resilience must be determined in the planning period for the system, years in advance an event occurs. Increasing the degree of resilience of IT & C systems is a technological and economic problem, as well as a problem of insuring security and compatibility of operation with other systems.

A system that shows persistent dysfunction after a hazard is non-resilient, or hazard-*vulnerable*. The vulnerability may be quite different depending on the type of hazard – flood, fire, strong winds, or earthquake.

The roles of IT&C in improving robustness and resilience include:

- Improving preparedness:
  - create models of hazards and disaster dynamics (see, for example, in this volume, the chapters by Bouder, Biliaev et al., Rothkrantz, Kandel et al., Lungu et al.);
  - improving the predictability of hazards (see, for example, in this volume, the chapters by Himoto & Kimata, Moreau et al., Panktarova et al.);

  - determining vulnerabilities (see, for example, in this volume, the chapters by Noenning & Schmiedgen, Watanabe, Damiani et al., Reinhardt, and Tsereteli et al.);
  - planning for disaster situations (improving preparedness);
  - creation and keeping of databases and knowledge bases required for dealing with disasters and their effects;

- Improving the efficiency of emergency response and disaster management (see, for example, in this volume, the chapters by Gaindric et al., Damiani et al., Guasch et al., Cojocaru, and Amarandei & Hagan);
- Improving the recovery (robustness) and adaptation to post-disaster circumstances (resilience), and alleviating the disaster effects (both robustness and resilience).

1.5.1 Roles of IT&C in Disaster Mitigation and Related Requirements

The main role of IT modeling applications is to determine if and how communities, companies, and society are robust, or at least are able to pass over disasters in a resilient manner. Data collection systems, e.g., sensor networks, decision-support systems, expert systems and other IT tools are aimed to insure robustness or at least increase resilience. To fulfill this goal, IT needs reliable telecommunication systems, that is, systems able to transmit the information. According to [1] (page 8, article 15) "Telecommunications" does not include means of information processing and new information generation. These are functions of IT. Unfortunately, many recent disasters have shown that both telecommunications means and IT systems are highly vulnerable.

In the remarkable study already cited, a committee of the National Academy [5] arrived at similar conclusions as the study performed for IFRC [15]. Namely, the Committee recognized as "key areas of IT-enabled capability" the need for information gathering and processing in view of creating a comprehensive, integrative picture of the situation ("Better situational awareness and a common operating picture"), and the need of IT tools to support decision-making ("Improved decision support and resource tracking and allocation", [15]). Also, the committee emphasized the need of "enhanced infrastructure survivability".

Especially in the brief period of time after a disaster (as earthquakes) strikes and sometimes in the hours prior and after a disaster (as flooding, hurricanes), the information flow in the disaster area much surpasses the one under normal conditions. This surge in information exchange is especially valuable for gathering information and for relieving the community (family members who are not in the region, family members in the region who lost trace of their kin also in the region). Special plans must be put in place to maintain the communications, moreover to increase the communication capacity during these critical lapses of time. Instead, what was witnessed in several disasters (see, e.g., [16–18]) was a dramatic or total loss of communication in the disaster area. Also, the diminishing or lack of decision-making capacity is a major issue in disaster area, as disaster mitigation requires highly efficient and fast decision-making on a vast scale and in a broad range of decisions.

Disaster mitigation requires in the first place gathering of data on the nature of the disaster and its dimension (local, large area, or widespread, comprising several large areas, such as several cities, in case of floods and earthquakes.) The next step is to collect data on (i) state of the population; (ii) the housing; (iii) the transport; (iv) the communication system; (v) and the IT system. Further data to be collected regard supplies, deployment of rescue etc. Data collection is useless if it is not organized, synthesized and made available to the decision-makers and to the operations. To make data useful, it should be organized, either manually on paper, or electronically, in databases. Either one method requires large specialized personnel. Considering the case of databases use, the needed personnel are, in the first place, computer operators and experts in this field, using for example applications that connect to the databases, installed on iPhones, tablets, and laptops. The planners should prepare decisions from where to bring the large number of equipment operators and data collectors, because the large amount of collection points and the number of personnel on site affected by the disaster will not allow the use only of the on-site operators. Moreover, planners should provide for the preparation in advance of the needed databases, and the training of the personnel at the national level for the use of the data collecting equipment and the databases.

It should be noted that special databases and software applications must be prepared to deal with problems occurring in disaster conditions. For example, biometric applications and databases must be deployed to acquire and preserve biometric identification data for the dead not having an ID on them, or for the displaced people, included wounded, who had not the possibility to carry IDs, but who need identification for bank operations, among others. Specific procedures and applications are also required to perform unusual operations in these circumstances, for example for allowing bank clerks and automata to deliver money based on biometric data. Also, databases and related applications should be tailored to the various types of disasters. While all databases should be interoperable and exchange data between them, actually constituting a single database composed of a constellation of particularized databases, the users wish to see only the database they need to. For example, in the complete database, one needs to have inbound population and personnel fluxes as well as outbound fluxes from the disaster region, but in case of nuclear disaster there is little use for the operations to know the number of population going in the disaster area, as that number must be zero. Presenting to the user the category of inbound population flux would be a distraction.

Databases must be supplemented with mechanisms that support automatic data collection from sensor networks and from various ad hoc sources, e.g., number of operating cellular phones at a specified location in the disaster area, at a given time moment, as reported by cellular phone companies/operators. For facilitating data collection, simple but powerful interfaces must be developed for the field operators and for the users. For example, the interfaces should show only the geographic region of interest (where the collector operates) and allow voice data collection. They also need to use internal checks for the data (e.g., not allowing errors as "age is about 355"), and provide for automatic corrections using language-specific knowledge (for example, in Romania, a name input as 'Amdrei' should either be automatically corrected into 'Andrei' or elicit a question like "Is it Amdrei or Andrei?").

Designing and developing the required databases and the various applications for use in data collection for disaster areas should preferably be done at a supra-state/international level, for interoperability, especially for trans-border disasters and for use by international help asked for at the disaster area. Multi-national development requires, in turn, standardization of procedures, databases and applications. No conventions or standards have been issued until now. This is one of the most urgent needs to be addressed at the regional (e.g., NATO, EU) and international level.

Data collected in the databases must be supplemented by applications that process and synthesize data (data mining tools, data analytics) in almost real-time and automatically send appropriate syntheses at the operative, planning, and decision-making levels. These reports may also be the subject of standardization. The reports should use appropriate language for the addressed users and keep the minimal information needed. For example, while quantities and specific names and form (e.g., capsules) of the prescriptions in a transport are required for the medical personnel, the transport planners should receive the information as "medications, ballot of about 1 m^3, 60 kg, not fragile".

Concluding, IT tools, as databases, data mining tools, and decision-support systems for disaster situations must be carefully conceived for increased efficiency, taking into account the urgency of the situation and the lack of time for human-made corrections and interpretation. In addition, the tools must provide capabilities for making sense of the uncertainty in data (see chapter by Kandel et al., Pankratova et al., and Kirschenbaum & Rapaport, in this volume, and a rich literature in NATO series, for example [19, 20]) There is a vast literature supporting the notion that under disaster situations, information is vague and has a random character (see also chapters by Bouder, Rothkrantz, Guasch, Damiani etc., in this volume; various papers specifically investigate the probability distributions of events under disaster circumstances, e.g. [21]).

1.5.2 Resilience and Robustness of IT&C

Katrina disaster was a wake-up signal in many respects, including the role of local communications and their fragility. Katrina inflicted a harsh lesson, which was at least partly learned by companies and US authorities. That led to several technical progresses. When Sandy shored, the results of those lessons were in place, but were not enough to tame all negative effects. For example, it was reported that some mobile data centers have been deployed, but could not operate before the hurricane passed, while ¼ of the towers were down when the hurricane passed and 1/3 of the telephony towers were down before the commercial power supply was reestablished. While under normal operation the equipment works under "in-door" conditions, during disasters it may face outdoor-like, even rough conditions. Therefore, not only the equipment, but also the cabling in the networks, the connectors (e.g., molded connectors), and the receptacles must withstand harsh environmental conditions, that is, must be rugged hardware, it they were to be available under disasters.

In a bleak picture, a well-structured lesson from Katrina disaster was summarized in [17] as

> Unavailability [of] public communication and other infrastructural facilities. Switching station, cellular antennas and other essential elements damaged. [17]

and the authors of that report conclude,

> Keep data and datacenters out of the harm's way; Don't assume the public infrastructure will be available; Plan for civil unrest; Assume some people will not be available; establish backup role takers for key personnel . . . [17]

Large companies and organizations, agencies involved in disaster protection, as well as states may consider backing up their IT&C infrastructure with mobile data centers. Such mobile datacenters, according to the published literature, proved highly useful in the Sandy hurricane hazard. There are several companies in US and UK, now extending in Asia, that offer datacenter backups and recovery services. Most of these specialized companies offer various solutions and full support, from disaster recovery planning to business continuity full support. Some of them also have mobile units to support data centers and even mobile data centers on large trailers. Combinations of cloud, mobile datacenters, and recovery center solutions may be the best answer in terms of both cost and flexibility.

It should be noted that IT networks, software supports, and computers are less "gracefully degrading" than traditional means to store information. For example, a map on paper can be soaked in water and still totally useful, while a computer cannot; the same map partly burned remains partly useful – but not a computer; a paper document resists tens of g acceleration, while any hardware will break into pieces at more than a few gs. Especially for people responding to disasters in the field, rugged hardware supplemented with older-type information sources, as maps and booklets are needed. In addition, IT&C equipment needs reliable sources of power, which often go depleted during lasting disaster conditions. According to anecdotic evidence gathered during Hurricane Sandy, using fuel cells as backup power sources for IT&C systems may have significant benefits when disasters strike. Further issues related to hardware efficiency and robustness are dealt in this volume in the chapters by Hlousek & Konecny, and by Mottock et al.

1.6 Conclusions

A partial list of conclusions of the discussion in this chapter and in the cited literature is:

1. IT&C has a recognized, central role in disaster response and mitigation.
2. While intimately relying on communications, IT has a distinct set of roles, so far not fully understood and only partly tapped in disaster mitigation.
3. There is a yet unsolved need for increasing the reliability of operation of communication means and IT equipment in various types of disasters, including chained disasters. "Graceful degradation" under disaster conditions, no "single point of failure", high redundancy, and equipment roughness must be imposed by regulations for all essential IT&C equipment and networks operating in disaster-prone areas.
4. The diversity of IT tools and the effectiveness of use of IT in disaster mitigation must be increased.
5. More advanced IT tools for modeling, data collection, and data processing and mining are needed.
6. IT industry should produce tools, like models, specific databases, decision-support systems, expert systems, resource allocation and management systems, and planning and task allocation tools that are acceptable under all legislations and national authorities.
7. Training the operative personnel for using the IT&C in disaster areas, and having available supplementary personnel for deploying, maintaining and operating IT&C equipment and software applications were weakness links signaled by several studies from past disasters.
8. Security of the deployed IT&C services in the disaster areas remains a largely unsolved problem.
9. IT is only a mean in a chain of means at the disposal of the management and society; therefore, overemphasizing IT role in the detriment of the fundamental issues of the society and of the disaster management may bring severe consequences.

References

1. Tampere convention on the provision of telecommunication resources for disaster mitigation and relief operations. http://www.ifrc.org/docs/idrl/I271EN.pdf
2. Strengthening of the coordination of emergency humanitarian assistance of the United Nations, Draft resolution submitted by the Vice-President of the Council, Fernando Arias (Spain), on the basis of informal consultations, Substantive session of 2012, New York, 2-27 July 2012, Agenda item 5, Special economic, humanitarian and disaster relief assistance. http://www.un.org/en/ecosoc/docs/adv2013/sg_report-adv_strengthening_coordination_of_%20humanitarian_assistance.pdf

3. Boğaziçi University, Istanbul Technical University, Middle East Technical University, Yildiz Technical University, Earthquake Master Plan for Istanbul. Published by Metropolitan Municipality of Istanbul, Planning and Construction Directoriat, Geotechnical and Earthquake Investigation Department, 7 July 2003. http://www.uni-kassel.de/fb14/stahlbau/eartheng/downloads/Istanbul_Earthquake_Masterplan.pdf
4. Hanfling D, Altevogt BM, Viswanathan K, Gostin LO (eds) (2012) Crisis standards of care: a systems framework for catastrophic disaster response, vol 1. National Academies Press, Washington, DC. http://www.iom.edu/Reports/2012/Crisis-Standards-of-Care-A-Systems-Framework-for-Catastrophic-Disaster-Response.aspx
5. Rao RR, Eisenberg J, Schmitt T (eds) (2007) Improving disaster management: the role of IT in mitigation, preparedness, response, and recovery. National Academy of Sciences Press, Washington, DC
6. FEMA, http://www.ready.gov/natural-disasters
7. FEMA, http://www.ready.gov/hazardous-materials-incidents, http://www.ready.gov/terrorist-hazards
8. Province of Alberta, Office Consolidation, Revised Statutes of Alberta 2000, Chapter E-6.8, 13 May 2011, Emergency Management Act, http://www.qp.alberta.ca/documents/Acts/E06P8.pdf
9. Emergency Management 323.02. August 12, 2013. Published and certified under s. 35.18. 2011−12 Wisconsin Statutes updated though 2013 Wis. Act 45 and all Supreme Court Orders entered before 12 August 2013. Published and certified under s. 35.18. Chapter 323. Emergency Management. http://docs.legis.wisconsin.gov/statutes/statutes/323.pdf
10. San Diego Municipal Code, Chapter 6: Public works and property, public improvement and assessment proceedings (10-2011), Ch. Art. 7 Div. 38 Article 7: Water System, Division 38: Emergency Water Regulations. http://docs.sandiego.gov/municode/MuniCodeChapter06/Ch06Art07Division38.pdf
11. Interview with Maurits R. Jochems, by Toni Pfanner. International review of the Red Cross. 89(866), June 2007. http://www.icrc.org/eng/assets/files/other/irrc_866_interview.pdf
12. Teodorescu HN (2013) SN Voice and Text Analysis as a Tool for Disaster Effects Estimation – A Preliminary Exploration. In: Burileanu C, Teodorescu HN, Rusu C (eds) Proceeding 7th conference on speech technology and human – computer dialogue (SpeD), IEEE, Cluj Napoca, Romania, 16–19 Oct 2013, IEEE 2013
13. Bankoff G (2010) No such thing as natural disasters. Harv Int Rev. Harvard University, Boston. http://hir.harvard.edu/no-such-thing-as-natural-disasters
14. Bankoff G (2007) Living with risk; coping with disasters. Hazard as a frequent life experience in the Philippines. Educ Asia 12(2):26–29
15. International Federation of Red Cross and Red Crescent Societies, Geneva (2012) The long road to resilience. Impact and cost-benefit analysis of disaster risk reduction in Bangladesh. http://www.ifrc.org/Global/Publications/disasters/reducing_risks/Long-road-to-resilience.pdf
16. NOAA reports on Hurricane Katrina. http://www.katrina.noaa.gov/reports/reports.html
17. Deepti A, Soumya P Business continuity at Northrop Grumman, http://www.slideshare.net/anshuman82/business-continuity-atnorthropgrumman
18. Miller LM (2012) Controlling disasters: recognising latent goals after Hurricane Katrina. Disasters 36(1):122–139
19. Teodorescu HN (2003) Information, data, and information aggregation in relation to the user model. Systematic organisation of information in Fuzzy Systems. Book Series: NATO science series, Sub-Series: Computer and systems science, vol 184. pp 7–10
20. Teodorescu HN (2003) Self-organizing uncertainty-based networks. Systematic organisation of information in Fuzzy Systems. Book series: NATO science series, Sub-Series: Computer and systems science, vol 184. pp 131–159
21. Teodorescu HNL, Cojocaru VP (2014) Experimental investigation of the reliability of reception of ultrasound signals in fire conditions. Fire Saf J 66:25–34

Chapter 2
Organizations Under Siege: Innovative Adaptive Behaviors in Work Organizations

Alan (Avi) Kirschenbaum and Carmit Rapaport

Abstract By examining the actual behavior of both managers and employees in work organizations during a crisis, we were able to better understand conditions facilitating an organization's ability to maintain operational continuity. Building on theories of organizational and disaster behavior, a working model was developed and tested from evidence acquired from work organizations that were subjected to massive Katyusha rocket bombardment of Northern Israel in 2006. The results support the notion that organizational response to a disaster includes a social process of innovative behavioral adaptation to changing and threatening conditions. Based on both perceived and actual financial performance levels during the crisis, we discerned that on the one hand, the organization's managers react within the administrative constraints of their organizations according to their perception of its performance. The day-to-day operations, however, are maintained as employees' adapt their own behavior to the changing demands of the situation. The analysis further shows that although plans, drills and emergency regulations are important for performance behaviors during the emergency, it was employees' innovative adaptive behaviors that contributed to maintaining actual organizational performance. These adaptive work behaviors depended on a series of social process predictors such as the levels of emergent and prosocial behaviors as well as the densities of social networks at the workplace. The findings both support and focus on the role that external organizational disruption can have on innovative organizational adaptation and change.

A. Kirschenbaum (✉)
Faculty of Industrial Engineering and Management, Samuel Neaman Institute for Advanced Studies, Technion – Israel Institute of Technology, Haifa 32000, Israel

Kirschenbaum Consulting Ltd, Haifa, Israel
e-mail: avik@tx.technion.ac.il

C. Rapaport
Faculty of Industrial Engineering and Management, Samuel Neaman Institute for Advanced Studies, Technion – Israel Institute of Technology, Haifa 32000, Israel

H.-N. Teodorescu et al. (eds.), *Improving Disaster Resilience and Mitigation - IT Means and Tools*, NATO Science for Peace and Security Series C: Environmental Security,
DOI 10.1007/978-94-017-9136-6_2,

Keywords Organizational continuity • Performance under crises • Adaptation • Social networks

2.1 Introduction

Social responses to disasters have been portrayed in terms of behaviors that emerge from groups during a crisis; being variously described in terms of innovation [1], situational adaptation [2], and evolving into disaster subcultures [3]. For the most part, such studies have focused on individuals, groups and agencies within the larger community [4, 5]. Sparse attention had been given to these same phenomena within work organizations with even less attention on actual behaviors [6, 7]. Like communities, work organizations represent structured social phenomenon embedded in society. And like communities, organizational continuity during disasters is critical for societal maintenance [8]. Therefore, we would expect similar types of social response behaviors in organizations. This, however, may not be the case as work organizations are far more constrained than communities in terms of their administrative and social structures. Our aim here will therefore be to explore if similar responses are indeed found within work organizations and if these responses have an impact on the continuity, maintenance and viability of such organizations.

Until now, efforts at explaining organizational continuity continue to be dominated by its crises management roots, including integrating strategic management with crises management [9]. This approach is reinforced by practitioners and governed by an international inventory of "standards and guidelines" (for example, [10–12]). As such, the literature is biased toward business aspects of organizational continuity, providing numerous "what to do" lists and specific case studies. Unfortunately, there are few confirmatory studies that such 'standards' are effective in maintaining continuity of operations. In a broad sense, the process of organizational continuity within work organizations remains an anomaly. This is troubling because of the practical implications that research can bring about to enhance or dampen an organization's ability to maintain operation of continuity.

To better understand this phenomenon, we will take the view describing community responses during disasters, namely that continuity also involves a social process of innovative adaptation. This means actual organizational behaviors that emerge during emergencies contribute toward the maintenance and continuity of operations. This strategy, we argue, provides two major benefits: it provides a framework for an interdisciplinary approach cementing organizational and disaster behaviors into a coherent framework and it contributes to organizational theory by focusing on ways innovative adaptation occurs during the sudden onset of environmental disruptions.

2.1.1 Unusual Events

Several approaches have been put forward that link macro-level organizational and disaster behaviors, all based on the notion that organizing and adapting is the prime set of behaviors associated with group survival [13]. The two general systematic approaches, contingency theory [14, 15] and ecological theory [16], emphasize the important, or even critical need in organizations for structural flexibility so as to fit the demands of a changing environment that may threaten an organization's operational maintenance. As organizations do not exist in a vacuum, they tend to adapt to environmental changes, such as disruptive disasters or emergencies, as a way to assure their survival. Therefore, any irregular event in the organization's internal or external environments such as a crises or full blown disaster might cause a disruption to the routine operations. As a consequence, actions are taken by members of these organizations to bring about changes in goal designation through operational means to achieve the original objectives.

The implication is that the reaction of an organization and its members to a disaster as a non-routine socially based event [17] may well be nontraditional, out of the ordinary and based on adaptation to the short-term challenging demands of the new environment. For example, work organizations may temporarily move operations, or have reciprocal agreements with other companies or supply chain alternatives. Typically, those involved in research related to organizational crisis management have put emphasis on short term changes in structural characteristics such as realigning organizational communication networks [18–20], organizational flexibility in the decision making process [21], improvisation [22, 23] and coordination and task integration [24]. In addition, some have argued that increasing organizational resilience during emergencies can be attained by enhancing an organizations social capital and networks thereby improving the ability to exchange information and foster the achievement of shared goals [25].

This pattern of response to disasters from both contingency and crisis management perspectives are expected to drive organizations to immediately adjust and adapt to the abrupt changes in the environment as a means to ensure operational continuity. However, repeated business collapse due to disasters, even among well established firms, [26], suggests that this may not be the case. Even with "pre-planning" for such emergencies, there is no consistent perception of what preparedness entails [27] and usually an underestimation of victims' ability to cope with disaster and the extent of expected disruption [28]. All these perspectives touch on organizational characteristics, but only rarely include those that are inherently associated with employee behavior within the organization. This raises an issue that perhaps can be resolved by examining more closely the concept of organization continuity and its antecedents.

2.1.2 Organizational Continuity

The present thrust of organizational continuity, especially with its emphasis on "business", reflects a Taylor like perspective of work organizations as rational administrative structures which can be easily manipulated; a prospective that frames continuity as a system engineering exercise rather than a social process. The present definition defines "business continuity" as "business specific plans and actions that enable an organization to respond to a crisis event in a manner such that business functions, sub-functions and processes are recovered and resumed according to a predetermined plan based upon their criteria to the economic viability of the business" ([29], p. 8) The "what to do list" differs by agency source. For example, the most common acronym is COOP sometimes defined as 'Continuity of Operations' [11, 30] or 'Continuity of Operations Planning' [31]. In general, the emphasis in all is on functions, operations, facilities, equipment and records, mentioning managerial leadership as an integral part of making the plan work. It assumes that intervention by following a 'do list' will assure continuity of operations. This perspective of organizational continuity leaves a lot to be desired as it fails to provide us with an understanding of what this construct represents in terms of organizational survival. It also assumes that organizational continuity is the end product of purposeful intervention rather than an integral part of the adaptation and operation maintenance that takes place during disasters and emergencies within organizations. It should also be stressed that guidance documents ("checklists") are primarily employed as criteria for judging the output of continuity plans, and do not provide explanations regarding the process by which such plans are effective in enhancing organizational continuity. A typical example of this, as previously noted, is the FEMA guide for business and home safety [32].

What we contend is that organization continuity (OC) is not only the outcome of a work organization's members coping with an emergency, but rather is a social process within the organization leading to operational maintenance and resilience. In this framework, OC is a social construct composed of multiple facets. As organizations are social units, we also suggest that social factors and processes inherent in disaster situations found to affect, for example, other types of organizations such as communities, may also be appropriate as guidelines in understanding the continuation of operation in work organizations. The literature on emergencies and disasters includes such factors as emergent behavior [33, 34], preparedness [35, 36], pro-social behaviors such as mutual help and volunteering [4, 37], social networks and information flow [13]. All are basic behaviors displayed in a broad range of emergency and disaster situations and can potentially be utilized in understanding the process of organizational continuity.

2.1.3 Disasters and Work Organizations

Most of the conceptual and empirical studies of work organizations' ability to maintain operational viability under threat and disaster have focused on work organizations' characteristics such as size, age, and ownership as determinants for increasing chances of survival [7, 26, 38]. Other studies have been conducted in terms of crisis management; mainly focusing on managerial decision making processes [21] and/or communication networks [18–20]. Not surprisingly, the majority of studies linking disasters and organizations have centered on the major public sector service agencies dealing with disaster management, primarily in the hope of increasing their effectiveness [36, 39]. This emphasis has, it seems, diverted attention from investigating innovative adaptive processes in a wide range of private and public organizations that have, or are undergoing the effects of disasters. In addition, the small numbers of studies that have looked at work organizations continuity have relied primarily on a post-disaster examination of (small) business that survived (Gordon and Richardson [40] in [7]). To overcome this gap, given the critical importance of resilient economic-based organizations in contrast to the sparse evidence enabling such resilience, it becomes incumbent upon us to explore disaster related behaviors during crisis situations. In addition, by being able to reasonably predict such behaviors, preparedness actions can be intelligently initiated leading to minimizing loss of life and financial costs.

The basis for such behaviors is the ability to overcome and mitigate the consequences of disasters and is mainly rooted in the collective emergent behavior of those affected [41]. This includes behaviors found both inside and outside the work organization. Overall there is strong empirical evidence that resilience and recovery is predominantly accomplished through group social processes primarily within the family and community [41, 42]. This can be seen, for example, in household and community preparedness behaviors [35], emergence of disaster subcultures [3] and the spontaneous appearance of prosocial helping "emergent groups" in disasters [42]. In addition, it can also be seen in the broad range of adaptation behaviors of individuals and family units to prolonged terrorism [36], increased workplace comradeship and helping behavior among employees in crises [43], and improvisation of inter-organizational coordination [44].

Underlying adaptive disaster behaviors such as improvisation lays an assumption that this type of disaster behavior is developed through a particular type of organizational culture (or climate) that enhances better individual adaptive behaviors.

Taken together, these fundamental disaster-related behaviors are also likely to be found in an organizational framework as they reflect social behaviors that are universal in character. What can be expected is that the structural framework of organizations will have an impact on how such disaster behaviors will be acted upon.

2.1.4 Behavior in Organizations During Disasters

As both managers and employees are inextricably bound within an organization, both are also involved in emergent behaviors that arise during a disaster. Research in the field of collective coping responses to stressors at work, for example, found that mutual help [45], increasing communication [46], reduced status differences [47], task-related communication and integration among employees [48] were the main behavioral collective mechanisms to external and internal pressures on the organization. Drawn primarily from contingency theory, we also note innovative adaptive behaviors as a response to disasters or emergencies that are based on factors outside the organization that might have an effect on the individual's behavior within her/his work place. Adaptive disaster behavior, however, cannot be expected to appear during routine periods of organizational activity as they are directed at survival during life threatening emergencies. We already know that family, community and social networks have a strong influence on an individual's behavior, especially in emergencies and under circumstances of uncertainty. These include information diffusion, social norms and risk perception [13, 36]. Therefore, employees might experience role conflict as a result of two conflicting obligations: on one hand, employees are subject to the organizations' administrative decisions, and behave according to their job commitment, professional status and the organizational culture. On the other hand, as social networks members (e.g., family, community, team mates), they are influenced by social pressures calling upon them to avoid the danger and not to go to work, sometimes with feelings of fear and anxiety. For example, the role conflict of family responsibilities in contrast to work commitments.

From a managerial perspective, the ability to implement rapid changes and adjust to a new environment due to a crises will differ from organization to organization, according to its market position [26], centralization of decision making process, the ability to transfer information quickly [49], organization's size and age [50], and previous experience with disruption [51]. These different managerial-administrative characteristics, especially in the way they are implemented, may have a positive or negative effect on the informal social processes in the organization, and consequently impacting the organization's ability to survive and recover. What is not clear, however, is the impact of managerial decisions during a crisis on the actual disaster behaviors of employees.

2.2 Working Model

To better understand the process of organizational continuity during disasters, we developed a working model (See Fig. 2.1). It focuses on employees' and different ranked managers' actual behavior as well as the organizational and administrative components in this process. Both encompass a broad range of activities and operations during a disaster within an organizational setting. As for the employees, we examined behavioral variables such as adaptive behavior, social networks and pro-social behavior found to enhance resilience in various social settings in past researches [2, 13]. In addition, as employees behavior is embedded in an organizational setting, we related to a managerial or administrative framework within which employees could act. These included both managerial and employee

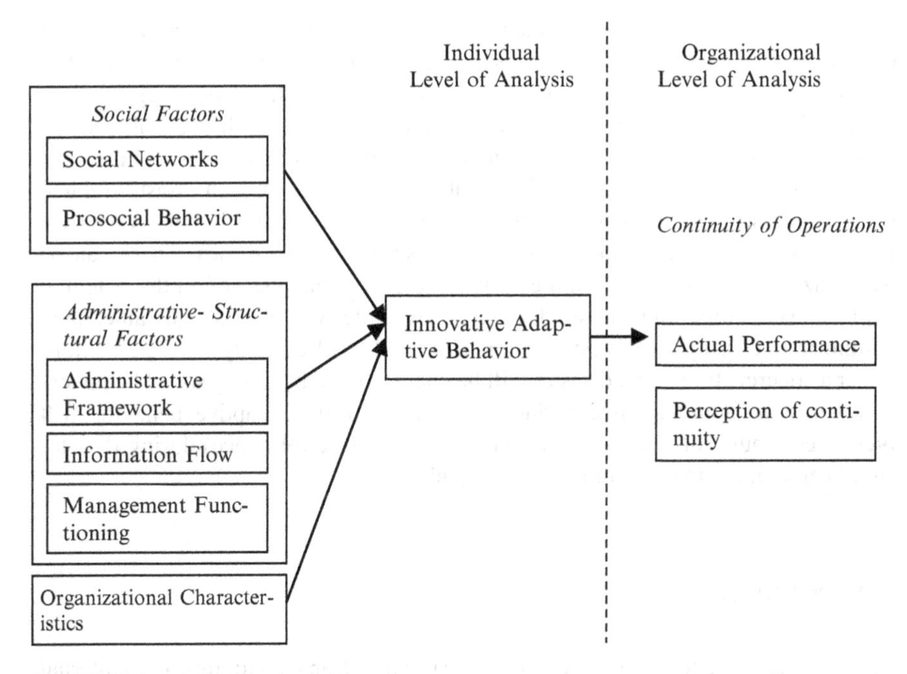

Fig. 2.1 Theoretical research model

responses to the crisis, such as information flow, and perception of management functioning. As our model suggests, an organization's operational continuity will be achieved through a combination of both managerial and employee actions. Common "business continuity" planning guides emphasize an organizational-administrative perspective as the key to maintaining operational continuity in response to an uncertainty and/or crisis. We, however, argue that a more broad process, based on social and administrative factors arises within work organizations during a crisis situation which is a critical determinant for operational continuation and resilience.

To support this argument, we divided the actions being carried out during a disaster into two main groups: social factors and administrative-structural factors. The first includes the actual behavior of employees and managers, through social networks and prosocial behavior. The second refers to distinct organizational-administrative actions and perceptions by organizations' members of actions that enable, in a work-organization setting, the appropriate behavioral adjustment to new crisis situations. Included are also the socio-demographic work related characteristics of the employees that provide an insight into how such background profiles can affect disaster related decisions. By combining these elements, we anchor the members' response directly to a social process carried out by managers and employees within the framework of the organization. In this way, we contend that organizations exhibiting an enhanced set of such social processes will perform better in times of disasters.

Maintaining levels of performance that can contribute to the ability of an organization to continue its operations, however, can alternatively be measured in terms of actual financial levels of performance or, as studies have suggested, be measured in terms of perceptions of performance [52, 53]. The theoretical working model provides an opportunity to examine both these measures. In addition, we have taken into account the possibility that the emergence of such disaster related social processes within organizations may not appear immediately but evolve over time. From the working model several propositions can be drawn. These can be generalized as follows: (H1) At the individual level of analysis, of all the potential explanatory factors, innovative adaptive behavior derives from social rather than administrative factors. (H2) At the organizational level of analysis, organizations under a concrete threat of a disaster will be characterized by strong social behaviors in contrast to administrative variables. (H3) Innovative adaptive behavior will predict continuity of operations (in means of actual performance) during disaster significantly more than administrative variables.

2.3 Strategy

In this research we focus on conceptualizing organizational continuity as an internal organizational social process that increases chances for maintaining operational continuity after a disaster. To understand this social process, we employed a field study to examine the actual behavior of employees and managers in different sector organizations who were in operation prior to, during and after the Second Lebanon War that occurred in northern Israel during July-August 2006. During 33 days of war, more than 3,600 Katyusha missiles were fired on northern Israeli civilian targets, disrupting civilian life in that area with lost revenues estimated at 1.4 billion dollars. Approximately one-third of the northern population left their homes for safer areas with warning alarms of immanent missile impact averaging three times per day in most urban and suburban areas.

To develop and test the concept of organizational continuity as a social process we extended the literature in crisis management that focus on managerial decision to include employees disaster related behaviors. To do so we made use of two sources of data: (a) weekly data from the employees and managers about their actual behavior before, during and after the missile bombardment; collected by a structured questionnaire given directly to employees and managers and (b) objective business data about performance (e.g., before, during and after the war sales, production levels, attendance, cash flow, etc.). By dividing the data into stages, 3 months before the war, at the beginning of the war, later on during the war as well as financial data 2 months after the war, we were better able to examine patterns of organizational learning, decision making and adaptation to the changing demands of the environment.

2.4 Methods

2.4.1 Data Sources

The core data set was based on purposeful sampling of private and public sector organizations from manufacturing, retail service providers to public institutions. Of the 40 organizations originally approached just after the missile bombardment ceased, 24 agreed to allow us to question employees and managers. However, we had a final sample of 17 organizations that had provided us with all information, including performance data. All the organizations were in full operation before the war and were all located in areas under bombardment. We randomly sampled 50 % of the employees in each of the companies. The overall response rate was 20 % with the final sample composed of 421 individual responses. A basic explanation for this response rate may lie in the fact that we surveyed only those who worked during the war days, and who could assess the organizational operation. The business organizations surveyed included seven branches of a nation-wide retailer services firm, located in city centers and shopping malls. Also included were four factories manufacturing medical equipment, steel, textile and chemicals. All these factories export their products and employed 70–250 workers. In addition, the public/governmental organizations included an academic educational center and an institution for disabled and mentally challenged adults. Finally, we surveyed various service-sector organizations, including a large service garage located in the heart of an industrial zone, sports center, cellular-phone firm, and a logistics service company. The data was collected 2 weeks after the war had officially ended (cease-fire was declared). However, the situation was still insecure and volatile. Although life got back slowly to normal, the emergency situation was still in place.

Before randomly distributing the anonymous questionnaires to employees and managers, a pilot study assessed the robustness of the measures. Little in the way of prior scales or measures were available for replication leading us to devise measures based primarily on theoretical variables proposed in the working model that could potentially contribute to our understanding of processes involved in organizational continuity. The initial result of the field survey reveals that respondents reported either direct-hits, near misses or missiles falling nearby. Nearly half stated that work routine remained as it had been prior to the Katyusha attacks. About a third reported that the work routine had been to a large extent been maintained while the remainder described a definite reduction in operational capacity.

Of the 421 employees that were sampled, there were slightly more women than men (57 % vs. 43 %). The average age of the employees was 41.8 (S.D = 11.49 years), 70 % were married and 70 % defined themselves as religiously secular. About 30 % (28 %) stated their income as average or above average (26.5 %) with 20 % below average (19.7 %). Most of the employees live near their workplace, close to 80 % (78 %) mentioned that commuting time ranged between a minute to half an hour. It is important to point here out that we surveyed three factories in three different "Kibbutz's" (the Israeli cooperative settlements) and in these cases most of the employees live in the Kibbutz itself.

2.4.2 Measures

2.4.2.1 Dependent Variable

Continuity of Operations

Continuity of Operations was measured on the basis of performance through two factors: (1) "actual performance": objective data based primarily on financial and production performance prior to, during and after the bombardment; and (2) "perception of continuity": the managers and employees' self perceptions of the extent to which the organization had managed to maintain its routine before, during and after the missile attack.

Actual Performance Measure

In order to compare performance levels of the varied work organizations, we chose to contrast pre-war with during (and shortly after) performance levels. For example, in manufacturing organizations we used actual data received from finance managers at each factory regarding the number of units produced during the war in contrast to pre-war levels. For service businesses and public organizations we used self reported data on the "readiness" to provide services measured in terms of time that passed until the customer was served in comparison to the pre-war period. This measure for service continuity reflected our concern that decreases in income may imply low customer flow but not necessarily reflect adapting to the situation. We therefore contrasted employees and managers readiness to provide service rather than sales itself as a means of reflecting the organization's efforts to achieve normal standards in non-routine times. Therefore, on the basis of these comparative measures we propose the following formula that provides the (percentage) degree of change over the war period compared with periods of routine operations. In short, "Performance" measured the decrease in income during the war compared to pre-war levels of performance.

$$P(t) = 100 - \left(\frac{WA}{BW}\right) * 100 \tag{2.1}$$

where: P(t) is Time Specific Decreases in Performance during war Contrasted to pre-war performance levels; 100 is Standardized pre-war performance level; WA is War Average Rates based on time specific performance rates (beginning + during war); BW is Pre-war performance rate.

2.4.2.2 Perception of Continuity

We used the variable "Perception of continuity" to evaluate organizational continuity as perceived by the employees and managers. Numerous social-psychological studies [52, 53] have looked at perception in an effort to explain organizational behavior. Building on this, we asked employees and managers about the extent to which the organization had managed to maintain its routine. This included the following four questions (answers ranging from "1" = "do not agree" to "4" = "I completely agree"): "to what extent do you agree with the following sentences:" "the routine was kept as much as possible", "my team achieved it weekly goals", "the situation caused disruption to my work routine", "there was a feeling of uncertainty regarding what I have to do", The Cronbach's Alpha index for these items is 0.68.

2.4.2.3 Independent Variables

Innovative Adaptive Behavior

Innovative Adaptive Behavior was measured by a series of questions regarding the non-routine disaster behavior that was carried out by employees during the bombardment. Referred to as "emergent phenomena" [41], such behaviors occur during a disaster and emergency with individuals and organizations replacing their traditional behaviors, structures and functions with new ones. Other researchers noted 'creativity' and improvisation to be the main and important characteristic of human behavior during crises [44]. Building on these notions, we employed the term "innovative adaptive behavior" to describe the appearance of new behaviors and activities that could be best described as adaptive, innovative, creative actions. Therefore, we expect that during disasters or emergencies, organizations, as with other kinds of social units, will experience the emergence of new behaviors that will enable organizational continuity of operations. This variable is composed as a sum of answers to a series of questions. Among these questions are (on a Likert scale ranging from 4 = to a very large extent, to 1 = to very small extent): "I took extra responsibilities in contrast to previous regular times"; "I worked overtime without being asked (more than in my usual shift)"; Cronbach's Alpha for this variable is 0.77.

Social Networks

Social Networks have been found to be particularly important in explaining a varied number of disaster behaviors at the community [36, 54] and individual levels [55]. While we will focus on the structural characteristics of network interaction, other facets are also involved such as norms, culture, symbols and values. It is for this reason we sought out social ties and networks among the employees that would partially reflect the broad band of such ties before, during and after the missile attack. The variable measure is a sum of the respondents' agreement on a Likert type scale (4 = to a very large extent, to 1 = to very small extent). Items included: "Social ties among team mates were strong before the war" and "It was important for me to consult with colleagues what to do when the alarm begins"; Cronbach's Alpha score for these items is: 0.79.

Prosocial Behavior

Prosocial Behavior refers to the social response of the employees and managers, in terms of helping behavior, leadership, actions of empathy and mutual help. This variable reflects a social process being carried out by employees during the disaster, and is not particularly related to the work itself. This concept expands the idea of 'convergence' found in the disaster literature when individuals (usually volunteers) seek to improve conditions created by a disaster due to physical proximity or abilities to do so [37, 56]. In our case such prosocial helping behavior by employees focuses on what occurred when seeking shelter and emotional support from workmates at the workplace. Such convergence should increase when "official authorities" do not necessarily fill in the personal and emotive needs of their employees during such crises. Here, we also employed measures based on a sum of answers to six questions on a scale ranging from 4 = "to a very large extent" to 1 = "to a very small extent". Among the questions were: "I provided mental support for my mates who need it in the shelter" and "I helped mates to reach the shelter during the alarm". The Cronbach's Alpha score for these items is: 0.85.

Administrative Framework

As in any social unit, employees and managers, as social players, are subordinate to the organizations, constrained primarily by administrative directives and reward systems. Understanding these constraints might shed light on behaviors when operational continuity was severely disrupted. For this reason we asked employees about their expected work obligations. This included eight questions that inquired of employees if "the management allowed employees to be able to work part-time", "the management forgave absences", "the management provided a safe shelter", "management reacted with understanding to parents with children", "management moved employees to other sites (if possible)", "management let employees take part in decision making process", "management took care for employees welfare", and "management arranged transportation to the workplace and back home (if not exists usually)". All these questions were based on a four levels Likert –type scale ranging from 'I disagree with this notion"(=1) to "I completely agree"(=4). The Cronbach's Alpha for these items is: 0.73.

Management Functioning

This variable measures how employees assessed the behavior of management in response to the changing situation. It includes three questions regarding if "the management made decisions quickly", "management changed goals according to the changing conditions" and the degree to which "management worked as usual". This variable was sum of the answers to these questions, where answers range from "I disagree with this notion"(=1) to "I completely agree"(=4), with a Cronbach's Alpha of 0.76.

Organizational Characteristics

We chose two fundamental characteristics to differentiate the organizations on the basis of output, namely manufacturing and non-production type organizations. Production units were coded as "1", and "non-production" (service, governmental and public organizations) was coded as "0". In addition we examined each respondent's work status level within the organization as a dummy variable. Here, managerial level employees (="1") included senior and junior levels shift managers, team leaders etc. while employees who did not define themselves as managers were coded as "0".

Information Flow

An important possible explanatory variable in the organizational continuity process outlined in the Model includes how information was disseminated within the organization and its impact on disaster related behaviors. Both in organizational and disaster studies, information plays a vital role in how decisions are made and what behaviors can be expected [57]. Within organizations, information can flow vertically as well as horizontally; from management down or among employees. To capture these variables and assess its impact on adaptation within an organization's maintenance of operations, we put forward three questions measuring the degree of information flow during the disaster. These included, among others (on the following scale (4 = to a very large extent, 3 = to a large extent, 2 = to some a extent, 1 = to very small extent)): "The management initiated informative talks to all employees about the situation"; "I quickly got information regarding the situation at work"; "My direct manager contacted me on a regular basis". The answers were summed into this variable. Alpha Cronbach's is 0.76.

Overall, Table 2.1 summarizes the descriptive statistics of the models' variables.

Table 2.1 Descriptive statistics for main variables (at the individual level)

Max.	Min.	S.D	Mean	Variable
16	4	3.50	16.20	Perception of continuity
16	4	3.31	9.67	Innovative adaptive behavior
24	6	4.36	17.27	Prosocial Behavior
24	6	4.32	14.14	Social Networks
32	8	4.71	19.50	Administrative Framework
12	3	2.35	8.65	Management Functioning
12	3	2.44	8.25	Information Flow

2.5 Results

2.5.1 Individual Level Analysis

2.5.1.1 Innovative Adaptive Behavior

An initial correlation matrix clearly demonstrated how these separate items were highly and significantly inter-correlated, reflecting the complexity and number of social behaviors interacting within the organization during crises (See Table 2.2).

Given the suggestion in the disaster literature that adaptive behavior emerges during disasters and in crisis situations, we set about determining which of these organizationally based social behaviors could best predict the adaptive ability of employees. An initial linear regression model revealed (See Table 2.3) that social network densities significantly explained the emergence of innovative adaptive behaviors, reinforcing H1. This hypothesize stressed that innovative adaptive behavior will have social roots rather than administrative ones, and as the results show, dense social ties stand as the basis of innovative adaptive behavior In the second iteration we inserted into the model organizational oriented variables, finding that social networks remain robust predictors of innovative adaptive behavior. Furthermore, for production organizations, adaptation was significantly higher than for service organizations.

Linear regression predicting innovative adaptive behavior by social and organizational variables (at the individual level).

2.5.2 Organizational Level Analysis

Our next concern was to assess the degree that such behaviors affected the overall performance of the organization experiencing the month long bombardment. To do this we first tracked over time how, and if, the crises had in fact affected the performance of the sampled organizations. To do so we first traced factual based performance records based on sales, cash flow, production and budget on a weekly basis. We opted for a mix of actual outputs but transcribed them all into

Table 2.2 Correlation matrix of main variables (at the individual level)

	N	1	2	3	4	5	6	7
1. Innovative adaptive behavior	392	1						
2. Prosocial behavior	381	0.33^{**}	1					
3. Social networks	383	0.35^{**}	0.52^{**}	1				
4. Administrative framework	382	0.21^{**}	0.31^{**}	0.32^{**}	1			
5. Management functioning	393	0.14^{**}	0.30^{**}	0.20^{**}	0.30^{**}	1		
6. Information flow	394	0.23^{**}	0.39^{**}	0.33^{**}	0.46^{**}	0.50^{**}	1	
7. Organizational type – service	421	−0.13	0.10	0.06	-0.15^{**}	−0.00	-0.11^{*}	1
8. Work level – managers	402	−0.00	−0.02	−0.00	−0.07	-0.10^{*}	−0.08	0.02

$^{**}p < 0.01$
$^{*}p < 0.05$

Table 2.3 Linear regression predicting innovative adaptive behavior by social and organizational variables (at the individual level)

Variable	Model 1	Model 2
Social oriented variables:		
Social networks	0.245^{**}	0.267^{**}
Prosocial behavior	0.077	0.093
Information flow	0.066	0.00
Administrative framework	0.064	0.066
Management functioning	0.032	0.08

$^{**}p < 0.01$

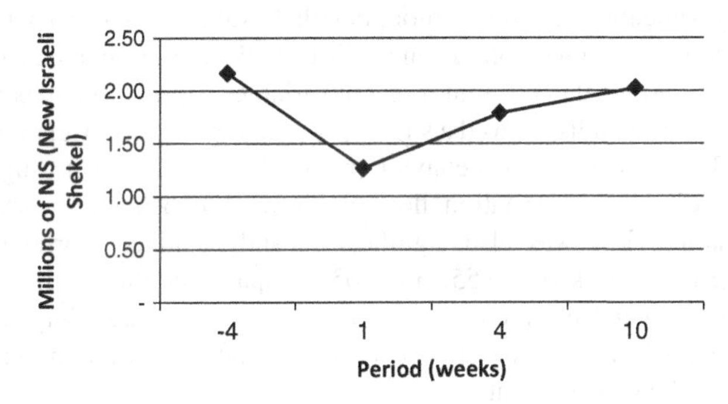

Fig. 2.2 Average sales rate of the sampled private (Organizations before, during and after missile attacks)

a single common denominator, namely revenues. As Fig. 2.2 shows, the initial start of the war led to a sharp dip in actual performance measured by objective output data during the first week followed by a slow but steady increase towards its end; but never-the-less reduced capacity performance after the war ended. These changes were found to be significant in a within-subjects repeated measures analysis $((F_{3,60}) = 101.067, p < .0001)$.

Table 2.4 Correlations among different variables (at the organizational level of analysis)

	N	1	2	3	4	5	6	7
1. Continuity of operation-performance	17	1						
2. Innovative adaptive behavior	17	0.56*	1					
3. Prosocial behavior	17	0.26	0.35	1				
4. Social networks	17	−0.33	0.08	0.45	1			
5. Administrative framework	17	−0.32	0.07	0.36	0.30	1		
6. Management functioning	17	0.27	−0.10	0.25	−0.52	0.10	1	
7. Information flow	17	−0.14	0.00	0.63**	0.33	0.55*	0.12	1
8. Perception of continuity	17	0.12	0.56*	−0.03	.10	−0.46	−0.33	0.00

*Correlated at 0.05 level
**Correlated at 0.01 level

The initial dampening of production and its steady increase suggested that some form of organizational change occurred in performance levels as the frequency of the bombardment did not abate during the entire war period. During this same period over a third of the Northern residents being bombarded left the area [58] depleting manpower resources and potentially affecting the maintenance of production and services. All this strongly suggests that after the initial "shock" of the disruption during the first week of the crisis, there began a process which led to changes within the organizational structure that affected continuity of operations. We thus contend that this resurgence in performance was due primarily to the emergence of adaptive organizational behaviors, a contention to be evaluated below.

To test this contention required a more detailed examination of social processes occurring within the organizations during a disaster. We therefore first examined the intercorrelations among the different research variables and also actual performance (See Table 2.4). The results showed a strong positive significant correlation between information flow and prosocial behavior ($r = 0.61$, $p < 0.01$). This suggests that in a concrete disaster, information flow and prosocial behavior work hand-in-hand. Information flow was also significantly and positively correlated to the administrative framework ($r = 0.55$, $p < 0.05$). Apparently, during the emergency information was shared between managers and employees regarding what is the expected behavior from the workers and what conditions (welfare, protection etc.) are provided by the management.

Furthermore, examining the two measures of continuity of operation, actual performance and perception of continuity intercorrelations with the research variables, has revealed interesting results. First, there was no significant correlation between actual performance and perception of continuity. This suggests, that at organizations where employees perceived the routine as continuing during the disaster, there is no evidence supporting actual continuation ($r = 0.12$, N.S).

In addition, perception of continuity was significantly correlated to innovative adaptive behavior, suggesting that higher levels of innovative adaptive behavior contributed to perception of continuation of routine. The results also show that

Table 2.5 Variables predicting continuity of operation (at the organizational level of analysis)

Variable	Model 1	Model 2
(Constant)	139.577	79.925
Administrative framework	−0.246	−0.344
Information flow	0.060	0.219
Innovative adaptive behavior		0.579*
Adjusted R^2	0.01	0.23

*p < 0.05

none of the variables, except for innovative adaptive behavior was significantly correlated to the level of actual performance. Here too, we see that organizations with high levels of innovative adaptive behavior were also more productive during the acute crisis period ($r = 0.56$, $p < 0.05$). These findings partially support H2 as we expected to find significant correlations between prosocial behavior and social networks reflecting ongoing social process within organizations during the crisis. This lukewarm support for H2 led us to examine in more detail social processes occurring within the organizations during a disaster.

Investigating this further, we employed regression models. Due to the relatively small number of participating organizations, we were constrained in the number of independent variables inserted into the regression (Table 2.5). Our intention here was to examine if "managerial" or "behavioral" variables would positively affect actual performance. We therefore ran a two-step linear regression. In the first model we regressed two "managerial" variables: administrative framework and information flow. In the second model we added innovative adaptive behavior. We used only innovative adaptive behavior, as we demonstrated earlier in the paper that this unique behavior has social roots, and therefore, it reflects the extent of the organizations' ongoing social process, although it is measured as an individual behavior. The results in this first model showed no significant effect on performance. But when we added to the model, in a second step, innovative adaptive behavior, we found it to be significant. In this second model managerial decision making made within an administrative framework and information flow were insignificant in predicting actual performance as in the first model. Rather, innovative adaptive behavior, performed by employees and managers, was found to improve performance. Therefore, H3, arguing that organizational performance will be effected by innovative adaptive behavior was supported.

2.6 Summary and Conclusions

The ability of managers and employees to maintain organizational continuity during a major crisis is a litmus test of an organizations ability to survive and develop. How and what is involved in maintaining continuity of operations is, however, still stymied by the lack of interdisciplinary consensus of what constitutes the core social processes involved. We have argued that examining disaster behaviors found in different social settings such as the family and community can be utilized in an organizational context and should provide a theoretical framework from which to better understand what constitutes organizational continuity and the role that managers and employees play in this process. To do so we developed a working model which views organizational continuity as consisting of a complex set of organizational social processes and focusing on disaster/emergency issues that would have a direct impact on organizational performance. In our view, the "bottom line" for maintaining continuity of operations should be measured in terms of a survival quotient based on actual financial performance. To this end we devised a research strategy that examined both employees and managers' behavior in Israeli public and private sector organizations that had been subjected to a massive missile attack over a month that severely disrupted daily life.

As a first step, we focused on the performance levels of the sampled organizations over time. The results of objective performance data showed that the initial onset of the disaster disrupted performance, leading to a major decline in the first week of the missile attacks, followed by a continuous steady increase in performance during the remaining 3 weeks of the war. This recovery period suggests that, despite the continuing bombardment and depletion of manpower, something occurred that acted to fortify the organizations operational ability.

Employing the theoretical working model as a benchmark, we first examined the extent that disaster related behaviors occurred for managers and employees within their respective organizational framework. Not unsurprisingly, we noted a complex number of socially based organizational behaviors that were directly applicable to the crisis situation. These included innovative adaptive behaviors that acted to fill in gaps in disrupted service and production systems and significantly affected continuity of operation in terms of actual performance. In addition, social behaviors such as pro-social helping and support activities among workmates, an intensification of social network interactions among employees as well as greater sensitivity to the flow of information form the basis of innovative adaptive behaviors, which, in turn lead to improved organizational continuity during and after the emergency.

The implications of these results also provided substantive clues as to the mechanisms through which organizational change occurs. By viewing organizational continuity as part of a larger theoretical framework focusing on organizational change, crises, in its broadest sense, opens a window of opportunity to gain a better sense of how change takes place. From the extreme case we examined here of missile attacks to the less discerned processes initiated by "management" to make their organization more competitive in the market place, all exhibit organizational behaviors that in one form or another relate to differing levels of crises. All relate to minimally maintaining continuity of operations, be it is due to potential physical destruction or being under priced by competitors. In these cases, crisis forms a common base line that would likely affect change as the bottom line objective is to heighten continuity. In our case, internal social processes enhancing innovative adaptive behaviors and crisis focused managerial behavior made a significant contribution to explaining bottom line performance Therefore, while the level of the "crises" may differ, the organizations' objective is to set in action both structural and group changes that imprint on the organizations ability to maintain or increase its performance and through this its continuity. Examining organizations that face major disruption to their continuity therefore provide a unique access to these processes.

Acknowledgments This research was made possible by a research grant from the Technion. We gratefully appreciate the assistance of the late Dr. Uri Ben-Nesher and Dr. Sharon Link for their initiative and organizing the field study. We would also like to sincerely thank the managers and employees of organizations who under difficult circumstances agreed to participate in this study.

References

1. Kendra JM, Wachtendorf T (2006) Community innovation. In: Rodríguez H, Quarantelli EL, Dynes R (eds) Handbook of disaster research. Springer, New York, pp 316–334
2. Kirschenbaum A (2006) Terror, adaptation and preparedness: a trilogy for survival. J Homel Security Emerg Manage 3(1):1–18
3. Granot H (1996) Disaster subculture. Disaster Prev Manage 5:36–40
4. Rodriguez H, Trainor J, Quarantelli EL (2006) Rising to the challenges of a catastrophe: the emergent and prosocial behavior following hurricane Kathrina. Ann Am Acad Pol Soc Sci 604:82–101
5. Stallings RA (2002) Weberian political sociology and sociological disaster studies. Sociol Forum 17:281–305
6. Mileti D (1999) Disasters by design: a reassessment of natural hazards in the United States. Joseph Henry Press, Washington, DC
7. Tierney K (1997) Business impacts of the Northridge earthquake. J Conting Crisis Manage 5(2):87–97
8. Peacock WG, Dash N, Zhang Y (2006) Sheltering and housing recovery following disaster. In: Rodriguez H, Quarantelli EL, Dynes R (eds) Handbook of disaster research. Springer, New York, pp 258–274
9. Preble JF (1997) Integrating the crisis management perspective into the strategic management process. J Manage Stud 34(5):669–791
10. British Standards Institution (2007), in P Woodman, Business Continuity Management; London: Chartered Management Institute, London, UK
11. Federal Emergency Management Agency (2008) Federal continuity directive (FC1) federal executive branch national continuity program and requirements. http://www.fema.gov/pdf/about/offices/fcd1.pdf
12. National Fire Protection Association (2007) Standard on disaster/emergency management and business continuity programs (1600). http://www.nfpa.org/assets/files/pdf/nfpa1600.pdf
13. Kirschenbaum A (2004) Chaos organization and disaster management. Marcel Dekker, Inc., New York
14. Burns T, Stalker GM (1961) The management of innovation. Tavistock, London
15. Harvey E (1968) Technology and the structure of organizations. Am Sociol Rev 33:247–259
16. Hanan MT (1988) Organizational population dynamics and social change. Eur Sociol Rev 4:95–109
17. Kreps G, Drabek TE (1996) Disasters are nonroutine social problems. Int J Mass Emerg Disasters 14:129–153
18. Quarantelli EL (1988) Conceptualizing disaster from a sociological perspective. Int J Mass Emerg Disasters 7:243–251

19. Seeger MW, Sellnow TL, Ulmer RR (2001) Public relation and crisis communication: organizing and chaos. In: Robert L (ed) Health public relations handbook. Sage, Thousand Oaks, pp 155–166
20. Sellnow TL, Seeger MW, Ulmer RR (2002) Chaos theory, informational needs, and natural disasters. J Appl Commun Res 30(4):269–292
21. Torrieri F, Concilio G, Nijkamp P (2002) Decision support tools for urban contingency policy. A scenario approach to risk management of the Vesuvio area in Naples, Italy. J Conting Crisis Manage 10(2):95–112
22. Harrald JR (2006) Agility and discipline: critical success factors for disaster response. Ann Am Acad Pol Soc Sci 604:256–272
23. Mendonca D, Wallace WA (2004) Studying organizationally-situated improvisation in response to extreme events. Int J Mass Emerg Disasters 22:5–29
24. Gittell JH, Cameron K, Lim S, Rivas V (2006) Relationships, layoffs and organizational resilience. J Appl Behav Sci 42:300–329
25. Sutcliffe KM, Vogus TJ (2003) Organizing for resilience. In: Cameron KS, Dutton JE, Quinn RE (eds) Positive organizational scholarship: foundations of a new discipline. Berrett-Koehler, San Francisco, pp 94–110
26. Gittell JH, Cameron K, Lim S, Rivas V (2006) Relationships, layoffs and organizational resilience. J Appl Behav Sci 42:300–329
27. Fowler KL, Kling ND, Larson MD (2007) Organizational preparedness for coping with major crisis or disaster. Bus Soc 46:88–102
28. Omer H, Alon N (1994) The Continuity principle: a unified approach to disaster and trauma. Am J Community Psychol 22:273–287
29. Shaw GL, Harald JR (2004) Identification of the core competencies required of executive level business crisis and continuity managers. J Homel Security Emerg Manage 1:1–14
30. Government Accountability Office (2006) Transportation security: efforts to strengthen aviation and surface transportation security continue to progress, but more work remains GAO-08-651T. United States Government Accounting Office. http://www.gao.gov/new.items/d08185.pdf
31. General Services Administration (2003) Emergency preparedness and continuity of operations (COOP) Planning in the federal judiciary. Congressional Research Services. http://www.au.af.mil/au/awc/awcgate/crs/rl31978.pdf
32. Federal Emergency Management Agency (2006) Open for business. Institute for business and home safety. www.ibhs.org/docs/openforbusiness.pdf
33. Auf Der Heide E (1989) Disaster response: principles of preparation and coordination. Mosby, St. Louis
34. Dynes RR (1994) Community emergency planning: false assumptions and inappropriate analogies. Int J Mass Emerg Disasters 12:141–158
35. Kirschenbaum A (2002) Disaster preparedness: a conceptual and empirical reevaluation. Int J Mass Emerg Disasters 20:5–28
36. Kirschenbaum A (2005) Preparing for the inevitable: environmental risk perceptions and disaster preparedness. Int J Mass Emerg Disasters 23:97–127
37. Wenger D, James TF (1994) The convergence of volunteers in a consensus crisis: the case of the 1985 Mexico City earthquake. In: Dynes RR, Tierney KJ (eds) Disasters, collective behavior, and social organization. University of Delaware Press, Newark, pp 229–243
38. Webb GR, Tierney KJ, Dahlhamer JM (2000) Businesses and disasters: empirical patterns and unanswered questions. Nat Hazards Rev 1(2):83–90
39. Von Lubitz DK, Beakley JE, Patricelli F (2008) Disaster management: the structure, function, and significance of Network-Centric operations. J Homel Security Emerg Manage 5(1):42
40. Gordon P, Richardson HW (1996) Beyond polycentricity; the dispersed metropolis. Los Angeles, 1970–1990. J Am Plann Assoc 62(3):289–295
41. Drabek TE, McEntire DA (2003) Emergent phenomena and the sociology of disaster: lessons, trends and opportunities from the research literature. Disaster Prev Manage 12:97–112

42. Quarantelli EL (1996) Emergent behaviors and groups in the crisis time of disasters. In: Kwan K (ed) Individuality and social control: essays in honor of Tamotsu Shibutani. JAI Press, Greenwich, pp 47–68
43. McKimmie BM, Jimmieson NL, Rebecca M, Moffat K (2009) Social support and fires in the workplace: a preliminary investigation. Work 32:59–68
44. Kendra J, Wachtendorf T (2002) Creativity in emergency management response after the World Trade Center attack. Preliminary paper #324 Disaster Research Center, University of Delaware, Delaware, USA
45. Anderson SE, Williams L (1996) Interpersonal, job and individual factors related to helping processes at work. J Appl Psychol 81:282–296
46. Orlikowski WJ, Yates J (1994) Genre repertoire: the structuring of communicative practices in organizations. Adm Sci Q 39:541–574
47. Wicks D (1998) Nurses and doctors at work: rethinking professional boundaries. Open University Press, Buckingham
48. Gittell JH (2008) Relationship and resilience: care providers responses to pressures from managed care. J Appl Behav Sci 44:25–47
49. Horwich G (1993) The role of the for-profit private sector in disaster mitigation and response. Int J Mass Emerg Disasters 11:189–205
50. Aldrich H, Auster E (1986) Even dwarfs started small: liabilities of age and size and their strategic implications. In: Staw B, Cummings L (eds) Research in organizational behavior. JAI Press, Greenwich, pp 165–198
51. Dahlhamer JM, D'Souza MJ (1997) Determinants of business-disaster preparedness in two U.S metropolitan areas. Int J Mass Emerg Disasters 15:265–281
52. Selvarajan TT, Ramamoorthy N, Flood PC, Guthrie JP, MacCurtain S, Liu W (2007) The role of human capital philosophy in promoting firm innovativeness and performance: test of a causal model. Int J Hum Resour Manage 18:1456–1470
53. Wall TD, Michie J, Patterson M, Wood S, Sheehan M, Clegg CW, West M (2004) On the validity of subjective measures of company performance. Pers Psychol 57(1):95–118
54. Hossain L, Kuti M (2010) Disaster response preparedness coordination through social networks. Disasters 34(3):755–786
55. Wellman B, Berkowitz SD (eds) (1988) Social structures: a network approach. Cambridge University Press, Cambridge
56. Kendra JM, Wachtendorf T (2003) Reconsidering convergence and convergence legitimacy in response to the World Trade Center disaster. In: Clarke L (ed) Terrorism and disaster: new threats, new ideas, vol 11, Research in Social Problems and Public Policy. Elsevier, Oxford, pp 97–122
57. Rutkowski AF, Van de Walle BA, Van Groenendaal WJH, Pol J (2005) When stakeholders perceive threats and risks differently: the use of group support systems to develop a common understanding and a shared response. J Homel Security Emerg Manage 2(1):1–17
58. Inbar E (2007) How Israel bungled the second Lebanon war. Middle East Q 16(3):57–65, http://www.meforum.org/1686/how-israel-bungled-the-second-lebanon-war#_ftnref22

Chapter 3
Risk Perception and Communication

Frederic Bouder

Many risk events whether natural or man-made have potentially devastating effects. For several decades, risk scholars, regulatory agencies and businesses have invested sustained efforts in developing and improving risk communication. Their attention has been focused on a number of risk situations from nuclear reactors [15, 19] to chemical plants [14] and radon [3]. Risk analysis has uncovered general drivers and patterns of communication that can be adapted beyond these distinct sectors. This considerable amount of research [6, 7] has helped to grasp with fundamental questions such as: are people rational or irrational when it comes to catastrophic events? Is there a way to develop sensible communication? And what role does perception play to inform communications? Evidence strongly suggests that effective risk communication is a crucial component of a robust strategy for improving disaster resilience and mitigation. Arguably, the major achievements of risk perception/risk communication have been therefore to:

1. Translate our understanding of the key factors that determine people's attitudes to risk into communications that are meaningful to those who are engaged in this communication process.
2. Identify key variables that influence the outcome of these communications in order to make them more effective.

Yet, in practice decision-makers have often neglected to take on board the lessons of risk communication science. This is especially true when they are confronted to emergency situations. This chapter offers a brief account of the latest development in risk communication science. This description is followed by the discussion of

F. Bouder (✉)
Department of Technology and Society Studies, Maastricht University,
Grote Gracht 90-92, Maastricht SZ 6211, The Netherlands
e-mail: f.bouder@maastrichtuniversity.nl

H.-N. Teodorescu et al. (eds.), *Improving Disaster Resilience and Mitigation - IT Means and Tools*, NATO Science for Peace and Security Series C: Environmental Security,
DOI 10.1007/978-94-017-9136-6_3,

a specific case where successful risk communication was put in place, i.e. the management of recovery process after the Buncefield explosion that took place in the United Kingdom in 2005.

3.1 What Is Risk Perception/Communication Science?

In the past 30 years risk communication has emerged as a distinct area of research. It is not the purpose of this paper to delve too much into the genesis of this discipline, as many excellent overviews are available on the subject (for instance: [8, 12, 13, 31, 37, 43, 44, 47–50]). Worth noting, however, Risk communication has its roots in risk perception studies which make risk perception and risk communication intrinsically linked.

3.1.1 Risk Perception Studies

Risk perception studies started in the 1940s with the work of the prominent geographer Gilbert White and his students at the University of Chicago [10, 11, 70, 71]. White was particularly interested in the relationship between natural hazards and human interventions. He famously argued that "floods are an act of good but flood losses are largely an act of man" [70] and in the 1960s conducted research for the American congress to improve flood protection. Around this time it also became clear that public concerns did not necessarily match the statistical estimates calculated by experts. Some risks seemed to be subject to less public scrutiny than others. For example most people tend to consider nuclear waste very risky [20, 26], while experts, on the other hand, consider that when properly managed the risk is low, almost negligible [67].

A traditional view is to blame the lack of data or the poor understanding from the public. Yet, since the late 1960s, major advancements in risk perception research have shown that this would be a highly simplistic interpretation. In particular studies initiated by Baruch Fischhoff, Paul Slovic and their colleagues shed a new light on the matter. In the late 1970s and early 1980s, matrixes comparing risk perceptions were designed, looking at the perceived levels of "unknown" and "dread" shaping attitudes towards risks [62–66, 68]. Many different risks were included in the research from car driving to nuclear plant explosions [59]. These so-called psychometric studies showed that "Unknown" and "Dread" factors varied considerably in the comparative scale. For example perceptions of chemical risks were much higher for DDT, PCBs or Pesticides than hair dyes, lead paint or auto lead. These studies went further. They uncovered a series of drivers of perception, such as whether the hazard is perceived naturalness, degree of control, catastrophic potential, and level of familiarity. Independently of the cultural context, individuals appear to fear events perceived as "natural" – e.g. volcano eruptions- much less

Table 3.1 Perception and level of concern (non-exhaustive list)

Leaning towards less concern	Leaning towards more concern
Voluntary: I use my mobile phone	Involuntary: they build a phone mast nearby
Control: I drive my car	No control: my spouse drives the car
Natural: a volcano erupts	Technological: a chemical plant blows up
High probability low consequences: I live near a river	Low probability high consequences: I live near a Nuclear plant
Familiar: I eat organic food	Non familiar: I eat GMO food
Adults receiving the yellow fever vaccine	Children receiving the MMR Vaccine
Male: I need smallpox vaccination	Female: I need smallpox vaccination
White: drugs against high blood pressure	African American: drugs against high blood pressure

Source: HSE [5]; adapted from various talks and presentations given by R.E. Lofstedt, as well as Slovic ([59], 1992)

than those perceived as "man made", such as a nuclear plant explosions. Similarly, people worry much more about rare events of potential high impact, i.e. a total breaking down of sea defences, compared to more frequent events of lesser impact, such as river flooding (see Table 3.1).

3.1.2 Risk Acceptance as Complex Phenomenon

Lessons from risk perception should not overlook the fact that risk acceptance is a complex phenomenon. For instance, when it comes to nuclear power, people are expecting the worst case scenario to happen [62]. The history of nuclear energy offers an interesting example of how acceptance may evolve over time. The human "atomic experience" started during Word War II. The A bombs of Hiroshima and Nagasaki, 60 years ago, were a shock to the world. Humankind became capable of developing mass destruction weapons, which could compromise its very existence. In contrast, only a few years later, the prospects of benefits from civil nuclear energy strongly emerged. This opened the door for the commercialisation of nuclear power. What was the impact on the public of these two, apparently contradictory, developments? Rosa and Freudenburg [56] looked at the US context, which is very interesting because of its pioneering nature. They highlighted that initial military developments were handled in secrecy. Before any commercial civil nuclear programmes were launched, nuclear issues were in fact barely discussed by the public.

The move towards the commercialisation of civilian nuclear energy modified this environment. As the number of debates grew, the public initially appeared supportive, and offered some degree of enthusiasm (Rosa & Freudenburg in Dunlap, Kraft and Rosa [56], pp. 32–63). Numerous surveys have assessed a wide degree of public attitudes towards nuclear energy [22, 42, 45, 46]. They all point into the same direction: gradually, the number of people expressing concerns about nuclear energy increased.

There is evidence that acceptance may evolve over time and that the same events may lead to different attitudes between different countries. The row that opposed Denmark to Sweden about the Swedish Nuclear plant of Barsebäck, built just opposite Danish shores, provides a good example of the complex nature of risk tolerance. As analysed by Löfstedt [5, 36, 38] the controversy centred around the Swedish nuclear (boiling reactor) plant located 20 km from both Malmö (Sweden) and Copenhagen, near the small harbour town of Barsebäck. The plant, built in the 1970s, is close to areas that are among the most densely populated in Scandinavia. On 28 July 1992, during the start-up of the second reactor at the Barsebäck plant, a safety valve became stuck, causing insulation material to fall into the reactor's water-cooling system, blocking the reactor's inlet filters. Within 20 min the incident was managed. Nevertheless, this event prompted the Swedish Nuclear Inspectorate (SKI) to force a shutdown of all similar filters ([38], p. 91). What was the reaction of the public? Barsebäck had an history of strong opposition from the Swedish Centre Party ([38], p. 95), which culminated in the mid-1970s and resulted in polarisation of "pro" and "anti" voices in the Swedish nuclear debate. However, opposition voices died out declined in the 1980s, after Sweden decided by referendum to continue producing nuclear energy. Since then, "Swedish opposition to nuclear power remained a marginal rather than a mainstream political issue" ([38], p. 96). By contrast, Danish opposition to the plant has grown over time. In the 1980s, the planning of a second reactor proved to be a sensitive issue between Sweden and Denmark ([38], pp. 96–97). Therefore, it is no surprise that Danish and Swedish opinions reacted differently to the 1992 incident. Danish opposition to the nearby Swedish facility was very fierce, 83 % of Copenhagen's population was against nuclear power and 82 % of the Danes wanted their country to pressure Sweden to close the nearby Barsebäck plant permanently [36]. According to polls, residents of Malmö, the nearby Swedish city, were by contrast overwhelmingly pronuclear (81 %). This was reinforced by the ability of SKI to ensure good management practices.

Therefore, it should be emphasised that the acceptance or the refusal of a risk is never straightforward or black and white. It would be too simplistic to portray the Barsebäck incident as a Denmark v/Sweden event. Individuals understand worries, and to some extent share them, but this does not necessarily mean that they will change their mind. For example, despite high levels of support, 72 % of Swedish respondents said they understood and appreciated Danish concerns. The Danish view remained quite static, however the Swedish opinion over the Barsebäck plant evolved over time [38]. This result suggests that regular checks on public views are also necessary.

3.1.3 Risk Communication as Emerging Discipline

Risk communication studies started in 1980s when academically trained communication consultants began to apply these lessons to concrete cases [16, 32, 48]. It is therefore fair to say that, risk communication is a relatively new offspring of risk perception studies. Risk communication has been best described as:

> (...) an interactive process of exchange of information and opinion among individuals, groups and institutions. It involves multiple messages about the nature of risk and other messages, not strictly about risk, that express concerns, opinions, or reactions to risk messages or to legal and institutional arrangements for risk management ([48], p. 21).

For some, a good communication strategy is about bringing the right data to people, for others it is about improving the "education" and the "transfer of knowledge" from technicians to laymen. In practice, distinct risk communications have tried to achieve unequally ambitious goals, from the modest objective of information-sharing, to more ambitious objectives- namely changing beliefs and inducing behavioural change [18]. What is then good risk communication compared to bad risk communication? According to the [48] definition, risk communicators should design strategies to assist the public in understanding the problem at hand. Fischhoff takes a similar view:

> By definition, better risk communication should help its recipients to make better choices. It need not make the communicators' lives easier... What it should do is to avoid conflicts due to misunderstanding... leading to fewer but better conflicts ([17], p. 31; also [16]).

Baruch Fischhoff [16] and William Leiss [32] have highlighted the evolutionary process that institutional risk communications went through, from ineffective top-down and one-way messaging towards a two-way communication process in which scientists, risk managers and various laypersons, engage in a social learning process. For Leiss three distinct strategies that scientists and government experts have used may be summarised as follows [32]:

(i) in its infancy risk communication has focused on conveying probabilistic thinking to the general public (...);
(ii) experts have then focused on persuasion and public relations to convince people that they should change their behaviour (...);
(iii) in the third and last phase, the focus has been on developing a two-way communication process in which scientists, risk managers, and various laypersons engage in a social learning process.

Looking back at 20 years of risk researches, Fischhoff identified eight development stages in the risk management discourses [16] (see Table 3.2). These stages could be used as a source of inspiration for an incremental risk communication strategy. We would summarise them as follows:

Stage 1: We need to get the number rights. Diligent analysts must do their best, gathering observational data, adapting it to novel cases with expert judgement, and trying to translate information into coherent models.

Stage 2: We need to tell the Public what the numbers are. A simple translation of numbers is not enough, as number do not speak for themselves. This implies an effort from risk analysts to present numbers in a way, which acknowledges

Table 3.2 Developmental stages in risk communication

All we have to do is get the numbers right
All we have to do is tell them the numbers
All we have to do is explain what we mean by the numbers
All we have to do is show them that they've accepted similar risks in the past
All we have to do is show them that it's a good deal for them
All we have to do is treat them nice
All we have to do is make them partners
All the above

Source: Fischhoff [16]

uncertainties, worries and values surrounding risk estimates. Experts have to admit one kind of subjectivity on the recipient side, as well as on their side.

Stage 3: We need to explain what we mean by the numbers. This process will be difficult and experts may face some opposition. In order to ease the process one should focus on numbers that really matter: telling much more than people need can be counter productive. This requires clear thinking about the scope of the message delivery.

Stage 4: We need to show that similar risks have been accepted in the past. Risk comparisons, which confront an unfamiliar risk with a more familiar one, can be useful, especially in the absence of a systematic analysis. However, one should be extremely cautious with risk comparisons, as risk decisions are not about risk alone. Risk comparisons are often misused to combat the alleged demand for zero risk: however, in contrast to a popular belief among scientists, most people do not expect zero risk and are perfectly capable of making their own risk/benefit decisions.

Stage 5: We need to show the public that it is a good deal for them. People need information about risks and benefits of activities that might affect them. Thinking about risks and benefits implies to overcome technical difficulties about these estimates and to provide acceptable trade-offs. The attractiveness of an action is highly dependent on how it is presented.

Stage 6: We need to treat people nicely. Getting the content of a communication right requires a analytical and empirical efforts, usually in the form of professionally trained risk communicators. The right language and attitude should be used. The smoothness of the message is of course no substitute for content.

Stage 7: We need to make them partners. Lay beliefs and experiences should not be undermined. Compared to expert knowledge, they operate in a different sphere but they do matter and impact on decisions. The right partnerships should therefore be explored in order to maintain a healthy "social contract" between those who create risks (as a product of other activities) and those who bear them.

Stage 8: All of the above

Risk communication implies to pay attention to all the preceding stages.

3.2 Devising Effective Risk Communications

Yet, one question that many have asked is: how can an organisation make it work in practice? For over a quarter of a century this question has been the subject of journal articles and books in both Europe and North America, highlighting good and bad experiences [7, 29, 33, 35, 53]. This literature has detailed a number of key messages such as the importance of taking on board key drivers of risk perception, the fact that proactive communication works better than reactive communication, and that persuasion is rarely a very effective approach to changing people's behaviour. Scholars have also captured the key variables that are essential to formulating sensible advice. This has led some organisations– from regulatory agencies to businesses as well as aware individuals to develop communication approaches that are known to work better, such as two-way, proactive and non persuasive communication. In practice, however, many other organisations have preferred to devise their own risk communications approaches with uneven success. This may be particularly damageable because practice has shown that too often failures in risk communication have played a decisive role in causing risk management failures [34]. Recently, a volume on "effective risk communication" was edited by Arvai and Rivers [2]. In their introduction they noted:

> Many colleagues felt that a reinvention of the risk communication wheel is becoming commonplace. Other shared the opinion that there has been alarmingly little introspection within the broader community of researchers and practitioners about what we have learnt from past experiences in risk communication (. . .) ([2], p. 1).

For examples the last of Fischhoff's risk communication developmental stages ("we need to make them partners") implies from institutional risk communicators that they refrain from adopting a patronising attitude vis-a-vis citizens. In practice, however, scholars and practitioners may interpret this imperative rather differently. Two contradictory solutions have been proposed to deal with this issue. One is to move away from the public and focus on stronger assessment and science, with the view that only rational benefit/cost analysis and risk/risk can produce reliable information, with the effect of reducing value driven conflicts [21, 24]. Others, however, have pointed out that strict risk assessment may only alienate the public and create greater distrust [9]. The proponents of the second approach, which call for public involvement, also point out that traditional expert knowledge may miss some of the issues, and that local knowledge could bring an essential dimension to finding locally appropriate solutions, a key aspect of waste management [52]. After all, experts themselves may be biased and are making value judgement in perceiving risks and benefits [23, 25].

3.2.1 Trust-Building

Trusted individuals and organisations will benefit from the advantage of being able to influence outcomes through persuasion [38]. On the other hand, they will not make much of an impact if they are perceived as untrustworthy. Once destroyed, trust is difficult to rebuild. According to P. Slovic much harder to rebuild confidence than to destroy it [60, 61]. The asymmetry between the difficulty of creating trust and the ease to destroy it, is resulting principally from four major factors:

- Negative (trust-destroying event) are more visible than positive (trust-building) events
- When events do come to our attention negative events carry much greater weight than positive events
- Sources of bad news tend to be seen as more credible than sources of good news
- Distrust, once initiated, tends to reinforce and perpetuate distrust

Solutions to re-establish confidence and trust, however, are not simple. One strong message that comes from risk communication science is that communicators will need to develop ways to work constructively, even in situation where we cannot assume that trust is attainable [27]. But what should exactly be done? What does trust-building actually imply? Trust has been analysed against three main criteria, which are fairness, competence and efficiency [38, 54, 55]. Efficiency is essentially about the use of resources. Competence has various facets. A central aspect is the professionalism of highly trained professionals. But it goes beyond this point. One attempt to improve the perceived competence of public authorities in the eye of the public has been, for example, to require authorities to develop explicit long term strategies subject to public scrutiny. Fairness implies three main components [1, 36]:

- Structural fairness, which refers to the make-up of the negotiation process (negotiating parties, the nature and the complexity of the issues etc.)
- Procedural fairness, which refers to the different types of negotiating tools (e.g. "tit-for-tat" solutions, fair chance solutions etc.)
- Outcome fairness, which refers to the allocation of burdens and benefits within the agreement itself, usually according to the three allocation principles, equality, equity and justice.

A fair process has often been interpreted as a transparent one. International bodies are advising countries to develop "independent, neutral, balanced and factual information" about issues related to safety ([51], p. 16). Praising efforts to create a transparent environment, this approach views information as a key function of regulation and, as a consequence instruments for the disclosure of information should be provided by the regulator [39]. However transparency and openness, access to information, is highly ambivalent [38, 39]. In a context of distrust, information will be misused and exploited by various interest groups and may lead to reactive and short sighted decisions. Although critical, trust is fragile and can be easily destroyed by ill prepared transparency [39].

Finally, this chapter ought to mention a crucial aspect that would require a separate discussion of its own: the management of the relation to the media is a key component of an effective communication strategy. Although the exact influence of the media may be contingent depending of the magnitude of the issue and its resonance among the public, the media play a critical role in shaping, amplifying or attenuating risks ([28, 30], pp. 202–229). This is why, depending on how well they are framed, public discussions about risk may contribute to building or destroying public confidence.

3.2.2 *State of the Art Risk Communication*

There is clearly an effort to be made on the part of key players to better communicate their decisions to the citizens and media and, if possible, develop partnership with them. Yet contextual differences suggest that one-size-fits-all recipes will be ineffective. What organisations need to develop effective risk communication is a "context based policy mix" involving a capacity to apprehend and understand public worries, the design of appropriate strategies, the sequencing of decision of action, as well as contingency planning and adequate follow-up communication. In 2010, Löfstedt formulated advice on "state of the art" risk communication. This advice conveyed eight recommendations based on an in-depth analysis of three seminal cases – i.e. a filter incident at the Barsebäck (Sweden) nuclear power station, the re-licensing hydropower dams on the Androscoggin river (Maine, US) and the siting and building of a waste tyre incinerator in Wolverhampton (UK) – (HSE [5], pp. 12–13):

1. Developing frequent dialogues between regulators, industry, media and key politicians. One key to successful risk communication is building relationships based on trust. This trust building can best occur prior to a crisis occurring;
2. Confrontations between the key parties in any dispute will not only destroy public trust in the key actors involved, but will also in more cases than not be socially amplified by the media;
3. Do not involve lawyers in risk communication disputes unless it is necessary. When it comes to disputes, lawyers in their very protective nature will inject more distrust in the process rather than less;
4. Risk communication will always be easier if the parties involved in a dispute are both competent and base their decisions on the best available science;
5. Involving highly trusted individuals as early as possible can help in solving risk communication disputes amicably;
6. Environmental NGOs, in the era of post trust, can significantly shape policy outcomes [38, 40];
7. The opinions of local policy makers are important particularly early on in a siting process. Where policy makers perceive a community benefit they may still push forward for a solution and convince their constituents even if some of the public oppose it;
8. Finally, the issue of taking responsibility for one's actions is often crucial.

3.2.3 *Buncefield Explosion Case*

The eight recommendations on "state of the art" risk communication were then tested on events that followed the explosion of the Buncefield Oil Storage depot. The incident took place on Sunday 11th December 2005 at an industrial installation of national importance based in Hemel Hempstead, Hertfordshire (UK). The site was the fifth largest oil-products storage depot in the United Kingdom, with a capacity of about 60,000,000 imperial gallons (272,765,400 l) of fuel. It supplied about 30 % of the consumption of Heathrow airport. Early in the morning, a number of explosions occurred at the facility:

> [They were] of massive proportions and there was a large fire, which engulfed over 20 large fuel storage tanks over a high proportion of the site. There were 43 people injured in the incident, none seriously. There were no fatalities. Significant damage occurred to both commercial and residential properties in the vicinity and a large area around the site was evacuated on emergency service advice. About 2,000 people were evacuated. Sections of the M1 motorway were closed. The fire burned for several days, destroying most of the site and emitting large clouds of black smoke into the atmosphere, dispersing over southern England and beyond (...). ([41]: vol. 2, 12).

Although no lives were claimed, the level of damage and disruption was impressive. To give a sense of the scale of the disaster it is worth referring to Hertfordshire's Chief Fire Officer's words, who described the explosion as "the largest incident of its kind in peacetime Europe" [57]. Some residents described the event as follows ([5], p. 19):

> We were in bed. It happened in the morning and we could feel the house shaking. We saw a big pillar of yellow flame. We also heard a very distinct noise that I still remember very clearly because it was very upsetting. We turned on the TV and we watched the news. Initially we thought a plane may have crashed. But Sky news picked up the story quickly and we received some information. They were very quick at reporting.

Words such as "lunar landscape" were used to describe the site. The fact that there were no casualties was described as "amazing" or "a miracle". On the other hand, damage was unevenly distributed. Distance was only a moderate predictor. Some buildings close to the oil depot appeared miraculously spared, while others further off were seriously damaged. It was also clear that the scale of the disaster would take months to be fully assessed and that there was a degree of hidden damage; for example some buildings may suffer structural damage and, possibly collapse at a later stage. Under such circumstances, risk communication was bound to be a delicate enterprise.

The response of the authority responsible for industrial safety oversight, the Health and Safety Executive (HSE), was proactive. The Executive used the possibilty that is granted under UK safety law[1] to authorise any person to investigate and make a special report about the circumstances of the incident on top of the criminal investigation. As a consequence, two streams of investigations were conducted

[1] Article 14 (2) of Health and Safety at Work etc. Act 1974.

in parallel, one to investigate the circumstances of the disaster and the other to establish facts and liabilities and, possibly, bring civil and criminal charges. The main advantage of this procedure was to restrict the involvement of lawyers to the necessary part, i.e. the criminal investigation. Unlike the criminal investigation which bound by tight confidentiality rules enforced by lawyers the MIIB was allowed to develop an open and proactive communication towards local citizens. In doing so the HSE met the 3rd recommendation of state of the art risk communication that suggests that too much juridicisation tends to undermined trust in the process.

A so-called *Buncefield Major Incident Investigation Board* (MIIB) was established within a month. It run from January 2006 until the end of 2008. The appointment of a Chair that would be perceived as neutral and independent from the Executive was of critical importance. The Chair of the HSE board, Sir Bill Callaghan, appointed the late Lord Newton of Braintree. Lord Newton had been a well-respected member of the House of Lords since 1997. He had previously been a Member of Parliament (MP) (1974–1997), minister (1984–1992)[2], and Leader of the House of Commons (1992–1997). This appointment clearly met recommendation 4 of state of the art risk communication, because Lord Newton had only been perceived as moderate and balanced. He was a trusted figure. At the initial stage of the investigation, however, HSE's credibility was challenged. Residents were not sure whom to blame: Oil companies who were perceived to have let it happen? HSE, who failed to protect the community? In the early days of the investigation MIIB was also regarded with suspicion. The composition of the Board was problematic because membership included three people from HSE and only two independent academics. Did this mean that the HSE was only paying lip service to establishing the truth? Did the "*HSE investigating itself*" as it was put by one local resident?

The proactive nature of the investigation launched by Lord Newton helped to dissipate fears.

One of the first actions taken by the Board was also to appoint a Community liaison officer to assist the Board. The Board was also minded to reassure local residents. The following process of two-way engagement was designed [5]:

1. Identify all people to work with
2. Attend public meetings and be visible
3. Organise meetings and make major announcements
4. Commit to provide information – within legal restrictions
5. Develop an understanding of who we need to target and make sure we have no surprises about who the audience is
6. Make sure people understand that we are "independent"
7. Be fully briefed

[2]Lord Newton has held various ministerial positions under conservative governments: Minister for Social Security and Disabled People (1984–1986), Minister for Health (1986–1988), Minister at the DTI (1988–1989), and Secretary of State for Social Security (1989–1992).

Conveying the right information to the public was also a central element of the strategy. MIIB produced a comprehensive body of 9 reports published in 10 volumes which were released in a timely fashion [5]:

- *21 February 2006.* First progress report by the Investigation Manager, describing the incident and the emergency response, the environmental, social and economic impact, as well as initial evidence on the causes of the explosion.
- *11 April 2006.* Second progress report by the Investigation Manager, describing the ongoing investigation.
- *9 May 2006.* Third progress report, describing the timeline of events at Buncefield.
- *13 July 2006.* Initial Report of the Board. The report reviewed what had been learned about the incident and set out the four areas of concern.
- *29 March 2007.* Report of the Board on the design and operation of fuel storage sites, the first of the four areas of concern identified in the Board's Initial Report.
- *17 July 2007.* Report of the Board on the on emergency preparedness, response and recovery; the second raft of recommendations to address the four areas of concern set out in the Board's Initial Report.
- *16 August 2007.* Report of the Board on lessons about the mechanism of the violent explosion at Buncefield and recommendations.
- *15 July 2008.* Report of the Board, into land use planning and societal risk (including recommendations).
- *11 December 2008.* Publication of Board's final report for the third anniversary of the Buncefield incident.

After the incident, media attention was very high and any communication vacuum could have triggered hostile coverage and steered a controversy. The release of reports at regular intervals kept the local community engaged. The first progress report was issued about 2 months after the incident, and only weeks after the first meeting of the Board on 24 January 2006. The two subsequent progress reports were issued shortly after (April and May). The Board did not only release timely information in a proactive fashion, but ensure that this information was well-articulated. Local residents highlighted that the initial progress reports used plain language and as such were easy to follow. This addressed an essential question – "how could this happen?" – where residents felt that they needed a straight answer, some of them having lost their homes or businesses. As Lord Newton put it:

> The (...) decision to establish this independent investigation was a significant move, highlighting the severity of the incident and the degree of concern for people living and working close to the Buncefield site, and the importance of the work to the wider industry (MIIB webpage[3]).

After the release of basic information about the explosion, the Board took a more systemic look at issues. In the fourth report, introduced the following areas

[3]Accessed at: http://www.buncefieldinvestigation.gov.uk/index.htm

of concern: (i) design and operation of sites; (ii) emergency preparedness; (iii) land use planning; and (iv) the Competent Authority's policies and procedures. Then, the Board formulated detailed recommendations to prevent future incidents. By the time subsequent reports, which became more technical, were issued the investigation team had already established a closer relationship with some residents, who could act as mediators within the community and turn to the team when clarification was needed.

To sum up, the MIIB showed a sense of responsibility – which is consistent with recommendation eight of state of the art communication, and a desire to positively engage with the public, which is consistent with recommendations two and three. As a result of this proactive management of the post-disaster context, MIIB was described as "very involved", "very approachable" and "willing to leave no stone unturned". The overall cost of the investigation was also perceived as appropriate, which is a known trust-building factor. The board did not rely on expensive public relation exercises. Instead, the investigation team used the Community Liaison medium to interact with the public through personal contacts, low-key venues such as town hall meetings, as well as networking techniques.

In addition the Board also identified and tried to develop frequent dialogues with key stakeholders, which is consistent with the first recommendation of state of the art risk communication. Board members highlighted that they targeted a large number of institutional and non-institutional players, some of them influential before the incident, others not, including:

- HRH Prince Charles
- MP Michael Penning
- Ministers
- Local politicians
- Staff of affected Local councils
- Local services
- PCT
- HSE
- Environment Agency
- MIIB
- Criminal investigation team

Private sector players mentioned include:

- Oil companies
- Other large companies (e.g. Kodak)
- SMEs
- Insurers
- Lawyers

Various stakeholders reacted differently to the call. Several residents who attended public meetings noticed that some of the most critical players, i.e. government officials, the police and the insurance industry did not always turn up to information meetings. However, this attitude damaged the reputation of the players themselves rather than that of the HSE. Some residents, for example, have complained about the passive attitude of the police force. Other key stakeholders were engaged but rather critical. Referring to the first meeting organised with the local business community, one member of the Board highlighted:

> We had a very hard time. People did not understand who we were, what the purpose of the meeting was.

At the time, the site operators – the terminal was owned by a joint venture between TOTAL UK Ltd. (60 %) and Texaco Chevron Ltd (40 %) – were also subject to criminal investigations lead by the HSE. Rather than co-operating with the Board, they decided to stay at bay. They launched their own public relation initiatives, for example distributing pamphlets and insisting they were also suffering from the damage. This approach, however back clashed. It was interpreted by the residents as somewhat disloyal and a sign that the companies could not be trusted. From the point of view of the HSE it had also the advantage of avoiding direct confrontations. Although potentially disastrous for the companies themselves this had another advantage from the point of view of HSE: no engagement meant no confrontation and less media attention. Unintendedly, the operators helped HSE build trust by avoiding "confrontation between the key parties" (second recommendation of state of the art risk communication).

The Board paid particular attention to opinions expressed by local policy makers, which is consistent with recommendation seven of state of the art risk communication. The local council did not conflict with the Executive and acted as a facilitator. At first, however, the local Member of Parliament (MP), Michael Penning expressed outrage about the explosion and skepticism about the work of MIIB. He had frequently called for a full parliamentary enquiry rather than an HSE investigation. Lord Newton managed to establish a working relationship with Michael Penning which strengthened the good reputation of the Board.

In a context characterised by trauma and shock, MIIB's risk communication helped to build a strong reputation as a source of trustworthy information. As trust grew between some local residents and the Board, local initiatives developed. For example engaged citizens worked closely with the liaison officers and the Board. As 'trusted third-parties' they set up an email account and were able to filter and contextualise discussions to inform less involved residents. A walkabout was organised to distribute leaflets with accurate information door-to-door. A taskforce was eventually established, with the effect of bringing more stakeholders on board, for instance the local police ([5], p. 24).

The review of the Buncefield case suggests that most of the eight recommendations for a successful "state of the art" risk communication were met despite a very difficult post-disaster environment. The conclusions presented in a report to the HSE indicated that:

- MIIB has developed frequent dialogue with regulators, industry and media as well as key politicians.
- MIIB has acted as a moderator to avoid confrontation between the key parties

- Although lawyers have been involved the separation between MIIB's investigation and the criminal case has pacified the debate on the root causes of the incident
- MIIB maintained a strong scientific dimension to the debate throughout the investigation process
- MIIB has created partnerships with independent members of the public trusted to in their community
- NGOs have taken little interest in the issue, which means that they have not shaped policy outcomes
- Opinions of local policy makers have been respected, even when they were critical of the investigation in principle
- The existence of a criminal investigation has been a critical factor to demonstrate that the issues of responsibility and liability are taken seriously ([5], p. 38).

3.3 Conclusion and Recommendations

The value of risk communication research remains largely unknown to key decision-makers, calling for more in-depth thinking about how best to communicate on risk communication [2]. The vast amount of work conducted in this area suggests that this deficit of knowledge is due less to a lack of proactivity on the part of risk communication experts than to the difficulty of policy-makers to revisit antiquated practices. The result is that, in too many cases, communications that intuitively "sound" right are still preferred to scientifically tested alternatives. This is how one can explain the fervour in regulatory, NGO and academic circles for ill-tested participatory and transparency techniques as an almost "automatic" way of regaining trust [40]. This development is worrying because regulators tend to loose trust after major safety scandals [38]. Yet, there is hope. There is much to learn from specific case studies, some of which are success stories.

Decision-makers often request policy recommendations Out of context recommendations, however, can be misleading. On the other hand "survivor's guides" when they are not misinterpreted as a cookbook can support a reflexive process leading to more effective outcomes. With this in mind, the author suggests to pay attention to five fundamental principles of risk communication to prevent failures [4]:

1. Assembling the evidence
   Demonstrate you have a credible basis for your position
2. Acknowledgement of public perspectives
   Understand how those affected understand the risk
3. Analysis of options
   Consider a broad range of options and the associated trade-offs
4. Authority in charge
   Define the nature of your involvement with the risk
5. Interacting with your audience
   Identify the audiences and the appropriate methods for communicating with them.

References

1. Albin C (1993) The role of fairness in negotiation. Negot J 9(3):223–240
2. Arvai J, Rivers L III (eds) (2014) Effective risk communication, Earthscan risk in society. Routledge, Oxon/New York
3. Bostrom A, Atman CJ, Fischhoff B, Morgan MG (1993) Evaluating risk communications: completing and correcting mental models of hazardous processes, part II 1994. Risk Anal 14:789–798
4. Bouder F (2009) A practical guide to public risk communication, the five essentials of good practice. Pamphlet for the risk and regulation advisory council. Department for business, innovation and skills, BIS, London
5. Bouder F, Löfstedt R (2010) Health and safety executive – HSE improving health and safety, an analysis of HSE's risk communication in the 21st century. Research report RR785, HSE books, Norwich
6. Bouder F, Löfstedt R (eds) (2014) Risk perception, critical concepts in the social sciences, vol 2. Routledge, Oxon/New York
7. Bouder F, Löfstedt R (eds) (2014) Risk communication, critical concepts in the social sciences, vol 3. Routledge, Oxon/New York
8. Breakwell G (2007) The psychology of risk. Cambridge University Press, Cambridge
9. Breyer S (1993) Breaking the vicious circle: towards effective risk regulation. Harvard University Press, Cambridge, MA
10. Burton I, Kates RW (1964) The perception of natural hazards in resource management. Nat Resour J 3(3):412–441
11. Burton I, Kates RW, White GF (1978, 1992) Environment as hazard. Oxford University Press, New York
12. Chess C, Salomon KL, Hance BJ (1995) Managing risk communication agency reality: research priorities. Risk Anal 15:128–136
13. Covello VT, McCallum DB, Pavlova M (1989) Principles and guidelines for improving risk communication. In: Covello VT, McCallum DB, Pavlova M (eds) Effective risk communication. Plenum, New York, pp 3–16
14. Covello VT, Sandman P, Slovic P (1988) Risk communication, risk statistics, and risk comparison: a manual for plant managers. Chemical Manufacturers Association, Washington, DC
15. Farmer FR (1967) Siting criteria – a new approach. In: Containment and siting nuclear power plants. International Atomic Energy Agency, Vienna
16. Fischhoff B (1995) Risk perception and communication unplugged: twenty years of process. Risk Anal 15(2):137–145
17. Fischhoff B (2012) Risk analysis and human behavior, Earthscan risk in society. Routledge, Oxon/New York
18. Fischhoff B, Brewer N, Downs JS (eds) (2011) Communicating risks and benefits: an evidence-based user's guide. Food and Drug Administration, Washington, DC, http://www.fda.gov/AboutFDA/ReportsManualsForms/Reports/ucm268078.htm
19. Fischhoff B, Slovic P, Lichtenstein S (1983) The 'public' vs the 'experts': perceived vs. actual disagreement about the risk of nuclear power. In: Covello VT, Flamm J, Rodericks J, Tardiff R (eds) Analysis of actual versus perceived risks. Plenum, New York
20. Flynn JH, Mertz CK, Slovic P (1991) The autumn 1991 Nevada state telephone survey. Nevada Nuclear Waste Project Office, Carson City
21. Freman AM (1993) The measurement of environment and resource values: theory and models. Resources for the future, Washington, DC
22. Freudenburg WR, Rosa E (eds) (1984) Public reactions to nuclear power: are there critical masses? Westview/American Association for the Advancement of Science, Boulder
23. Gould LC, Gardner GT, DeLucca DR, Tiemann AR, Doob LW, Stolwijk JAJ (1988) Perceptions of technological risks and benefits. Russell Sage Foundation, New York

24. Graham JD, Wiener JB (1995) Risk versus risk: tradeoffs in protecting health and the environment. Harvard University Press, Cambridge, MA
25. Kahneman D, Slovic O, Tversky A (1982) Judgement under uncertainty: heuristics and biases. Cambridge University Press, New York
26. Kasperson RE (1990) Social realities in high-level radioactive waste management and their policy implications. In: Proceedings, international high-level radioactive waste management conference, vol 1. American Nuclear Society, LaGrange, pp 512–518
27. Kasperson RE, Golding D, Tuler S (1992) Societal distrust as a factor in sitting hazardous facilities and communicating risks. J Soc Issues 48:161–187
28. Kasperson JX, Kasperson RE (2005) The social contours of risk: publics, risk communication and the social amplification of risk, vol 1. Earthscan, London/Sterling
29. Kasperson RE, Palmlund I (1987) Evaluating risk communication. In: Covello VT, McCallum D, Pavlova M (eds) Effective risk communication: the role and responsibility of government and non government organisations. Plenum, New York
30. Kasperson RE, Renn O, Slovic P, Brown HS, Emel J, Goble R, Kasperson JX, Ratick S (1988) The social amplification of risk: a conceptual framework. Risk Anal 8(2):177–187
31. Leiss W (1989) Prospects and problems in risk communication. University of Waterloo Press, Waterloo
32. Leiss W (1996) Three phases in the evolution of risk communication practice. In: Kunreuther H, Slovic P (eds) Challenges in risk assessment and risk management. Special Issue of the Annals of the American Academy of Political and Social Science, vol 545. May 1996, pp 84–95
33. Leiss W (2001) In the chamber of risks: understanding risk controversies. McGill-Queen's University Press, Montreal
34. Leiss W (2014) Learning from failures. In: Arvai J, Rivers L III (eds) Effective risk communication, Earthscan risk in society. Routledge, Oxon/New York, pp 227–291
35. Leiss W, Chociolko C (1994) Risk and responsibility. McGill-Queen's University Press, Montreal
36. Löfstedt R (1996) Fairness across borders: the Barsebäck nuclear power plant. Risk Health Saf Environ 7:135–144, Spring 1996
37. Löfstedt RE (2003) Risk communication: pitfalls and promises. Eur Rev 11(03):417–435
38. Löfstedt RE (2005) Risk management in post-trust societies. Palgrave, Basingstoke
39. Löfstedt R, Bouder F (2014) New transparency policies: risk communication's doom? In: Arvai J, Rivers L III (eds) Effective risk communication, Earthscan risk in society. Routledge, Oxon/New York, pp 73–90
40. Löfstedt R, Bouder F, Wardman J, Chakraborty S (2011) The changing nature of communication and regulation of risk in Europe. J Risk Res, forthcoming in special issue, 14(4):409–429
41. Major Incident Investigation Board (MIIB) (2008) The buncefield incident 11 December 2005. The final report of the major incident investigation board. http://www.buncefieldinvestigation.gov.uk/reports/index.htm#final
42. Mazur A (1981) The dynamics of technical controversy. Communications Press, Washington, DC
43. McComas KA (2006) Defining moments in risk communication: 1996–2005. J Health Commun 11:75–91
44. Morgan GB, Fischhoff B, Bostrom A, Atman C (2001) Risk communication: a mental models approach. Cambridge University Press, Cambridge
45. Nealey SM, Melber BD, Ranking WL (1983) Public opinion and nuclear energy. Lexington Books, Lexington
46. Nealey SM, Hebert JA (1983) Public attitudes towards radioactive waste. In: Walker CA, Gould LC, Woodhouse EJ (eds) Too hot to handle? Social and policy issues in the management of radioactive waste. Yale University Press, New Haven, pp 94–111
47. NRC (1983) Risk assessment in the federal government: managing the process. National Academy Press, Washington, DC
48. NRC (1989) Improving risk communication. National Academy Press, Washington, DC

49. NRC (1996) Understanding risk. National Academy Press, Washington, DC
50. OECD (2002) OECD Guidance document on risk communication for chemical risk management. OECD, Paris. http://www.olis.oecd.org/olis/2002doc.nsf/43bb6130e5e86e5fc12569fa005d004c/cb81407367ba51d5c1256c01003521ed/$FILE/JT00129938.PDF
51. OECD/NEA (2003) The regulator's evolving role and image in radioactive waste management, lessons learnt within the NEA forum on stakeholder confidence. OECD, Paris
52. Pidgeon N (1996) Technocracy, democracy, secrecy and error, accident and design: contemporary debates in risk management. In: Hood C, Jones DKC (eds), University College London Press, London, pp 161–171
53. Powell D, Leiss W (1997) Mad cows and mothers milk. McGill-Queen's University Press, Montreal
54. Renn O (2005) White paper on risk governance, towards an integrative approach. International Risk Governance Council, Genève
55. Renn O, Webler RT, Wiedemann P (eds) (1995) Fairness and competence in citizen participation. Evaluating new models for environmental discourse. Kluwer, Dordrecht/Boston
56. Rosa E, Freudenburg W (1993) The historical development of public reactions to nuclear power: implications for nuclear waste policy. In: Dunlap R, Kraft E, Rosa E (eds) Public reactions to nuclear waste, citizens' views of repository siting. Duke University Press, Durham/London, pp 32–63
57. Sky News (2005) Fire rages after blasts at oil depot. http://news.sky.com/skynews/Home/Sky-News-Archive/Article/20080641205711. Accessed 11 Dec 2005
58. Slovic P (1992) Perception of risk: reflections on the psychometric paradigm. In: Krimsky S, Golding D (eds) Social theories of risk. Praeger, New York, pp 117–152
59. Slovic P (1987) Perception of risk. Science 236:280–285
60. Slovic P (1993) Perceived risk, trust, and democracy. Risk Anal 13(6):675–682
61. Slovic P (1997) Trust, emotion, sex, politics, and science: surveying the risk-assessment battlefield. In: Bazerman M, Messick D, Tenbrunsel A, Wade-Benzoni K (eds) Environment, ethics and behavior. The New Lexington Press, San Francisco
62. Slovic P, Fischhoff B, Lichtenstein S (1979) Rating the risks. Environment 21(4):14–20, 36–39
63. Slovic P, Fischhoff B, Lichtenstein S (1980) Facts and fears: understanding perceived risk. In: Schwing R, Albers WA Jr (eds) Societal risk assessment: how safe is safe enough? Plenum Press, New York, pp 181–214
64. Slovic P, Fischhoff B, Lichtenstein S (1981) Perceived risk: psychological factors and social implications. In: Warner F, Slater DH (eds) The assessment and perception of risk. The Royal Society, London
65. Slovic P, Fischhoff B, Lichtenstein S (1982) Rating the risks: the structure of expert and lay perceptions. In: Hohenemser C, Kasperson JX (eds) Risk in the technological society, AAAS symposium series. Westview, Boulder
66. Slovic P, Fischhoff B, Lichtenstein S (1985) Characterizing perceived risk. In: Kates RW, Hohenemser C, Kasperson J (eds) Perilous progress: managing the hazards of technology. Westview, Boulder, pp 91–125
67. Slovic P (1992) Perception of risk: Reflections on the psychometric paradigm. In: Krimsky S, Golding D (eds.) Social theories of risk. Praeger, New York, pp 117–152
68. Slovic P, Layman M, Flynn J (1993) Perceived risk, trust, and nuclear waste: lessons from Yucca mountain. In: Dunlap RE, Kraft ME, Rosa EA (eds) Public reactions to nuclear waste: citizens views of repository siting. Duke, Durham
69. Slovic P, Lichtenstein S, Fischhoff B (1984) Modelling the societal impact of fatal accidents. Manage Sci 30:464–474
70. White GF (1945) Human adjustment to floods, Department of Geography research paper no. 29. The University of Chicago, Chicago
71. White GF (1961) The choice of use in resource management. Nat Resour J 1:23–40

Chapter 4
Establishing Social Resilience with (Public-Private Partnership)-Based BCM (Business Continuity Management)

Kenji Watanabe

Abstract The importance of interoperability among Business Continuity Management, BCMs, and economical incentives as drivers for a resilient society is analyzed.

Keywords BCM (Business Continuity Management) • PPP (public–private partnership) • Interdependencies • Interoperability • Resilience • SPOF (Single Point of Failure) • Financial incentive

4.1 Introduction

In addition to the long history of natural disasters such as earthquakes, floods, or typhoons, Japan has experienced several large-scale earthquakes in the recent few years and several wide-area disruptions in the critical social functionalities caused through interdependencies within supply chains [1]. In the modern networked society, it is very rare that products and services are provided by a single organization, regardless of organizational form—public, private, or NPO/NGO. Most of the processes to develop and deliver the products and services are divided vertically and horizontally throughout supply chains. This is very efficient and effective social structure in normal condition but at the same time it has wide range of interdependencies within the social system, which may cause wide-area chain disruptions of products and services in disaster situations. Considering the situation of the highly interdependent society, it is almost impossible for an organization to have substantial resilience [2] in operation just with BCM (Business Continuity

K. Watanabe (✉)
Graduate School of Social Engineering, Nagoya Institute of Technology,
Gokiso-cho, Showa-ku, Nagoya 466-8555, Japan
e-mail: kewatanabe@nifty.com

H.-N. Teodorescu et al. (eds.), *Improving Disaster Resilience and Mitigation - IT Means and Tools*, NATO Science for Peace and Security Series C: Environmental Security,
DOI 10.1007/978-94-017-9136-6_4,

Management) for its own limited scope. Thus, it is getting important to establish a community-based BCM with participations from all related organizations.

4.2 Quick Review of the Great East Japan Earthquake with Regional BCM Point of View

4.2.1 The Chain-Failures Through Dependencies

The Great East Japan Earthquake was a magnitude 9.0 undersea earthquake off the coast of Japan that occurred on 11 March 2011. It was the most powerful known earthquake ever to have hit Japan, and the fifth most powerful earthquake in the world since modern record-keeping began in 1900. The earthquake triggered powerful tsunami waves that caused over 15,000 deaths, 129,000 buildings totally collapsed, and nuclear accidents, primarily the level 7 meltdowns at three reactors in the Fukushima. Those chain disasters had resulted in wide-area and long-term spread operational disruptions through supply chains not only in the distressed areas, but also in all over Japan and overseas (Fig. 4.1).

4.2.2 Spread Damages Through Major Supply Chains and Realized Concentration Risks

The damages from the earthquake, tsunami and chain-disasters were too big and widespread to be managed with the PPP (Public-Private Partnership) frameworks, which had not worked well because of missing interoperability and joint-preparedness within private and public sectors [3].

Many supply chains that were specialized horizontally and vertically experienced large impact disruptions caused by the SPOFs (Single Point of Failures) in automotive, chemical, electric parts, paper, and energy industries. Typical SPOFs were SMEs (Small and Medium Enterprises), which have very special technologies or techniques and high market share in the specialized fields. Those SMEs were positioned in the lower-tier suppliers (i.e. under 3-tier supplier) and not in the day-to-day supply chain management activities of final user companies of the SPOFs' products and services. They can be called "hidden risks" of the major supply chains.

The municipal governments and the central government agencies tried o rescue those SPOFs in their jurisdictions but could not even recognize the damage situations of SPOFs because of the lack of contacts and communications i.e., not prepared. As a result, many SMEs with relatively high market share with very special technologies or techniques lost or reduced their business because of alternative production by competitors and product design changes by their user companies (Fig. 4.2).

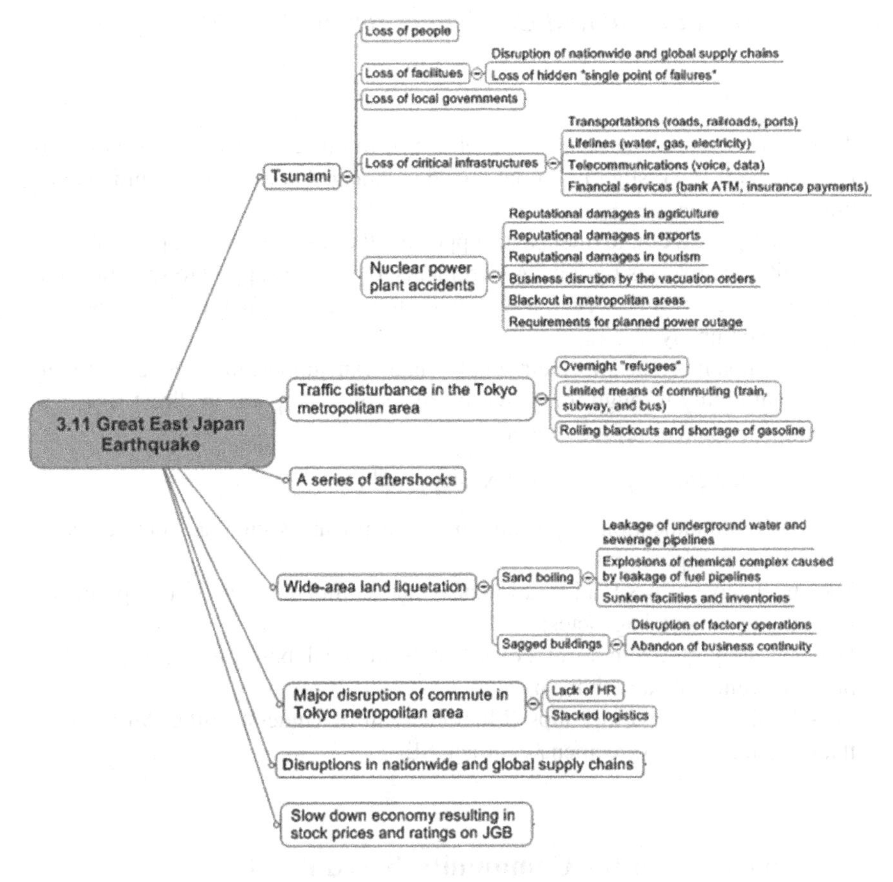

Fig. 4.1 The wider repercussions into the intangible social functionalities and values

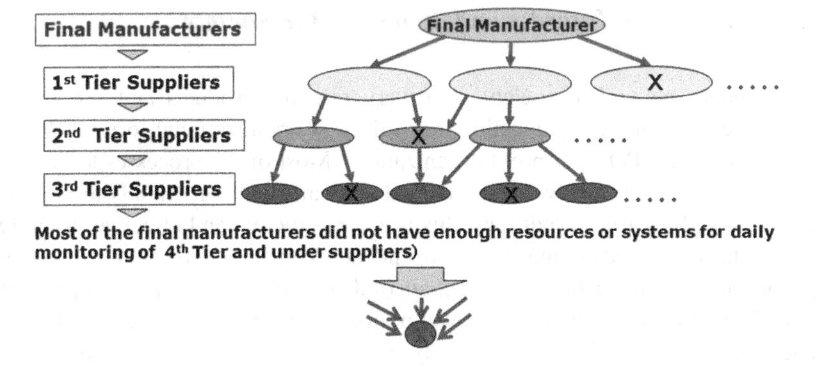

Fig. 4.2 Realized concentration risks

4.2.3 Lessons Learnt and Challenges for the Next Steps

The lessons learnt include:

- Although public sector support was observed in this case, its effectiveness was limited and it came slightly too late for the damaged companies to mitigate the impact of disaster;
- Maintaining local employment and supporting the economy are major objectives and legitimate reasons for local governments to provide support to specific companies for recovery. However, it is difficult to have accountability in supporting a specific company;
- Major companies that own their supply chain can support the recovery of their local community through the provision of recovery support to direct suppliers and the provision of refuge shelters.

Therefore, the challenges for the next steps include:

- Lack of communication, especially between private sector and public sector about business operations;
- Too much layers in situation awareness and command and control processes within governmental agencies;
- Delayed supports from local governments in local business recoveries (less priority against life and living);
- Less feasibility of partnerships (Memorandum of Understanding, MoU rather than Service Level Agreement/ contract, SLA).

4.3 Importance of the Community-Based BCM and PPP (Public/Private Partnerships)

4.3.1 Increasing Interdependencies of Our Society

In the modern networked society, it is very rare for products and services to be provided by a single organization, regardless of organizational form—whether public, private or NPO (non-profit organization). Most of the processes to develop and deliver products and services are divided vertically and horizontally throughout supply chains. In normal conditions, this is a very efficient and effective structure; however, due to the wide range of interdependencies that the system entails, disaster situations may result in a far-reaching chain of disruption to the delivery of products and services. As a result, while the power of disasters have not changed much, their impacts on our society have dramatically increased because of interdependencies with functional diversifications and concentration of specific functionalities of our society (Fig. 4.3).

In the highly interdependent society, the organization that fails to adopt a holistic approach to business continuity planning will find it almost impossible to have

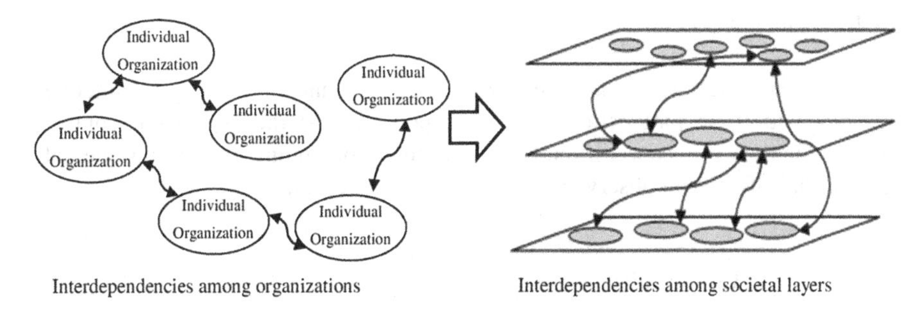

Fig. 4.3 Interdependencies among organizations and social layers

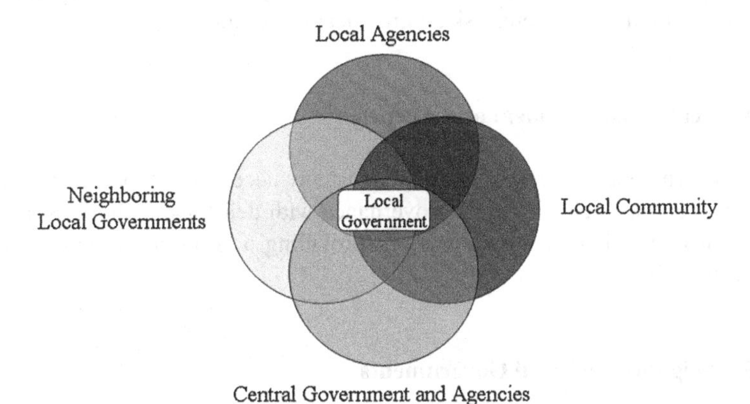

Fig. 4.4 Expanding BCM scope in the public sector (conceptual)

substantial operational resilience. It is therefore becoming increasingly important to establish a community-based business continuity plan (BCP) with participation from all related organizations.

4.3.2 Emerging Needs for Expanding the Scope of BCP and the Role of PPP

To assure the resilience of any organization that has interdependency with external organizations, it is important to expand the scope of its BCP to an appropriate level.

4.3.2.1 Expanding BCP Scope in the Public Sector

To ensure its resilience, a local government, for example, should consider interdependencies with external entities such as local agencies, local community, central government/agencies and neighboring local governments (Fig. 4.4).

4.3.2.2 Local Agencies

A local government must manage the central, regional, or local governmental agencies within its jurisdiction, and is responsible for the command, control and coordination of resources. Ineffective coordination among local governmental agencies has often been observed in previous disaster cases.

4.3.2.3 Local Community

A local government must monitor and take necessary action for residents and any organizations in the area. Here, risk communication is vital.

4.3.2.4 Central Government and Agencies

A local government must report to and request any necessary support from central government and agencies. National governance with delegated authorities to local governments sometimes causes delays in providing a coordinated national-level response to a disaster.

4.3.2.5 Neighboring Local Governments

A local government must coordinate with its neighboring local governments before, during and after the event, to protect and maintain resilience of the area and surrounding areas against shared risks or disasters.

4.3.2.6 Expanding BCM Scope in the Private Sector

To assure operational resilience, private sector companies should expand the scope of their BCP to involve other parties within the supply chain (Fig. 4.5).

4.3.2.7 Corporate Groups

To enhance group resilience, companies within corporate groups must share and leverage resources with related companies or offices in the group. Actual cases are recognized in several global enterprises, especially in manufacturing industries that share and leverage materials and production capabilities.

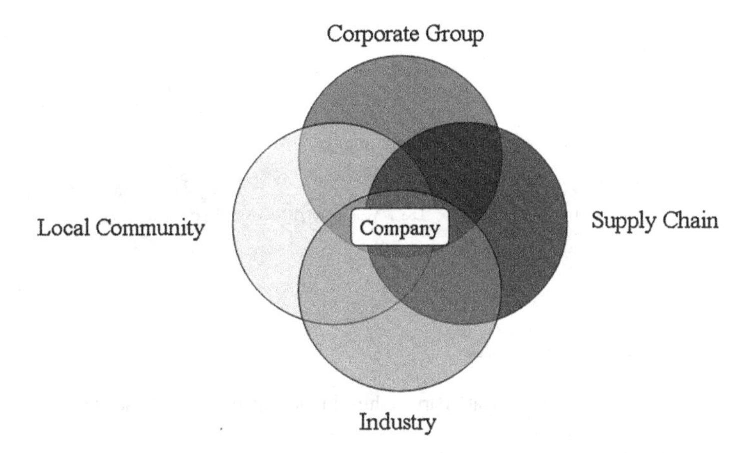

Fig. 4.5 Expanding BCM scope in the private sector (conceptual)

4.3.2.8 Supply Chain

A company must discuss with its suppliers and buyers to coordinate how to manage operations as a whole supply chain and cost allocations among participants to business continuity activities in any emergency situations. Guidelines and best practices will differ across industries depending on the workflow and supply chain; for example, manufacturing in the semi-conductor industry is a global process, divided into multiple sub-processes at different sites, while the automotive industry depends on a huge range of suppliers.

4.3.2.9 Industry

To gain intelligence for resilience and a structure for mutual assistance, a company must share its experiences with other players in the industry. Regulated industries, such as the financial industry, have an information sharing and benchmarking mechanism in some countries.

4.3.2.10 Local Community

A company must work with local communities including local residential and neighborhood associations, chambers of commerce, and local governments and agencies. Even large global enterprises have many local offices or factories which are physically located within local communities and which share local resources and infrastructure.

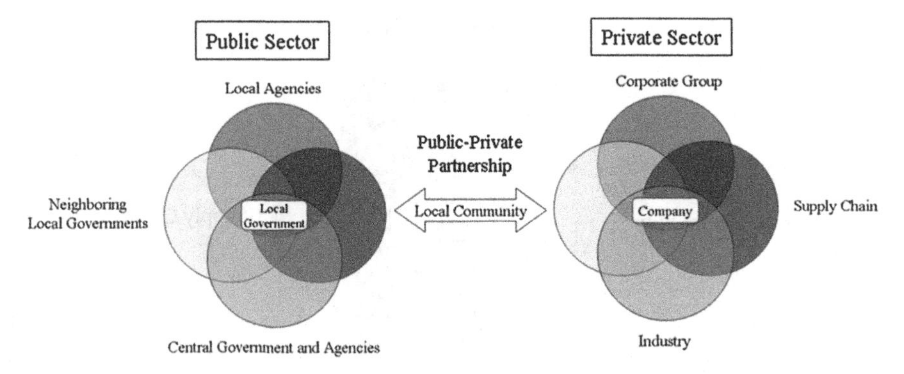

Fig. 4.6 The role of PPP (Public-Private Partnership) in the expanding BCM scope

4.3.3 Local Community as an Interface Between Public and Private Sector Business Continuity Management

In the expanding scope of business continuity management (BCM) in the both the public sector and the private sector, both sides share a common area: the local community. In this area, the public–private partnership (PPP) has an important role to promote BCM for both sides and also to enhance social resilience as a whole (Fig. 4.6).

4.4 Economical Incentives for BCM Promotion Driver

In order to promote individual BCM to each stakeholder in a local community, governmental promotions may not enough and some economical incentives will be necessary especially for the private sector. In this section, financial incentives with some cases in Japan are discussed.

4.4.1 Financial Incentives

The preparedness finance from the DBJ (Development Bank of Japan) has been well accepted in the financial market and other semi-governmental financial institutions and regional banks have introduced similar types of loan products to the market. In such cases, financial incentives work for both the lenders and borrowers. On the lender's side, they can reduce credit risk by focusing the purpose of loan on improving the borrower's resilience. On the borrower's side, they can enhance their resilience through loans with lower interest rates.

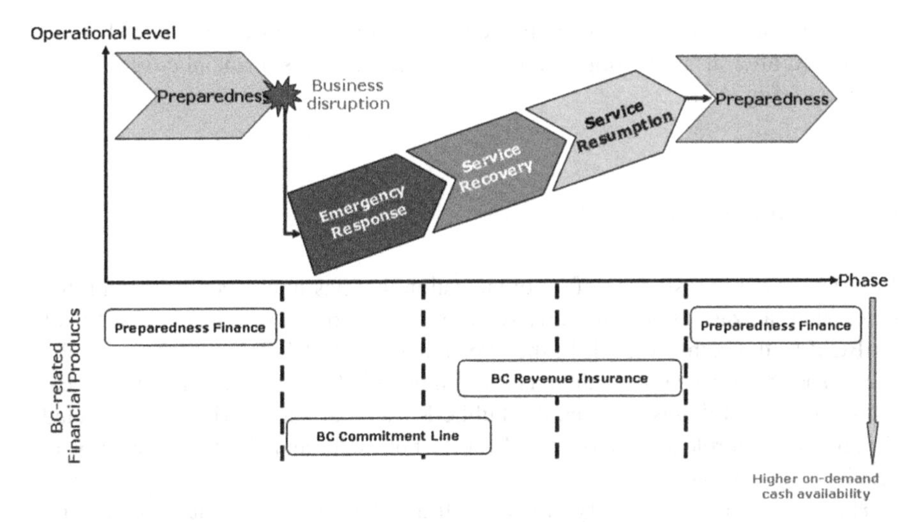

Fig. 4.7 Conceptual coverage of BC-related financial products

In addition to the above, some insurance companies have started reducing the premium for enterprise revenue insurance if the DBJ has rated the company to be insured. Considering the success of the above schemes so far, the promotion of financial incentives to the capital market represents a very effective and sound approach, providing a 'win-win' situation for both public and private sectors. This model offers the public sector sustainable tax income and employment in the event of disaster. Several financial products have been already developed in the Japanese financial market, and the market is trying to support private sector efforts to enhance resilience to contribute to social resilience.

Products include:

- Preparedness finance: The DBJ has developed a preparedness rating as a new dimension in the decision process for corporate finance. The rating measures are based on the guidelines and self-assessment matrix provided by the Cabinet Office. This financial product mainly covers the pre-disaster phase (Fig. 4.7);
- Business continuity commitment line: The DBJ has also developed disaster-triggered backup line for recovery from expected large-scale disasters. It has functional similarity with the catastrophe (CAT) bond and provides standby available cash. This product mainly covers the 'emergency response' phase (Fig. 4.7);
- Business continuity revenue insurance: Sompo-Japan, one of Japan's major insurance companies, has developed revenue insurance with a lower rate if an applying company has a preparedness rating from the DBJ. The discount rate is between 5 and 10 %.

If the financial market could combine these products to cover all phases before, during and after disasters, organizations could get seamless financial coverage at a lower rate (Fig. 4.7).

4.5 Conclusions

In order to assure resilience of an organization that has interdependency with the external organizations or third-party vendors, it is important to expand the scope of its BCM to the comfort level. In the expanding scope of BCM in both the public sector and the private sector, there is a common area for the both sides. It is "local community" and this is the area the Public-Private Partnership (PPP) is expected for an important role to promote BCM for the both sides and also to enhance social resilience as a whole.

However, in the long history of natural disasters in Japan, the wide-area natural disasters have forced every organization in public sector and private sector to work together, sometimes with resource conflicts in the process of the community recoveries. In order to enhance social resilience, more economical incentives should be developed to promote community-based BCM especially for SMEs that have not enough resources for BCM by themselves. The approach to motivate mutual supplement by both of public and private sectors should be conceptualized to be applied to other countries or discussions in international standardization to coordinate the public and private sectors in the same platform to achieve social resilience in the PPP scheme.

References

1. Oke A et al (2009) Managing disruptions in supply chains: a case study of a retail supply chain. Int J Prod Econ 118(1):168–174
2. Sheffi Y (2005) The resilient enterprise. MIT Press, Cambridge
3. Watanabe K (2009) Developing public-private partnership-base business continuity management for increase community resilience. J Bus Contin Emer Plan, Stewart H (ed) 3(4):335–344

Chapter 5
A Model for the Spatio-temporal Distribution of Population using Country-Wide Statistical Data and Its Application to the Estimation of Human Exposure to Disasters

Keisuke Himoto and Jumpei Kimata

Abstract Extent of fatality due to natural disaster depends largely on spatial distribution of population at the moment disaster occurs. In this study, a computational model for estimating day-long spatio-temporal distribution of commuters (workers and students) was developed using a number of country-wide statistical data of Japan such as "population census", "survey on time use and leisure activities", and "economic census". The model estimates behavior of individual commuters by considering their attributes including gender, occupation, place of work (or school), and place of residence. As a case study, the proposed model was applied to the Keihanshin (Kyoto-Osaka-Kobe) metropolitan area, one of the largest metropolitan areas in Japan. Estimated maximum number of commuters unable to return home in an expecting earthquake scenario was between 1.1 and 1.9 million depending on the assumptions on traffic disruption following disaster and the maximum walkable distance of each commuter.

5.1 Introduction

Appearance of disaster considerably changes due to the time of its occurrence. Kobe earthquake (1995) occurred at 5:46 in the early morning of 17th January. Although the area of severe seismic shaking was relatively small, the most severe shaking hit the center of Kobe city, one of the largest cities in Japan. As a consequence, number of collapsed houses in the affected area counts 104,906 and fires started in the city

K. Himoto (✉)
Building Research Institute, Tachihara 1, Tsukuba, Ibaraki 305-0802, Japan
e-mail: himoto@kenken.go.jp

J. Kimata
Construction Division, Aichi Prefectural Government, San-nomaru 3-1-2, Naka-ku, Nagoya, Aichi 460-0001, Japan

H.-N. Teodorescu et al. (eds.), *Improving Disaster Resilience and Mitigation - IT Means and Tools*, NATO Science for Peace and Security Series C: Environmental Security,
DOI 10.1007/978-94-017-9136-6_5, © Springer Science+Business Media Dordrecht 2014

area following the earthquake burnt 7,534 buildings [1]. Because most of people in the affected area experienced the shaking when they were asleep, majority of fatalities, which is reported to be 6,437, were caused mostly by collapse or burn of their own houses.

On the other hand, Tohoku earthquake (2011) occurred at 14:46 in the afternoon of 11th March. Estimated moment magnitude of this earthquake was 9.0, the largest among the recorded history of Japan. Wide range area of eastern Japan experienced severe ground motion and tsunami arrival subsequent to it. It is reported that the number of collapsed houses counts 126,574, and that of fatalities including missing persons counts 21,377 [2]. Contrary to Kobe earthquake, most of the physical and human loss was caused by the tsunami arrived on the Pacific coast. Because most of people were away from home for work or school when the earthquake occurred, family members were forced to evacuate from the tsunami by their own. In addition, they had to take a shelter for several following days separately, because the transportation and communication networks were fragmented in the affected areas.

Appearance of the above disasters might have been much different if the occurrence times were the other way round, i.e., if Tohoku earthquake were to occur at 5:46 and the tsunami were to arrive when people are asleep, number of fatalities would have been no less than the actual. This is attributed to the population distribution at the moment the earthquake occurred. This is to show that impact of disaster is substantially affected by the extent of human exposure to hazard.

The importance of human exposure evaluation on planning disaster mitigation strategy is widely recognized. Nojima et al. proposed PEX (population exposure to seismic intensity) index and conducted a retrospective assessment of five major earthquake events in Japan [3]. Dilley et al., on an attempt of identifying key hotspots of natural disaster, assessed the risk of disaster-related mortality by combining hazard exposure with historical vulnerability for gridded population for six major hazards: earthquakes, volcanoes, landslides, floods, drought, and cyclones [4]. Peduzzi et al. proposed DRI (disaster risk index) for monitoring the evolution of global risk to natural hazards, in which human vulnerability was measured by crossing exposure with selected socio-economic parameters [5]. Aubrecht et al. provided an overview on available multi-level geospatial information and modeling approaches from global to local scales that could serve as inventory for people involved in disaster-related areas [6].

One of the possible approaches for the human exposure evaluation is to use the population census data. The Center for International Earth Science Information Network (CIESIN) provided the Gridded Population of the World (GPW) v.3 representing the residence-based population distribution across the globe. The Global Rural–urban Mapping Project (GRUMP) v.1 was built on GPW increasing its spatial resolution by combining census data with satellite data [7]. While the spatial resolution of population distribution is of critical importance for disaster impact assessment, enhancing temporal resolution of population distribution poses another great challenge. Dobson et al. developed the LandScan Global model and database which represents an “ambient” population distribution over a 24 h period by integrating diurnal movements and collective travel habits into a single measure [8]. Bhaduri et al. extended the LandScan Global model and developed the LandScan USA model which allows the creation of population distribution data at a spatial resolution of 3 arc sec (~90 m). The model contains both a night-time residential as well as a baseline day-time population that incorporates movement of workers and students [9]. Freire, by combining census data and mobility statistics with physiographic data, developed a model to allocate day-time and night-time population distribution in Portugal [10]. Unlike the above mentioned census-based models, Osaragi developed a model for estimating the travel behavior of individuals calibrated with questionnaire- and person-trip surveys, and delineate the overall population distribution by integrating the output travel behavior of individuals [11].

From the viewpoint of disaster mitigation planning, several requirements could be imposed on the output of human exposure estimation, e.g., (1) spatial and temporal resolution of the output should be fine enough to accurately capture the appearance of disaster; (2) the output should contain not only distribution of population, but also its attribute such as gender or age for the better estimation of human loss from hazard; and (3) the model should be scalable and applicable to wide-range areas because affected area of disasters generally crosses over administrative boundaries. To add, (4) it is preferable that the model is extensible, i.e., the model is not designed for a specific area, but is applicable to an arbitrary area. However, to date, existing models have only partly fulfilled these requirements.

Census data and person-trip survey data are two major types of base data for the estimation of population distribution. The advantage of census data is its completeness and periodicity as the base data. To add, data of equal quality are available all over the country. On the other hand, because the census data can only provide population distribution at two time-points a day, enhancement of temporal resolution has been a challenge for the models using data of this type. The advantage of the person-trip survey data is its thoroughness which contributes to the modeling of detailed travel behavior of individuals. On the other hand, because person-trip surveys are conducted only at large cities and their surrounding areas in Japan, the models of this type is not applicable to an arbitrary area.

In this paper, a model for estimating the day-long population dynamics of people is presented. The model is a census-based model, but has incorporated features of person-trip-based models at the same time. It simulates travel behavior of individuals using multiple country-wide statistical data to enhance its temporal resolution. In this paper, the model is further applied to the Keihanshin (Kyoto-Osaka-Kobe) metropolitan area as a case study, and the impact of disorder of transportation network in an anticipated earthquake scenario is studied.

5.2 Outline of the Model

The estimation model is developed using "population census" data which provides static population distribution at two time-points at relatively high spatial resolution [12]. The model simulates the day-long travel behavior of individual commuters

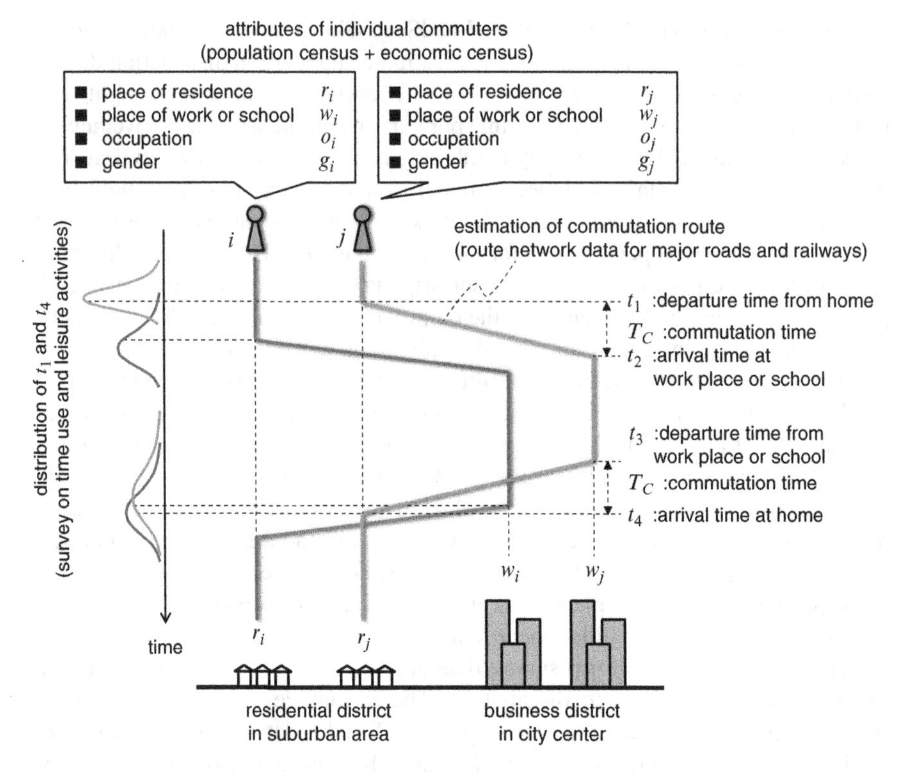

Fig. 5.1 Schematic of the population dynamics model. Spatio-temporal distribution of population is estimated by simulating travel behavior of individual commuters

by interpolating the population distribution of the two time-points spatially and temporally using "economic census" [13] and "survey on time use and leisure activities" [14] in an integrated manner. Figure 5.1 shows the schematic of the model. Although behavior of commuters is complex in general, we assumed that all commuters behave routinely in a day, i.e., they take routine actions in an order of,

1. leave home (t_1),
2. arrive at work place or school (t_2),
3. leave work place or school (t_3), and
4. arrive at home (t_4).

Note that signs in the parentheses are the times for the respective action. Location of a commuter at an arbitrary time can be traced by determining the times of actions t_1, t_2, t_3, and t_4, and commutation route; a commuter is at home during the period $t_4 - t_1$, on the commutation route during the period $t_1 - t_2$ and $t_3 - t_4$, and at work place or school during the period $t_2 - t_3$. However, assuming that the commutation times to work place or school and that on the way back are the same, t_2 and t_3 can be expressed as,

$$t_2 = t_1 + T_C \text{ and } t_3 = t_4 - T_C$$

where T_C is the commutation time. Thus, the four parameters necessary to determine the time of the routine action can be reduced to three, i.e., t_1, t_4, and T_C.

The times of the routine action t_1 and t_4, commutation time T_C, and commutation route are determined probabilistically by considering attributes of individual workers, which are,

- place of residence (r)
- place of work or school (w)
- occupation (o)
- gender (g)

These attributes of individual commuters are extracted from the cross tables of "population census" aggregated for each municipality within the target area. However, because the cross tables are available only at the level of municipalities, extracted place of residence r and place of work or school w of individual commuters are further processed to enhance their resolution by the Monte Carlo sampling method using "economic census".

5.2.1 *Times of Routine Action t_1 and t_4*

Times of routine action t_1 and t_4 for an individual commuter are determined from the distribution functions of departure time from home $f_1(t)$ and arrival time at home $f_4(t)$, which are both derived from "survey on time use and leisure activities". These distributions are obtained for every combination of occupation o and gender g, which are considered influential on t_1 and t_4.

However, a drawback of this approach is that the times of actions t_1 and t_4 are obtained from mutually independent distribution functions of $f_1(t)$ and $f_4(t)$. Thus, time spent at work place or school T_W and time spent at home T_H derived from times of routine action t_1, t_2, t_3, and t_4 may be inconsistent with generally expected time durations. In order to accommodate such an inconsistency, additional constraints are set in obtaining the times of routine action t_1 and t_4, i.e.,

$$t_3 - t_2 = t_4 - t_1 - 2T_C > T_W \text{ and } 24 - (t_4 - t_1) > T_H \text{ (hr)}$$

where T_W is obtained from the distribution function $f_W(T)$ derived from "population census". T_H is the sum of the mean hours of sleep T_S and mean hours for personal care T_{PC} which are both derived from "survey on time use and leisure activities".

5.2.2 *Commutation Route and Time T_C*

According to the 4th person trip survey of the Keihanshin metropolitan area conducted in 2000 [15], major means of commutation can be grouped either into

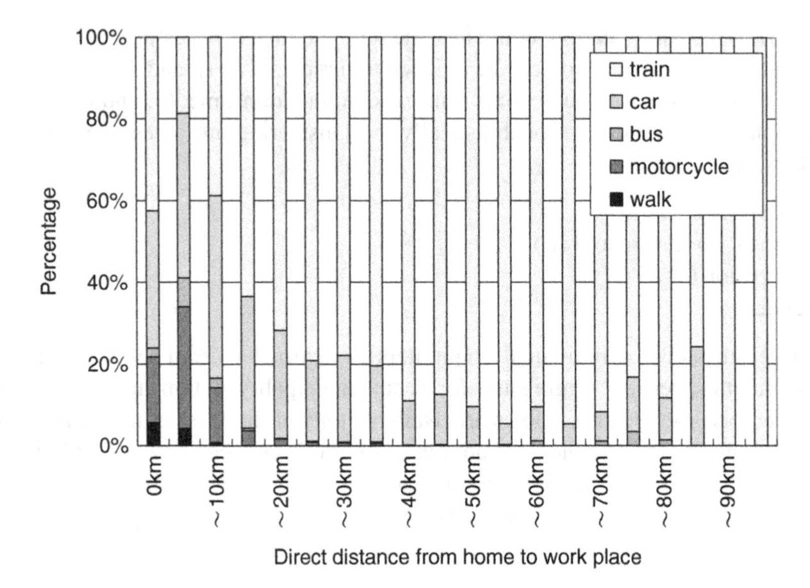

Fig. 5.2 Means of transportation for commutation in the Keihanshin metropolitan area with regard to direct distance between home and work place [15]

Table 5.1 Means of commutation and its assumed travel speed [16]

Means of commutation		Travel speed (km/h)
Walk		4.0
Bicycle		11.0
Car, Bus, and Motorcycle	Highway	71.1
	Limited highway	41.7
	National road	37.4
	Prefectural road	33.9
	Municipal road	17.9
Train		40.0

"railway", "car", "bus", "motor cycle", and "walk". Figure 5.2 shows proportion of the means of transportation with regard to the commutation distance. On simulating travel behavior of a commuter, means of transportation is probabilistically and individually determined using this proportion.

Commutation route of a commuter is selected by seeking the shortest route from the relevant route network individually. However, effect of fare or number of connections on one's route choice is neglected for simplicity. To add, only major roads shown in Table 5.1 was used for the route choice of commuters using other than "train", because the size of road network data is substantially larger than that of railway network data in general. Commutation time T_C for the selected commutation route is calculated by assuming a uniform travel velocity for each means of transportation as shown in Table 5.1 [16].

5.3 Day-Long Movement of Commuters in the Keihanshin Metropolitan Area

The model is applied to the Keihanshin metropolitan area for the estimation of day-long movement of commuters. The target area is the second largest metropolitan area of Japan with Kyoto, Osaka and Kobe as the central cities. In addition, the area includes 148 other cities and towns with the number of people commuting to the central cities exceeding 1.5 % of all the population. Note that any of area of these municipalities is not isolated, but adjoins to each other. Total population of the Keihanshin metropolitan area is 18.7 million living in an area of 11,600 km^2. Target population of the present model is 8.3 million of workers and 3.6 million of students; the total 11.9 million is equivalent to 63.6 % of the overall population.

Estimated distributions of commuters at four time points of 2:00, 8:00, 14:00, 20:00 of a weekday are aligned in Fig. 5.3. The model depicts a day-long movement of commuters between the suburban area and city centers; people dispersedly distributed in the suburban area at night gathers to the city centers such as Kobe, Osaka, and Kyoto in daytime and return home again at night.

In order to validate the model, estimated number of commuters is compared with that of the 4th person trip survey of the Keihanshin metropolitan area [15]. The comparison result at four time points of 2:00, 8:00, 14:00, and 20:00 of a weekday is shown in Fig. 5.4. Each data point represents the population in enumeration districts of the person trip survey. Although there are certain discrepancies, data points gather around the diagonal line on which the estimation result is identical to the person trip survey result. Such scatter of the data points may be attributed to the out-of-routine actions which are not considered in the present model.

5.4 Effect of Traffic Disorder in the Event of an Anticipated Earthquake in the Keihanshin Metropolitan Area

Japan is a country around which four continental plates meet each other, i.e., Pacific plate, North American plate, Eurasian plate and Philippine plate. The Tohoku earthquake which stroke eastern Japan was induced by a source fault at the boundary of two continental plates among the four. Such a gigantic earthquake may occur at any part of boundary of these continental plates. Figure 5.5 shows a map of western Japan with location of plate boundaries indicated. The situation of western Japan resembles that of eastern Japan; the plate boundaries closely lie along the coastline exposing the coastal area with the risk of severe ground motion and tsunami.

The fatal experience of the Tohoku earthquake in 2011 triggered the debates on the possibility of an earthquake which will occur along the Nankai, trough which is a plate boundary between Eurasian plate and Philippine plate. This has repeatedly caused earthquakes with the moment magnitude of over 8 at intervals between 100 and 150 years. The cabinet office of Japan recently released an estimation report

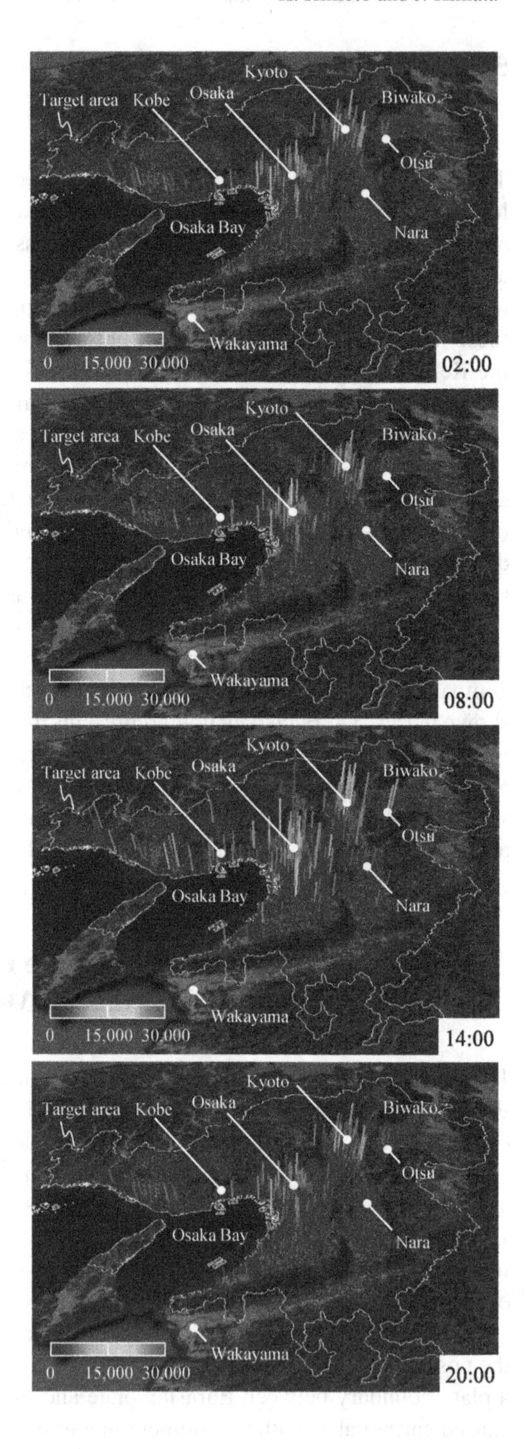

Fig. 5.3 Estimated spatial distribution of commuters in the Keihanshin metropolitan area in a weekday

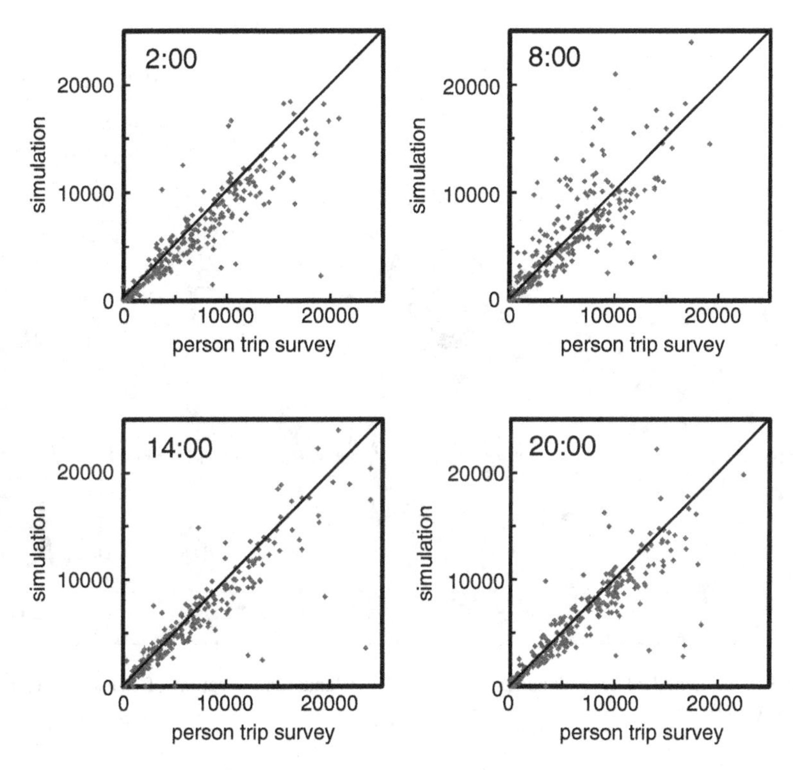

Fig. 5.4 Estimated and observed number of workers at each unit areas at four time point in a weekday

on the losses due to the anticipated Nankai trough earthquake [17]. As shown in Fig. 5.6, Keihanshin metropolitan area is within the extent of impact of this gigantic earthquake. Note that the JMA (Japan Meteorological Agency) seismic intensity is a scale which represents the level of ground motion with the ascending order of 0, 1 (slight), 2, 3 (weak), 4, 5−, 5+ (strong), 6−, 6+, 7 (severe). It is empirically known that certain minor damage may be caused to old wooden houses with the ground motion of level 5−. Such a severe ground motion will lead to huge physical damages involving collapse of buildings and infrastructures, occurrence of fires, landslide in mountainous area, etc.. However, in addition to this, Tohoku earthquake shed light on another type of consequences which lead especially metropolitan areas into disruption. Although physical damage was relatively small in the Tokyo metropolitan area, full suspension of railway services due to the earthquake caused millions of commuters gathered in the city center to become stranded. Similar disruption is expected in the Keihanshin metropolitan area in the event of the anticipated Nankai trough earthquake. In this section, number of commuters unable to return home due to the traffic disorder is estimated using the simulated results of population distribution in the previous section.

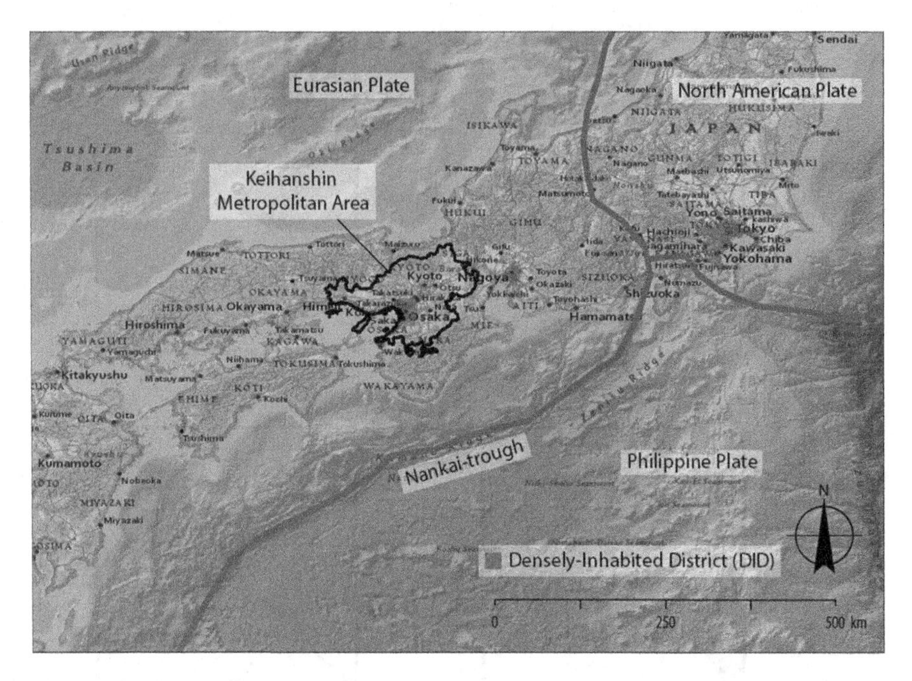

Fig. 5.5 Continental plate boundaries around Japan islands and location of densely-inhabited districts (*DID*)

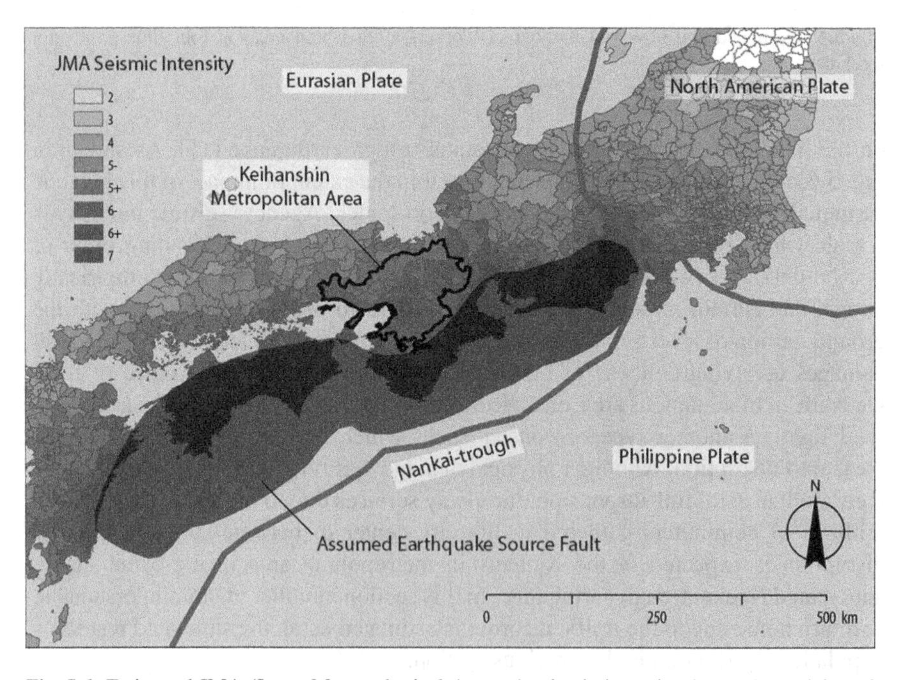

Fig. 5.6 Estimated JMA (Japan Meteorological Agency) seismic intensity due to the anticipated Nankai trough earthquake

Table 5.2 Cases considered for the estimation of commuters unable to return home

	Unavailable transportation network		Maximum walkable distance
	Rail network	Road network	
I	✓	✓	A
II	✓	✓	B
III	✓		A
IV	✓		B

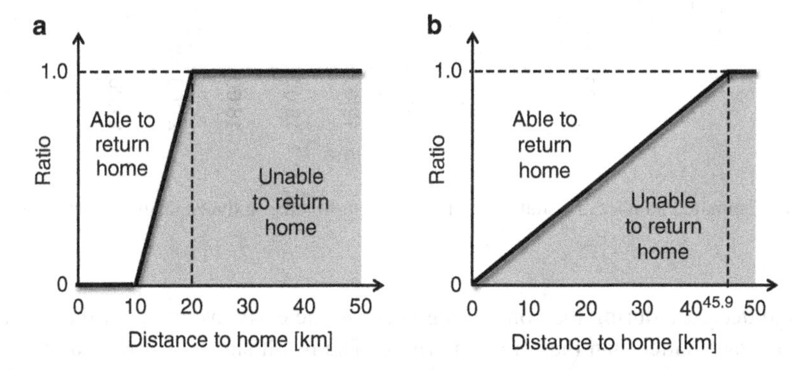

Fig. 5.7 Assumed maximum walkable distance of commuters in case of traffic disorder due to the earthquake **a**) the Miyagi-oki earthquake (1978) observarion; **b**) the Tohoku earthquake (2011) observation

5.4.1 Cases Studied

Four cases are considered on estimating the number of commuters unable to return home. As shown in Table 5.2, they are in combination of two conditions, i.e., "availability of transportation network" and "maximum walkable distance of commuters". In case the means of commutation of a commuter is unavailable, this commuter considers returning home on foot. Whether this commuter can actually return home on foot or not is probabilistically determined using the criteria in Fig. 5.7. There are two criteria which discern the possibility of a commuter to return home on foot as functions of one's location from home at the time the earthquake occurs. These criteria are based on the evidence of past earthquakes, i.e., (A) Miyagi-oki earthquake in 1978 and (B) Tohoku earthquake in 2011 [17].

5.4.2 Estimation Results

Estimated number of commuters unable to return home is displayed in Fig. 5.8. In all the cases, the number rapidly increase during the commuting hours in the morning, followed by a period of relatively small change in the daytime. It then

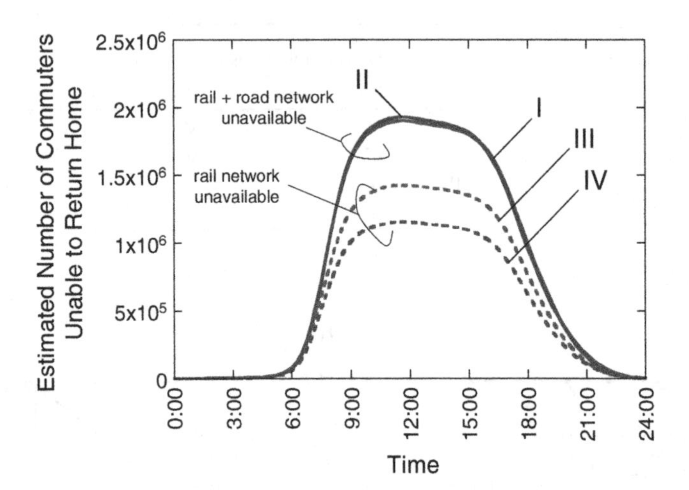

Fig. 5.8 Estimated number of commuters unable to return home in the Keihanshin metropolitan area

starts to decrease during the commuting hours in the early evening, yet the rate of change was smaller than that in the morning. The result shows that the number of commuters unable to become home is the maximum between 11:30 and 11:45 for all the cases. However, there are certain discrepancies in the maximum estimated number: 1.88 million for case (I); 1.91 million for case (II); 1.41 million for case (III); and 1.15 million for case (IV). The cabinet office has also conducted similar estimation to the present analysis using the person trip survey data [17]. They estimated the number of commuters unable to return home to be between 2.2 and 2.7 million in the Keihanshin metropolitan area. This is larger than that of the present analysis. However, note that there are 6.76 million non-commuters in the Keihanshin metropolitan area which are not considered in the present analysis. Nevertheless, an extension of the model will be required in order to conduct a more adequate estimation.

In cases (III) and (IV) at which rail networks are assumed to be unavailable, the estimated number of commuters unable to return home is smaller with the criteria (B) for the maximum walkable distance. On the other hand, in cases (I) and (II) at which both rail and road networks are as-summed to be unavailable, the estimated numbers resemble each other regardless of the criteria for the maximum walkable distance. This is because the proportion of commuters unable to return home is estimated higher with criteria (A) in comparison with (B). A major difference between the two criteria is that while all commuters can return home between $0 \sim 10$ km with criteria (A), a portion of commuters cannot return home with criteria (B). That is to say that there are many commuters who use cars, buses, and motorcycles as means of commutation between $0 \sim 10$ km in the Keihanshin metropolitan area.

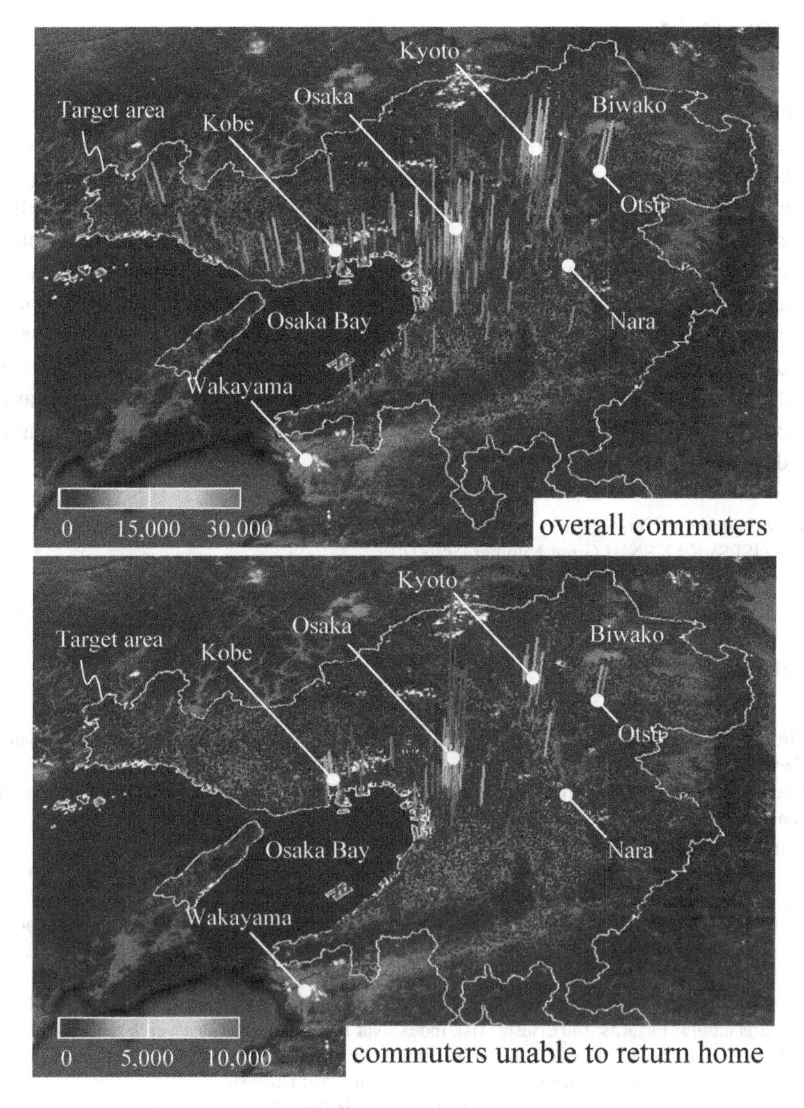

Fig. 5.9 Estimated spatial distribution of overall commuters and those unable to return home in the Keihanshin metropolitan area at 11:45 in a weekday

Snapshots of estimated distribution of overall commuters and that of commuters unable to return home at 11:45 are shown in Fig. 5.9. This figure shows that overall commuters concentrate around major cities such as Kyoto, Osaka, and Kobe. However, those unable to return home are concentrated at even more specific areas in these major cities. This implies that there are many facilities and services which attract commuters from distant locations in these areas.

5.5 Conclusions

A model for the day-long population movement of commuters is proposed in this paper using several country-wide statistical data such as such as "population census", "survey on time use and leisure activities", and "economic census". The model is applied to simulate travel behavior of 11.9 million commuters including workers and students in the Keihanshin metropolitan area, one of the major metropolitan areas in Japan. As a result, maximum number of commuters unable to return home due to traffic disorders in case of the anticipated Nankai-trough earthquake is estimated to be between 1.15 and 1.91 million within the cases considered. Development of the model is still ongoing. Major points of refinement include: (1) estimation of travel behavior of non-commuters including infants, housewives, or elderly people; and (2) estimation of population movement in non-weekdays.

Acknowledgement This ongoing work is supported by Japan Society for the Promotion of Science (JSPS) KAKENHI Grant Number 21360291.

References

1. Fire and Disaster Management Agency (FDMA) (2006) Final report of 1995 Hyogo-ken Nanbu Earthquake, 2 pages
2. Fire and Disaster Management Agency (FDMA) (2013) 148th report of 2011 off the pacific coast of Tohoku Earthquake, 40 pages
3. Nojima N, Kuse M, Sugito M, Suzuki Y (2004) Population exposure to seismic intensity for assessment of seismic disaster potential. J Jpn Soc Nat Disaster Sci 23(3):363–380 (in Japanese)
4. Dilley M, Chen RS, Deichmann U, Lerner-Lam AL, Arnold M, Agwe J, Buys P, Kjekstad O, Lyon B, Yetman G (2005) Natural disaster hotspots: a global risk analysis. The World Bank, Washington, DC
5. Peduzzi P, Dao H, Herold C, Mouton F (2009) Assessing global exposure and vulnerability towards natural hazards: the disaster risk index. Nat Hazards Earth Syst Sci 9:1149–1159
6. Aubrecht C, Ozceylan D, Steinnocher K, Freire S (2012) Multi-level geospatial modeling of human exposure patterns and vulnerability indicators. Nat Hazards
7. Balk D, Yetman G, de Sherbinin A (2010) Construction of gridded population and poverty data sets from different data sources. In: Proceeding of the European forum for geostatistics conference, Talin, 5–7 October 2010, pp 12–20
8. Dobson JE, Bright EA, Coleman PR, Durfee RC, Worley BA (2000) LandScan: a global population database for estimating populations at risk. Photogramm Eng Remote Sens 66(7):849–857
9. Bhaduri B, Bright E, Coleman P, Urban ML (2007) LandScan USA: a high-resolution geospatial and temporal modeling approach for population distribution and dynamics. GeoJournal 69:103–117. Springer, http://link.springer.com/journal/10708
10. Freire S (2010) Modeling of spatiotemporal distribution of urban population at high resolution – value for risk assessment and emergency management. In: Konecny M et al (eds) Geographic information and cartography for risk and crisis management, Lecture notes in geoinformation and cartography, Berlin, pp 53–67

11. Osaragi T (2012) Modeling a spatiotemporal distribution of stranded people returning home on foot in the aftermath of a large-scale Earthquake. Nat Hazards
12. Ministry of Internal Affairs and Communications (MIC) (2007) Report of the 2005 population census, http://www.stat.go.jp/english/data/kokusei/index.htm. Last date accessed: Dec 2013
13. Ministry of Internal Affairs and Communications (MIC) (2007) Report of the 2006 economic census, Tokyo. http://www.stat.go.jp/english/data/jigyou/index.htm. Last date accessed: Dec 2013
14. Ministry of Internal Affairs and Communications (MIC) (2007) Report of the 2006 survey on time use and leisure activities, Tokyo. http://www.stat.go.jp/english/data/shakai/index.htm. Last date accessed: Dec 2013
15. Ministry of Land, Infrastructure, Transport and Tourism (MLIT) (2000) Report of the 4th Keihanshin Metropolitan Area Person Trip Survey, 35 pages
16. Ministry of Land, Infrastructure, Transport and Tourism (MLIT) (2010) Report of the 2010 Census on Traffic in Major Metropolitan Areas, 157 pages
17. Cabinet Office (2012) Second report of the WG on the countermeasures against the gigantic Nankai trough earthquake, http://www.bousai.go.jp/jishin/nan.kai/nankaitrough_info.html. Last date accessed: Dec 2013

Chapter 6
From Noise to Knowledge: Smart City-Responses to Disruption

Jörg Rainer Noennig and Peter Schmiedgen

Abstract Based on the assumption that so-called "Smart Cities" ought to have an ability to detect and properly respond to disasters, the article presents a descriptive model for urban disasters. The discussion of how urban disasters evolve leads to an inquiry into the systemic mechanisms of cities, and the dynamics of urban disasters. Cities can be described as amplification systems with the capacity to magnify input values through positive feedback. This stimulates the core hypothesis of the paper: Incompatibilities between urban systems generate friction ("noise") as input variable to the urban amplification mechanism which eventually triggers the occurrence of urban disasters. Disaster can be thus defined as positive feedback of negative input values to a scale of complexity beyond control. However, low-level noise shall be regarded as a natural feature of cities, or any other complex systems as it plays a central role in indicating risks and building up disaster resilience. The model describes potential paths of disaster build-up and mitigation, and induces three knowledge-based strategies for urban disaster resilience: (1) "Noise intelligence", which calls for an awareness of urban frictions in order to supply early information on potential disaster build-up; (2) "Disaster Creativity", which holds that creative ad hoc problem solving is a key ability in the case of disaster, apart from inflexible principles of disaster management; (3) "Noise to Knowledge", which is a learning concept based on the assumption that creative disaster response must be trained with the "natural noise" as emerges in any complex urban system.

J.R. Noennig (✉) • P. Schmiedgen
Department of Architecture Junior Professorship for Knowledge Architecture, Wissensarchitektur, Technische Universität Dresden, Mommsenstraße, 9, Dresden, Germany
e-mail: joerg.noennig@mailbox.tu-dresden.de

H.-N. Teodorescu et al. (eds.), *Improving Disaster Resilience and Mitigation - IT Means and Tools*, NATO Science for Peace and Security Series C: Environmental Security,
DOI 10.1007/978-94-017-9136-6_6,

6.1 Smart Cities as Disaster-Resilient Cities

"Smart" and "intelligent" have become popular terms in recent times. Whatever is being produced and put to market will most certainly be labelled "smart" or "intelligent", given it being on a certain technological level. Most of the descriptions and definitions of "smart things" refer to some kind of intelligent behavior, to the application of information technology and communication networks, to some form of strategic knowledge management, and to a certain effectiveness and efficiency in operations. However, these "smart" notions show a stunning incongruity, and often a circular argumentation: smartness refers to intelligence, and vice versa. Thus, not much meaning is added. This may be due to lacking reflection on the epistemic and systemic background of the very matter at stake: rapidly evolving complex systems. What is more, "smart" and "intelligent" are especially popular in the context of urban planning, development, and management. As it comes to the "City of the Future", they are among the most preferred attributes, along with terms like "sustainable", "innovative", or "ecological" [1]. However, this ubiquitous call for smartness and intelligence should not be totally dismissed only for it's undefined terminology and vague application. What is the actual background dilemma that urges the permanent application of this somewhat outworn notion?

The call for "smartness" may be interpreted as a counter-reaction to dissonances resulting from fast technological development and networked systems. Accelerated evolution in societies and technologies – before all urban ones – leads to asynchronous and disharmonious systems. The incongruence and incompatibility can be termed "friction". When escalating, such frictions are sensed as "painful" on the side of human individuals resp. society: they turn into stress, discomfort, and confusion. If further escalating beyond controllability, a state of disaster is reached. This obviously should be prevented by systems termed as "smart" and "intelligent": they are supposed to interact smoothly and effectively.

For the further discussion, it will be helpful to distinct between the notions of "intelligence" vs. "smartness". Generally, intelligence has been defined as the ability of problem solving independent of circumstances, or context [2]. A person's intelligence, or IQ, does not change substantially when the situation changes. Intelligence is a rather constant property, acquired through genetic predisposition, long-term study, and learning. "Smartness", in contrast, is less a property but a behavioral quality based on adequate response to environment. Persons with a defined IQ thus can still behave more or less smartly depending on their reaction to a situation: sometimes response is quick and witty, sometimes slow and dumb. Smart is what comes close to optimal reaction to a certain context. Here, "optimal" means a minimal misfit and maximum adaptation to situation. The very quality of being "smart" is the avoidance or reduction of friction.

Simply speaking: "smartness" works like a painkiller. Smart devices, environments, or technologies etc. help to solve and avoid problems in complex situations. Thus "smartness" is highly relevant also in the context of urban disasters as situations of highest complexity which cause painful effects for individuals as well as for society. Smartness, before all, is needed to cope with such qualities of disaster. The argument works also the other way round: disaster resilience is a necessary core ability of smart cities which depend on other smart qualities like effective transportation, networked information technology, or communication networks.

The following paragraphs will explain how the establishment of "disaster smartness" is based on intelligence about potentially disastrous frictions ("noise") that arises between conflicting urban systems, and on creative problem solving in the face of unforeseeable crisis situations. For this ends, two basic conditions will be discussed: the functioning of urban systems on the one hand, and the nature of disaster on the other. Both form the cornerstones for a model of urban disaster intelligence and smartness.

6.2 Cities as Systems of Systems

A city can be described as a system composed of many systems. Any city consists of a multiplicity of infrastructures. Technical infrastructures are e.g. transportation (road, rails, mobility services), communication networks (internet, telephone), or energy supply (electricity, gasoline). Social infrastructures include e.g. educational system (schools, universities) or health care (hospitals, rescue workers). Commonly these components are described as distinct and specialized systems that have evolved into their own forms of practice and theory. It is important to understand how each system develops along its own path of evolution, based on its specific pretext, background and rationale, which in turn largely determines how it will develop in future. In other words: To a large extent, systems are railed, or locked-in, on their specific tracks. Science-and-technology-studies (STS) call this condition "path dependency" [3, 4].

Yet, a brief glance at urban practices (e.g. planning, construction, management, marketing) shows that in reality these systems overlap and coincide. In many cases they are heavily conflicting; friction appears. It is before-mentioned path dependency of the various individual systems in addition to the increasing complexity of the urban conglomerate that leads to disharmonious and asynchronous development.

The incongruence of evolving systems can be described from an informational point of view. Composed of a variety of other systems, cities are big systems and producers of big data. As it is hard to harmonize the path-dependent individual systems with each other, it is hard as well to integrate the heterogeneous mass of information produced by them. As the most probable case is that urban systems are not fully integrated and function rather asynchronously, it can be assumed that large amounts of disintegrated data will be produced: "big noise". Here, noise is understood as data produced by a system that is relevant to others, and which cannot be interpreted by the other systems. From this, the core hypothesis of this paper derives: In the urban realm, big data production by heterogeneous and uncoordinated systems creates "big noise" (see Fig. 6.1). Big noise – if further amplified and escalating – is a potential trigger for urban disasters. The following paragraphs will explain this argument in detail.

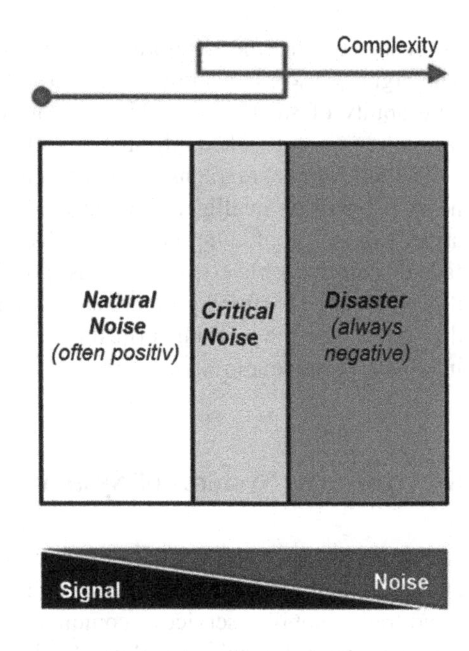

Fig. 6.1 Three-stage disaster model

6.3 Cities as Positive Feedback Systems

Throughout the history of human culture, urban agglomerations have featured two specific qualities:

1. They function as amplifiers for human affairs, as accelerators for almost any vital issue relevant for mankind. From a systems perspective, cities with their high density of actors, information, and interaction can be understood as feedback systems where output products are quickly being re-inserted into the same production loop as fresh input, thus creating a dynamic amplification cycle which cybernetics has termed "positive feedback". Thus cities function as amplification mechanisms regardless of what is being maximized or multiplied.
2. In their iterative cycles of amplification and acceleration, cities also add new attributes to the items at stake. Tight networks of communication and exchange as well as complex processes of self-organization lead to the emergence of new qualities. Cities do not merely reproduce their "input" on larger scale, but also mutate, transform, and alter their quality: they are mechanisms for sophistication and complexification too.

The rise of urban civilization towards an "Urban Age" can be attributed to the successful enhancement of desirable civil qualities by way of urban amplification and complexification. This formula of progress can be translated into systems terminology as "positive feedback of positive input". The term "positive input", however, refers to what a certain society regards as valuable (this is of course culture-dependent). When making positive values such as knowledge, security, wealth e.g. the basis of the iterative cycle of the urban generator, more positive output of similar kind shall result. As statistics show, a city provides higher chances for the transfer and multiplication of positive impulses. Researchers have indicated that the growth of urban agglomerations correlates to growth in economic opportunities, innovation and knowledge expansion [5]. As long as positive input is being amplified, such development dynamics may be called "under control". Arguably, there cannot be enough of "goods" like health, wealth, or security. This may explain why accelerative dynamics in urban systems such as industrial development, scientific progress, and expansion of health care are widely supported.

However, cities have the uncanny potential to magnify and complexify negative values too. Their amplification of "input variables" such as poverty, crime, or traffic leads to well-known effects: traffic collapse, environmental pollution, fast spread of epidemic diseases through higher infection probability. Even if not disastrous in themselves, conditions such as floating populations or mass squatting have high potential to trigger urban disasters like crime, violence, or drug abuse. An example is the explosion of crime in fast growing megacities in developing countries. According to WHO Global Health Observatory, OECD Statistics, UNDOC Organized Crime Division, only in Brazil more than 500,000 people died from gun violence in the past 30 years. Given that a society views such conditions as undesirable, the existence of these amplified negative values suggests that urban feedback systems have run out of control. When positive feedback of negative input escalates, the result may be properly termed "urban disasters".

As mentioned before, cities are highly overlapping systems of systems – spanning from social-political to technological systems, from collective to individual components. In this complex array, it is the frictions in, or between, the multiple systems that becomes a negative input value potentially amplified to the level of crisis. In other words: through the dynamic escalation of "noise" into "big noise", agglomerations become disaster-prone.

Poverty, for example, can be seen as a developmental misfit between social and economic systems, potentially triggering social riot. Another example, traffic collapses when transportation needs and infrastructural development diverge, thus leading to shortage of supply, be it food, drugs, or manufacturing items. Similarly, epidemics spread when hygienic and medical care does not match with number and structure of fast growing populations. – In all of these cases, there are critical mismatches between social, technological and other systems. Certainly, there can be no perfection in the total urban arrangement. There is a natural incompleteness and incompatibility in all these complex systems. In them, "natural noise" is even a vital condition for further development and progress. However, the amount of friction in relation to amplification power is crucial. The amplification and complexification of "natural noise" to "big noise" marks the point when positive feedback runs out of control. The fact that noise not only arises at the interfaces between the urban systems but also that the various systems themselves are transmitting and back-feeding noise from one to another, explains the high vulnerability of fast developing urban agglomerations.

6.4 Noise-Based Model for Disaster

Providing a descriptive scheme for before-mentioned features of urban systems, this paragraph is to present a generic model of how disasters emerge and proliferate. "Disaster" itself is understood here as an event unlikely to emerge in a certain environment, running out of control, and leading to problems hard to solve. Based on the assumption, that disasters are enhanced noise, the model distinguishes three stages, or thresholds, of noise amplification and complexification (Fig. 6.1).

1. The basic state, "Natural Noise" is a normal property of any complex system, be it a biological organism or a large city. Often it is a positive asset associated with creativity, self-organization, and robustness – the "small chaos" that is a pre-requisite for the emergence of innovation and resilience.
2. When "Natural Noise" increases in amplitude beyond normality level, it turns into "Critical Noise". This happens when the positive feedback mechanisms of amplification and complexification have kicked in, and noise becomes problematic and critical, a preliminary stage of disaster.
3. A level of "Disaster" is finally reached when the amplification of "Critical Noise" runs out of control, or turns destructive. Significantly, there is an increase of complexity, as disasters pose non-simple problems that appear difficult to solve. "Critical Noise" thus is not only magnified, but develops destructive order by chaining effects, problem cascades, self-organization, etc. Although it may seem paradoxical to attribute to disastrous events higher states of order, it is the very property making them so difficult to deal with.

To make the scheme more understandable, it can be specified for different types of systems (e.g. technology, society). For each system, concrete references for the various stages of escalation from "Natural Noise" over "Critical Noise" to "Disaster" can be assigned without difficulty (Fig. 6.2). For example, in social systems certain friction between groups and parties seems to be inevitable "Natural Noise". However, when escalating beyond a certain level, up-rises and riots ("Critical noise") result which may go out of control and potentially result in anarchy, chaos, and civil war. Global newspapers give full report on such socio-political dynamics

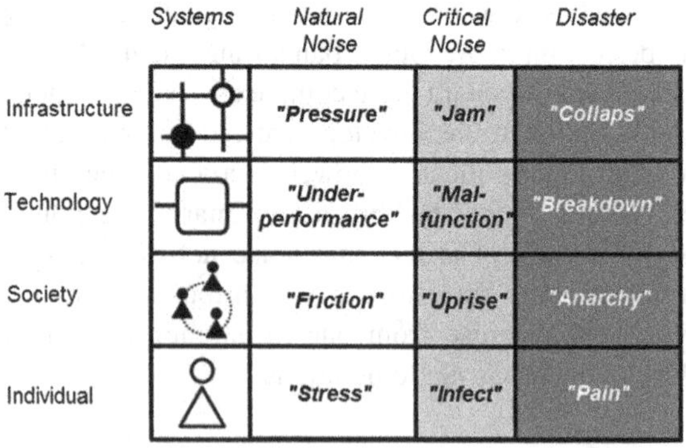

	Systems	Natural Noise	Critical Noise	Disaster
Infrastructure		"Pressure"	"Jam"	"Collaps"
Technology		"Under-performance"	"Mal-function"	"Breakdown"
Society		"Friction"	"Uprise"	"Anarchy"
Individual		"Stress"	"Infect"	"Pain"

Fig. 6.2 Examples of noise escalation for different systems

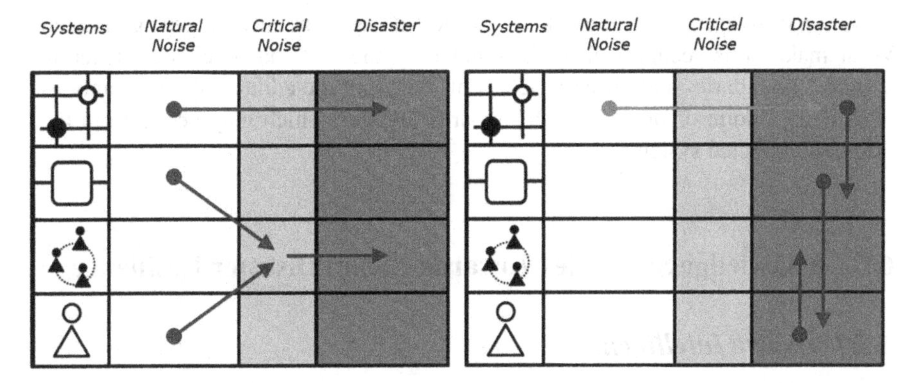

Fig. 6.3 Intra- and intersystemic disaster build-up (*left*); cascading disaster (*right*)

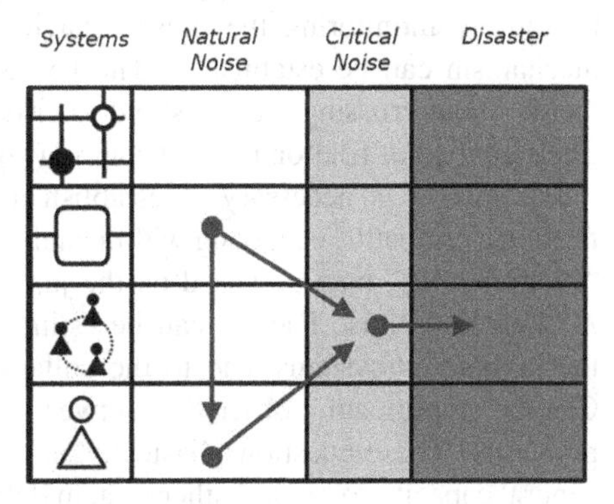

Fig. 6.4 Triggering disaster through cascading Natural Noise

every day. Another example, for the individual human being physical stress may be identified as "Natural Noise". Beyond a certain level though, stress may turn into mental defects as well as into bodily malfunctions ("Critical Noise") which – when going out of control – turns into serious pain or typical urban diseases.

In the formalized scheme, the build-up of disasters can be traced through the various stages of noise amplification, as well as across the various systems' boundaries (see Fig. 6.3). There are two generic types of disaster build-up: "intra-systemic" vs "inter-systemic". The earlier is characterized by the growth of noise within one system, e.g. increasing malware within a computer system leading to eventual breakdown. The latter is characterized by the convergence of noise originating from different systems, e.g. the coincidence of defective traffic infrastructures and unreliable signal systems leading to traffic collapse or accidents. The inter-systemic type is apparently of higher complexity and more difficult to identify, and thus more susceptible from the perspective of disaster research.

Further, the scheme allows the tracking of cascades of disasters as they skip from one system to another one (right scheme in Fig. 6.3). Once having reached overcriticality in one system, such cascades create an avalanche of complex problems that chain through various systems.

Noteworthy but not as easy to comprehend is another scenario: "Natural Noise" that cascades on low level (left column in Fig. 6.4) throughout various systems may amount a critical mass of noise that is sufficient for eventually triggering disaster. What makes this scenario difficult is not only the fact that it crosses different systems' boundaries and different levels of scale, but the additional uncertainty as regards the "point of breakthrough" towards disaster, which may be far from the originally affected systems.

6.5 Knowledge Strategies Towards Urban Disaster Resilience

6.5.1 Noise Intelligence

The first knowledge strategy draws intelligence from information ante disaster: by closely monitoring the noise-behavior in urban systems, an early warning-mechanism can be established. The key issue here is the identification of noise that is about crossing the threshold to critical amplification, that is: accumulating a certain load of friction beyond control (Fig. 6.5).

For this, it is necessary to establish a measure of amplification risk for the frictions, or misfits, occurring within, and between, systems. One approach for such "frictiometrics" may be based on the juxtaposition of critically involved systems. A "Potential Noise Factor" can be estimated by asking: Will the misfit between the juxtaposed systems lead to the collapse of either system? More specifically: Can the amplification of friction between the systems potentially trigger insoluble problems? These questions, basically, are to inform whether amplified noise has a general capacity to render either system defunct. Thus the robustness and particular importance of each system are to be compared probabilistically, as well as the interdependence of the separate systems. Especially "nervous" constellations will be identified, thus forming a risk map indicating points of closer investigation and scrutiny (Fig. 6.6).

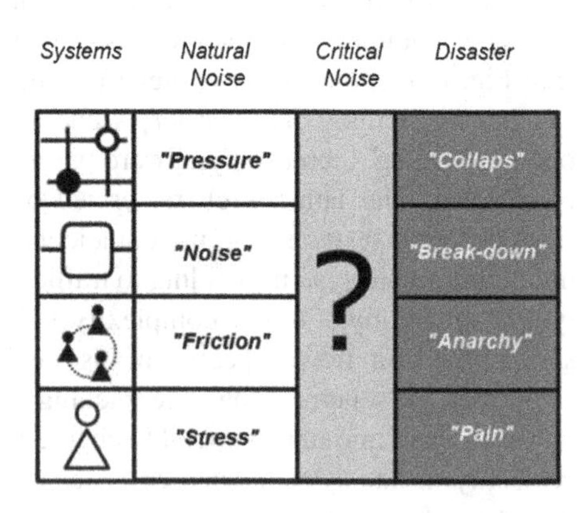

Fig. 6.5 Determining critical thresholds of amplification

		Misfit		
		Infrastructure misfits technology	↑	1
		Technology misfits infrastructure	↓	0
		Infrastructure misfits society	↑	1
		Society misfits infrastructure	↑	1
		Infrastructure misfits individual	↓	0
		Individual misfits infrastructure	↓	0

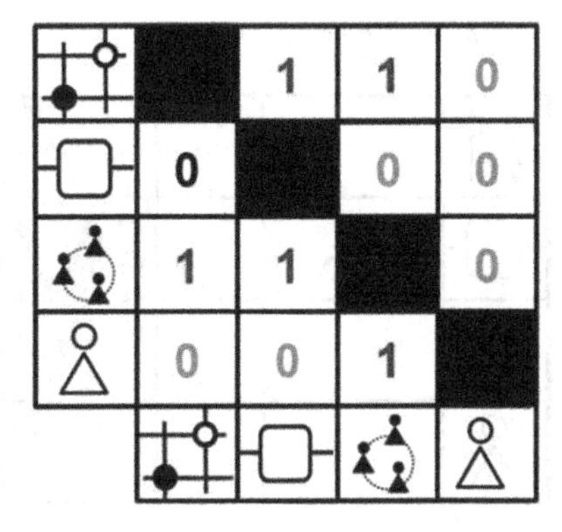

Fig. 6.6 "Frictiometrics". Determining critical misfits between systems

Fig. 6.7 Sensing and negotiating "Natural Noise"

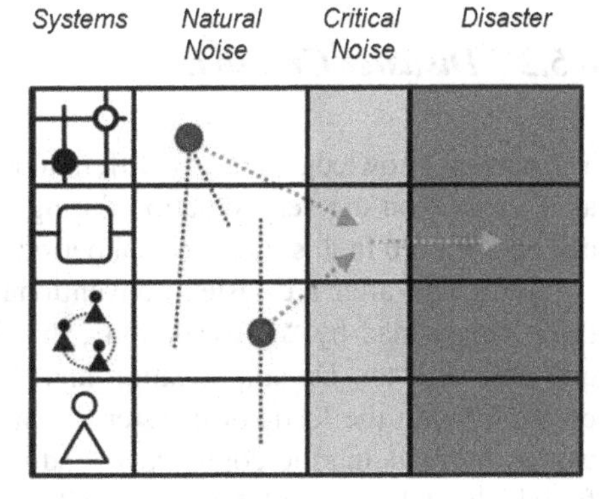

With the mapping of critically interlinked systems, a preventive mechanism can be schemed that is capable of sense the dynamics of urban noise on its way to disaster. A major component, however, is a comprehensive structure that identifies the various forms of friction e.g. pressure, tension, stress. Simply put, the task is to sense dynamic amplification of Natural Noise in "nervous" constellations. If amplification happens, quick negotiation of noise is needed before it reaches a critical level and develops into disaster (see Figs. 6.7 and 6.8).

As structures for sensing and negotiating the dynamics of noise in urban systems, two components are important: (1) As a major component of the sensing structure, not only sensor technologies (e.g. security surveillance, machine control, environmental monitoring) but also social structures (e.g. media, communities, opinions and memes) are important. For them, specific stress indicators, noise-factors, and friction signaling must be developed. (2) For the negotiation of critical noise on early stage, specific actors, agencies, and platforms are needed that function as noise brokers at the intersection of data analytics, insurance business, and urban management.

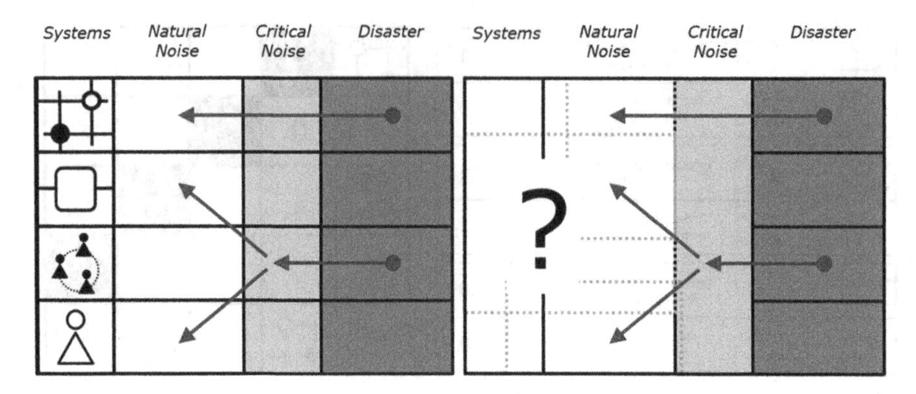

Fig. 6.8 Disaster management: de-amplification, de-escalation (*left*); design heuristics (*right*)

6.5.2 *Disaster Creativity*

The second knowledge strategy differs much from the intelligence-based first one, as it focuses on the very situation during disaster, and how appropriate responses can be achieved in this most critical period of time.

During and after a disaster, conventional disaster management seeks to reduce the problem-load by "de-escalation", that is: by a controlled return to the status quo ante disaster. De-amplification and de-complexification is mostly attempted by distributing the loads of disaster to various actors, societies, systems. In other words: by reducing the complexity load to partial problems of lower complexity, for which established solutions are available. In the schematic model (Fig. 6.8, left), this shows as a progress from the right column to lower orders of noise (central and left column). However, the deficit in this approach is indicated already in the term "management" which signals division of labor, routine tasks, and organization. Management is possible if there are reliable subsystems and defined scopes of action.

After serious disaster, however, situations have thoroughly changed. Before-hand conditions may have disappeared and be of no reference anymore. Institutions may have failed, infrastructures collapsed, key persons vanished. Instead of routine-based management, creative approaches are needed then. From a perspective of decision making, the return-path to de-escalation and consolidation resembles the heuristics of a design process. Here too, the route to be taken is unclear and without reliable structures. An unknown area ahead is to be systematically explored and organized. A time-graph of a disastrous event can picture this easily (Fig. 6.9).

Before the disaster strikes, the systems run steadily within in their usual range of volatility. After the strike, however, their functions are unsettled; they run unstable, if at all.. There is uncertainty and unsteadiness; chaos spreads. In order to establish order within this rapidly changing situation, an enormous capacity of ad hoc problem solving and creativity is needed. To be short: in many aspects, disasters

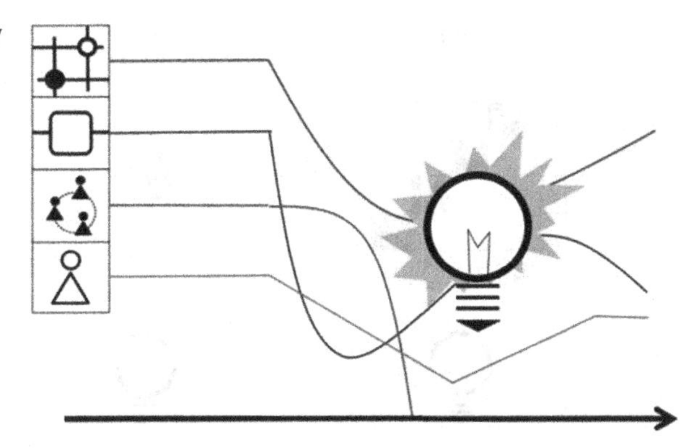

Fig. 6.9 Maximum creativity needed after disaster (*left*)

must be treated like complex design problems. That in turn implies the application of creative techniques and tools as well as methods for complex scheming such as improvisation, ideation, modelling, or prototyping.

The inversion of the argument holds true as well. Catastrophic failures of large-scale constructions and design projects e.g. Berlin Brandenburg International Airport make it plausible to treat complex design tasks as potential disasters too, as the far-reaching technological, economic, and socio-political avalanches suggest which were created by them.

The demand for ad hoc creativity can be supplemented with the cybernetic concept of "requisite variety" respectively System Theory's notion of "contingency": A system's adaptation to complex environment (which disaster is indeed) can succeed when the inner complexity of the system corresponds to the outer [6, 7]. System and environment are to be adjusted; their levels of organization must match. Then, in order to cope with the improbabilities of disaster, a struck system too is to behave with low probability: it has to be innovative and creative and match the scale of disaster with an adequate scale of inventiveness. Large-scale disasters are not properly responded to with common ideas and daily routine. From this a crucial demand infers: Especially under disaster conditions, the intrinsic capacities of invention, innovation, and creativity must get activated. Being creative in catastrophes is certainly a demanding task. Yet, there are few situations when creativity is more needed.

6.5.3 Noise to Knowledge

The third knowledge strategy attempts to establish a "learning cycle", as it combines the preventive strategy of "Noise Intelligence" with the reactive mode of "Disaster Creativity". In essence, it advocates a training scheme which transforms "information" and "creativity" into "knowledge" and "experience", thus forming a solid base for future disaster resilience.

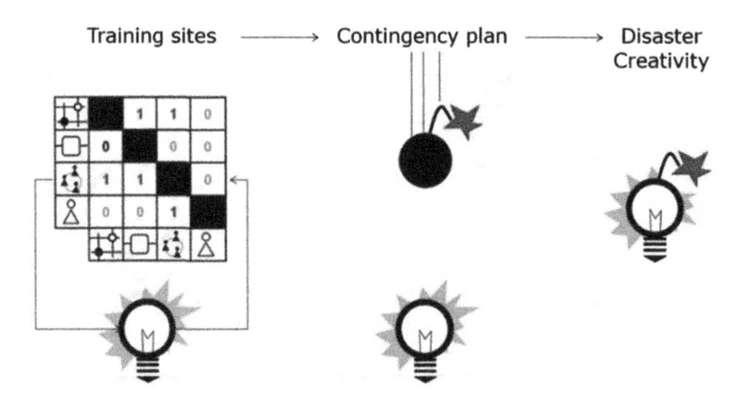

Fig. 6.10 From training of Natural Noise to Disaster Creativity

What has been called "Disaster Creativity" before can be hardly acquired in times of disasters striking. Firstly, because chances for training would be extremely rare. Secondly, because it would be too late then – the moment to possess disaster creativity has come already. The opportunity for training must be sought on the lower levels of Natural Noise, whose dynamics are described and analyzed by strategy 1, "Noise Intelligence". The noisy frictions and misfits between systems are not-yet disasters; however they provide testing grounds for disaster creativity. Natural Noise in the system, and awareness of it, are always a good strategy to ensure innovation, and creativity. This includes to the acceptance of new systems or system modifications introducing new noise to perfectly interacting systems. Especially for learning processes, Natural Noise is indispensable: to be "a bit disastrous" just enhances the capacities of reconstruction, recombination and reorganization. A major objective of strategy 1 "Noise Intelligence" therefore is to indicate the potential training grounds for strategy 2, "Disaster Creativity". The combination enables comprehensive preparation and contingency planning. By these forms of awareness and alertness, disasters will be viewed as rare events that give chance for radical innovation and for their faster acceptance (Fig. 6.10).

A creative and learning attitude to disaster, however, is a matter of cultural disposition and values. Different cultures have different notions of disaster. Not necessarily a positive yet inevitable issue, certain cultures welcome and accept disasters as chance for change and learning progress. In cases, they seem to create disasters in waiting – by confidently assuming that the most can be made from them. The "Urban Age" which is to a large extent "Asian Age" seems to be driven by such disaster creativity: the paramount challenges due to environmental pollution, overpopulation, low hygiene etc. are stimulating astonishing effort and inventiveness. The most advanced "Smart City"-concepts are in the making in Far East, and certainly include smart concepts for disaster resilience too.

6.6 Conclusions

As an indispensable part of their "smartness", so-called Smart Cities and communities should develop comprehensive disaster resilience. For this end, a sensitivity for frictions emerging from disharmonious urban systems is the basic parameter. In order to avoid their escalation to disaster scale, the early detection and negotiation of critical dynamics of frictions is necessary ("Noise Intelligence"). Coupled with the ability of ad hoc-problem solving and invention during and after crisis ("Disaster Creativity") an effective learning mechanism can be established. Disaster creativity must be trained on urban systems and interfaces which are most nervous and prone to escalation of noise.

References

1. Albino V, Berardi U, Dangelico RM (2013) Smart cities – definitions, dimensions, and performance. In: Proceedings IFKAD 2013, Matera
2. Spearman G (1904) General intelligence, objectively determined and measured. Am J Psychol 15:201
3. Arthur B (1997) The economy as an evolving complex system. Addison-Wesley, Reading
4. Arthur B (2009) The nature of technology: what it is and how it evolves. The Free Press, New York
5. Bettencourt L, Lobo J, Kühnert C, Helbing D, West G (2007) Growth, innovation, scaling, and the pace of life in cities. Proc Natl Acad Sci U S A 104:7301–7306
6. Ashby WR (1956) An introduction to cybernetics. Chapman & Hall, London
7. Luhmann N (1973) Zweckbegriff und Systemrationalität. Suhrkamp, Frankfurt a. M

Conclusions

[illegible]

References

[illegible]

Chapter 7
Challenges of Image-Based Crowd-Sourcing for Situation Awareness in Disaster Management

Guillaume Moreau, Myriam Servières, Jean-Marie Normand, and Morgan Magnin

Abstract One of the main issues for authorities in disaster management is to build clear situation awareness that is consistent and accurate in both space and time. Authorities usually have Geographical Information Systems (GIS) for reference data. One current trend of GIS is the use of crowd sourcing, i.e. gather information from the public to build very large databases that would be very costly otherwise. The OpenStreetMap project is one famous example of this. One can easily imagine that, provided some communications networks are available, photographs shot with smartphones could help to build knowledge about the disaster scene. Advantages would be that costly sensor networks would be less necessary, more objectivity would come from pictures than from the public interpretation and that a huge amount of data could be collected very fast. Yet this raises a number of challenges that we will discuss in this paper: the first is geo-referencing of pictures, i.e. locating the pictures received with respect to the current GIS. Even if GPS are more and more common on smartphones, not all pictures will have GPS coordinates and GPS accuracy might not be enough. The second one is the need for a spatio-temporal data model to store and retrieve redundant and uncertain data, the third one is the need for new visualization techniques given the amount of data that will be stored. Finally, legal and ethical issues are raised by the use of massive images.

G. Moreau (✉)
Computer Science and Mathematics Department, École Centrale de Nantes,
1 rue de la Noé, BP 92101, 44321 Nantes cedex 3, France
e-mail: guillaume.moreau@ec-nantes.fr

M. Servières • J.-M. Normand • M. Magnin
École Centrale de Nantes, 1 rue de la Noé, BP 92101, 44321 Nantes cedex 3, France
e-mail: myriam.servieres@ec-nantes.fr; jean-marie.normand@ec-nantes.fr;
morgan.magnin@ec-nantes.fr

H.-N. Teodorescu et al. (eds.), *Improving Disaster Resilience and Mitigation - IT Means and Tools*, NATO Science for Peace and Security Series C: Environmental Security,
DOI 10.1007/978-94-017-9136-6_7,

7.1 Introduction

When a natural (tornado, tsunami, earthquake, floods, etc.) or an industrial (nuclear, chemical) disaster happens it is essential for governments to be able to quickly delimit the spatial footprint of the disaster and evaluate the damages in order to react accordingly. The decision making process, based on the awareness of the disaster's consequences, has to prioritize where help has to be sent, both in terms of rescue forces (military and medical) and basic needs (water, food, medical care or shelter supplies) and as such is extremely an important aspect of disaster management. In this paper, we present how a crowd-sourced image-based for situation awareness system could be used to help in the decision making process. We define a crowd-sourced image-based for situation awareness system as a system based on geo-referenced images uploaded by users from the area where the disaster happened. These images should help understand which areas have been hit by the disaster as well as to estimate the importance of damages. As a consequence, in the following of this paper, we only discuss *localized* disasters (or catastrophes) i.e. that have a restricted spatial footprint like the ones named above. Questions arise as to which challenges such a system will have to face (uncertainty/liability of the geo-referenced images, etc.), what would be the limitations and advantages of a crowd-sourced image-based system compared to existing solutions? In the remainder of this paper, we will first present existing systems or approaches that address the problem of situation awareness for disaster management. Then we will describe the general architecture of a crowd-sourced image-based situation awareness system for disaster management before detailing the foreseen advantages. In a third part, we will focus on the challenges such systems have to face, in particular dealing with the spatio-temporal uncertainty of images, on data visualization and analysis issues, as well as on the legal and ethical aspects that might arise with disasters. Finally, we will draw a conclusion.

7.2 Related Works: Context

In recent years, satellite systems in combination with digital image-analysis technologies have proven to be very efficient tools for crisis or disaster (natural, industrial, humanitarian) management. Advances in sensing systems allowed for obtaining more and more complete (thermal, infra-red, radar, SAR, etc.) and precise (meter precision for optical and radar; decameter for thermal) images from the Earth while communication, network progress as well as cooperation agreements between countries significantly democratized access to huge databases of such satellite imagery.

As a consequence, satellite imagery has already been used as an efficient tool for disaster or crisis management in cases such as: oil spilling (e.g. [13]), earthquakes (e.g. [33]), tsunamis, forest fires, landslide, etc. (see [40] for examples).

NASA's Landsat, NASA's and Japan's ASTER [8] or ESA's ENVISAT programs are examples of successful satellite systems used for disaster or crisis management throughout the world.

Geographical Information Systems (GIS) allow to store, manipulate and analyze geolocalized data. Urban communities embedded in such spatial databases not only information about their networks (roads, water pipes, electricity and communication wires, etc.) but also risks cartography and risks mitigation plans. In the context of crisis or disaster management, GIS can be used in combination with crowd-sourcing Web technologies, to rely on inhabitants to provide or update information in the GIS. In [27], Roche et al. provide examples of how sharing photo, links, videos, etc. can provide meaningful information in time of crisis or disaster, such as the Katrina hurricane, the Christchurch earthquake, etc. Recently, within the TRIDEC European project Richter and Hammitzsch [26] proposed to develop a smartphone application dedicated to reporting and notifying natural disasters.

In a similar way, the OpenStreetMap project [20], shares georeferenced data and even allow people to contribute in case of crisis as after the Haiyan typhoon in the Philippines.

Of course, whenever humans are allowed to report or update sensitive crisis or disaster-related information, it becomes necessary to ensure that the information is reliable. This process should ideally be realized automatically, and be able to determine whether crowd-sourced data are reliable or not. As an example, Gupta et al. [10] showed that it was possible to automatically detect fake images of the Katrina hurricane aftermath that were posted on Twitter.

The use of GIS is very useful for risks prevention and was used for example on landslide prevention and management [3] or flood hazard management (in combination with remote sensing) [36] with a spatial description of risk zones. These simulations and geographical analysis allow a preparation and planning of rescues before the crisis if a part of it at least was predictable.

7.3 Proposal Overview

In this section, we will introduce what would a general images-based crowd-sourcing system look like. We will then detail the forecast advantages as well as the challenges that such a system would have to address before drawing a few assumptions for the remainder of the paper.

7.3.1 General System Vision

When disasters occur, situation awareness is obviously a key challenge for authorities in order to organize emergency response as well as resilience. It is sometimes difficult to obtain or consolidate data because of danger, access difficulties, lack of staff... Our key idea is to use images shot by the public to improve situation awareness.

We believe that being able to obtain information from the general public can be an asset in case a disaster happens as highly trained emergency professionals cannot be everywhere or numerous enough on a large space. Yet, it is common that people in distress situations overreact and they are by definition less skilled in describing complex and difficult situations than professionals. That is why we assume that using image-based contributions might be more objective than speech-based data. Therefore, our proposed system relies on:

- **Crowd-sourcing** which is according to Wikipedia, "the practice of obtaining needed services, ideas, or content by soliciting contributions from a large group of people". It has been successfully applied to large projects such as Wikipedia itself and OpenStreetMap [20] in the mapping domain.
- **Smartphones**: modern smartphones are equipped with a number of sensors, the most common one being a camera. The penetration rate of smartphones in developed countries is also very high. Most smartphones are also equipped with rough localization capabilities.

A potential usage scenario would be the following: In a disaster scene, many people would be able to take pictures of the scene, including damages and would send these images to a central system that would gather all information from various sources: recent images but also existing database such as digital maps, existing risk mitigation plans, simulations results... All this information would be processed by specialists of the command center who then would be able to make better decisions thanks to a more accurate and updated view of the situation. Associated with new visualization tools such as Virtual or Augmented Reality, better decision making could be achieved.

7.3.2 Foreseen Advantages

We believe our proposal has the following assets:

- The main advantage of our proposal is to extract the information from its origin, i.e. the location where the disaster took place. As the general public is the information source, we are likely to rely on massive data and thus coverage. Projects like OpenStreetMap have shown they have come to a very wide coverage. Moreover the variety of information sources should allow avoiding unreliable information to be trusted.
- We believe that real-time feedback is also possible because in recent disaster situations, social networks have been widely used to spread information as well as to report about personal issues in a timely manner. Today, most large accidents are reported on networks like Twitter earlier than on traditional media such

as television. Moreover, TV use those networks to get information, most news companies have emails or websites where they can be fed with news validating the fact that crowd-sourcing can be a reliable and real-time source of information.

- Images are more likely to be considered as *raw* (unbiased) data than text that might be transformed by users' affect and limited knowledge of the situation. Obviously an image can also have an affective value or a subjective composition that would give wrong ideas about the situation, but we argue that the number of images and their quite objective content can be of good value. Moreover, images are easier to geo-reference than speech. This topic will be addressed in the next paragraphs.
- Another advantage of our proposal is that it fully integrates existing data such as maps, risk mitigation plans or pre-computed simulation data as well as new information. This new information can help set the input parameters of simulation models. For example, the rise of the water can be roughly deduced from pictures and can be used as an input to determine a flood level over a territory. In a more general way, cross validation through various information sources is believed to be a good way of consolidating knowledge about the situation.

Yet, those claimed advantages raise a number of challenges that will be discussed in the next section.

7.3.3 Assumptions and Challenges

The first challenge raised by our proposal is image geo-referencing, i.e. how to determine the position and orientation of an image. As mentioned earlier, most smartphones are equipped with rough localization sensors such as GPS or magnetic compass. This is not enough in our case because GPS accuracy in an urban environment can be up to 100 m at 1 Hz. Furthermore, the accuracy of a magnetic compass is only up to 10 degrees, the two other rotations remaining undetermined.

The second challenge that needs to be addressed is the data model. Geo-referenced and time-stamped images must be stored into a database (that could be cloud-based) but also have to be mixed with other data including mapping information (GIS), simulation results, risk mitigation plans and inferred knowledge. The data model has to take into account heterogeneity of data in nature (thematic, 1D, 2D, 3D or even 4D), in scale and in time. It must also handle uncertainty as the inferred knowledge is also limited by missing or unreliable information. Multi-representation is also important.

The next challenge deals with information representation. Beyond the difficulties related to the various nature of data, the system must also be able to take into account massive data input as there can be numerous pictures of a single event.

At last, it has to be mentioned that collecting images that may include pictures of people raises a number of legal questions. Is it legal to distribute pictures of people who have not given their explicit consent, who is the owner of such pictures? Would disaster situations need a specific legislation such as what occurs in war periods?

Those challenges will be discussed in the next sections, yet we would like to lay the stress on an assumption that will remain true throughout the paper: we assumed that a minimum level of network connectivity would be available.

7.4 Challenges

7.4.1 Geo-Referencing Images

There is a great mass of geospatial data accessible from diverse origins and of different qualities but which may nevertheless serves as a support to on-site geolocalization. On the other hand, urban objects have specific geometries (most buildings have planar facades) that can facilitate computer vision tasks needed to compute localization.

To retrieve a precise localization from pictures taken in a disaster scene (complex scenes with presence of occlusions, illumination variations, etc.) it will be mandatory to mix pre-existing GIS data and data extracted from embedded sensors using a common mobile device. Various GIS data can be used: the first one is textureless, containing mainly geometric attributes (2D or 3D), while the second one contains textural information.

Localization based on pre-existent GIS data along with external information as images or videos is a registration problem. This kind of problem is usually achieved by solving the classical computer vision problem of pose estimation. Although it is a classical problem in mobile robotics or Augmented Reality that has received a lot of attention, it remains a scientific lock in the case of urban environments where the available 3D models are large, can only be partially observed under difficult conditions (lighting, occlusions, etc.) and where no artificial markers can be used conversely to traditional indoor Augmented Reality.

The 6D (position and orientation with respect to the real scene) camera localization problem, that will allowed to geo-reference images, is generally divided into two distinct phases. The first step is the initialization one, which corresponds to the (potentially inefficient) search of a first solution to the problem of the camera localization (typically a 3D position and 3D orientation). The second step is the tracking step, which consists of an incremental update of the 3D position and 3D orientation of the terminal in order to follow the camera movement. This is a difficult but largely addressed problem, including real-time calculation of camera pose at $t + dt$ as a function of the camera pose at time t and under a small movements assumption [41].

The pose computation must ensure six degrees of freedom, giving absolute measurements over a given reference, ensure high robustness, and operate in real time. This will be mandatory for AR visualization purpose but can take more time for information gathering.

Computing the position and orientation of a camera may rely on different techniques and could be classified as follows:

- Sensor fusion based: these techniques take advantage of multiple sensors (GPS, compass, inertial, acoustic) on a single device in order to compute the camera pose [15];
- Feature matching based: the most common techniques use feature points matching (also called keypoints or points of interest); those techniques heavily rely on feature detectors (see [34] for a review on existing detectors); other techniques relying on angles between facades [5] or skyline matching [45] also exist.
- Marker-based: the markers can be either "natural" (elements in the environment whose 3D models are known [28]) or artificial (e.g. the ARToolkit library [14]). They are mostly inapplicable in an outdoor context.
- Template-based matching: given a database of known images, these techniques try to find in the current frame some small patches of the database.
  Currently, well established methods are available for model based tracking, especially for multi planar environments [25, 29, 30] and open problems are common with the pose initialization problem: identifications of parts of the model which become visible, robust matching between the model and the image when the model has been acquired at distant time with different lighting and acquisition conditions, identification of good features to track for efficient pose estimation (texture, key-points, contours, sky lines).
- Texture-based methods: texture-based camera tracking proves to be efficient and robust when sufficient texture information is available in the scene. Planar surfaces [29, 30], CAD models [25] as well as free-form triangulated surfaces [38] have been considered. Existing methods rely on feature matching between subsequent video frames [30], video frames and keyframes [38] or video frames and synthetized images (tracking-by-synthesis) [25, 29].

In order to obtain a first rough localization of the device sending images, it is possible to use positioning sensors (GPS, accelerometers, etc.) that are available on many mobile devices. Nevertheless, data obtained by these sensors is relatively imprecise, especially in urban environments. It is thus necessary to rely on the use of the device's camera, even if the positioning sensors can also be used to compute a first rough estimate of the device's pose.

Challenges are first in the initialization process of geolocalization and then in the tracking phase (only for augmented reality based visualization). In fact, current approaches [28, 31] do not focus on the initialization phase and use assumptions or relatively hard prerequisites (panoramic images, complex 3D models of the environment, etc.). We believe that the initialization process leaves room for improvement. Particularly, we intend to take advantage of multi-sensor capabilities of today's mobile devices to improve the initialization phase and supplementing them with other interactive methods and leveraging GIS data input, even in a very degraded form (only 2D ground map as example).

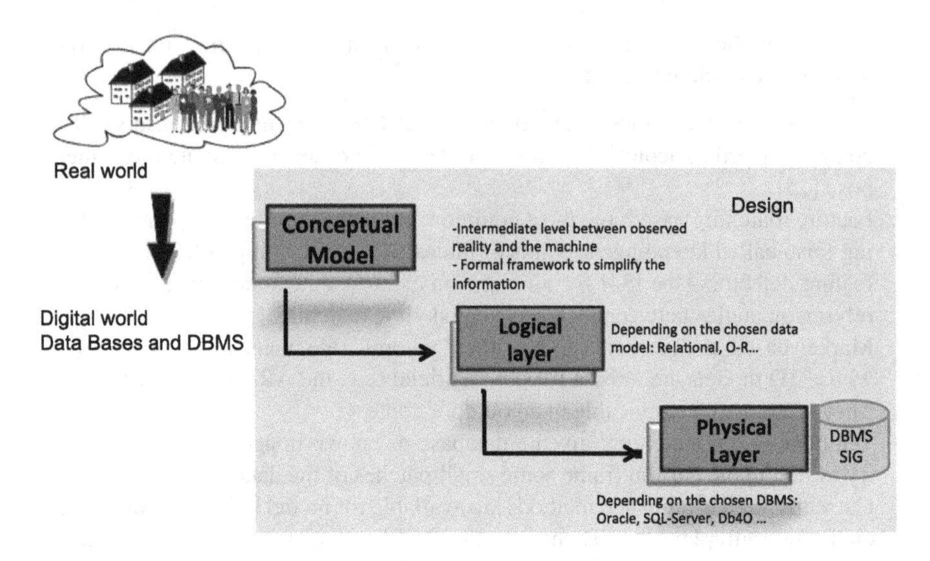

Fig. 7.1 Modeling stages [43]

7.4.2 Spatio-Temporal Models in Presence of Uncertainty

As pointed in Sect. 7.3.3, defining a data model capable of embedding all the complexity and heterogeneity of the geolocalized data we need is a challenge.

Modeling the data always starts with an observation step of the real world centered on the user needs. This will first provide with a conceptual representation of the data which is a formal framework to describe the information content. The next modeling steps after defining a conceptual model are the logical and the physical levels, as represented in Fig. 7.1. The logical scheme will be created from the conceptual scheme following the data model. Then the physical level will be the implementation of the data storage.

The key problem is to define the conceptual model that will meet our expectations. A conceptual model should provide user expressive means to express themselves and at the same time provide them readable and easy to understand data schemes. Regarding to our problem, the characteristics to model are diverse and complex. As presented in [43], himself based on [4, 19, 22, 23], requirements of conceptual models can be listed as follow:

- Compatibility with other modeling methods: data can be only spatial or only temporal but data can also be alphanumerical without any spatial or temporal characteristic. Also, composition, aggregation or generalization links between objects or more generally all complex structures should be modeled.
- Modeling of discrete and continuous views of space: such models should be able to represent at the same time localized objects considered as discrete or more global phenomena (as wind or pollution for example) considered as continuous.

- Orthogonality between model concepts: the user should be free to choose if the item he/she wants to represent will be an object, an attribute, an association between two objects, etc.
- Orthogonality between the structural, spatial and temporal dimensions of an object: a choice of representation in one dimension does not prevent from choosing other possibilities in another dimension.
- Events modeling: the conceptual model should allow the representation of events occurrences and their interactions.
- Spatial hierarchy: the conceptual model should provide a hierarchy of spatial concepts, to give the designer the ability to model simple objects (line, polygon, surface), complex objects and even objects without determined shape as well as to allow 3D modeling.
- Temporal hierarchy: simple and complex temporal concepts should be offered to model the whole item life-cycle.
- Modeling of historicity: keeping a record of some object change is mandatory for some study case.
- Spatial and temporal constraint on entity associations.
- Data multi-representation: allowing a different view on the data depending on the performed task (multi-scale for example).
- Uncertainty: some items' spatiality and/or temporality are known with an uncertainty that needs to be modeled.
- Interoperability of conceptual models: different actors use the same data with different modeling concepts. To work together they need easily readable diagrams.

Regarding these different criteria, the MADS model [21] is designated by various studies [9, 16, 18, 21] as the more complete model regarding design requirement for spatio-temporal data. It is a conceptual model that takes into account data multi-representation. The object-oriented structure can enrich it with multiple advantages as heritage or polymorphism. Zaki [43] has developed an extension of this model using an approach adapted to the characteristics of urban data, implementing a distinction between conceptual schemes to represent physical entities (continuants) and conceptual diagrams to represent events (occurents). But this model does not implement uncertainty yet.

Spatio-temporal models that handle data uncertainty have been mostly developed in the archaeology field [6, 7]. These data have the same imperfections as the ones we need to handle: uncertainty, inaccuracy, ambiguity and incompleteness. They can be used to complete the enrichment of the MADS model and as such, could be a viable solution to our data modeling problem.

7.4.3 Analysis and Visualization Issues

Being able to analyze and visualize geo-referenced data in real-time raises interesting challenges.

At first, data analysis for crowd-sourced systems has to address the problem of data reliability, namely can we trust data we receive and should it be stored in our GIS? Several algorithms have been developed to address this issue, including voting schemes, general consensus, outliers detection but as our group is mainly involved into imaging and GIS, we consider them as being out of the scope of this paper.

Once data is analyzed and can be trusted, there remains the problem of visualizing it. What should be displayed and how? We believe that a system for situation awareness for crisis or disaster management should be able to visualize both 2D and 3D data. Indeed, 2D data are easier to visualize and understand, but 3D conveys more information which can prove to be important.

As a consequence, we believe it is important for the system to let the user switch between allocentric (topview) and egocentric (also called first-person or immersive) views in order to display 2D and/or 3D information at different level of details. As a consequence, we will focus in the following on how to cope with various levels of details of information as well as on a type of visualization interesting for a situation awareness system, namely augmented reality.

Yet there remains a discrepancy between 2D and 3D maps. On the one hand, 2D maps include a lot of abstraction whereas 3D models tend to be as photo-realistic as possible. Therefore, some researchers among which Döllner and our group have proposed scenarios and techniques to include more abstraction in 3D city models. The works included the proposal of Urban or Geographic level-of-details (LoDs) that go beyond the concept of traditional Geometric LoDs, the latter techniques being dedicated to a small number of objects having large number of polygons whereas here we are facing a large number of low-polygons objects. Proposed techniques are based on the (well known in 2D) concept of generalization, i.e. adapting the amount of information to the scale of the map. This implies going beyond traditional simplification techniques multi-level building-level aggregation based on viewpoints for example. An example of Geographic LoDs can be seen in Fig. 7.2.

A further challenge is to be able to display adequate information on such 3D maps. It obviously involves the problem of legibility of such maps that could be decomposed into two sub-problems:

- **Labeling**: label placement is a more complex issue in 3D than in 2D because of the additional perspective problem and because it is view-dependent which requires real-time algorithms conversely to 2D maps.
- **Information density** as for geographic LoDs that take into account the specific features of geographic data with respect to traditional geometric models, information density and clarity must also be adapted to the users' need, hence the concept of semantic LoDs [44].

An interesting solution for decision makers while dealing with a crisis and/or disaster situation would be to be able to visualize real-time information directly on physical paper maps. Information can either be projected directly onto the physical map, or can be visualized through a medium (such as a smartphone or a tablet) that films the physical map and adds information when displaying the map on its screen (see example in Fig. 7.3).

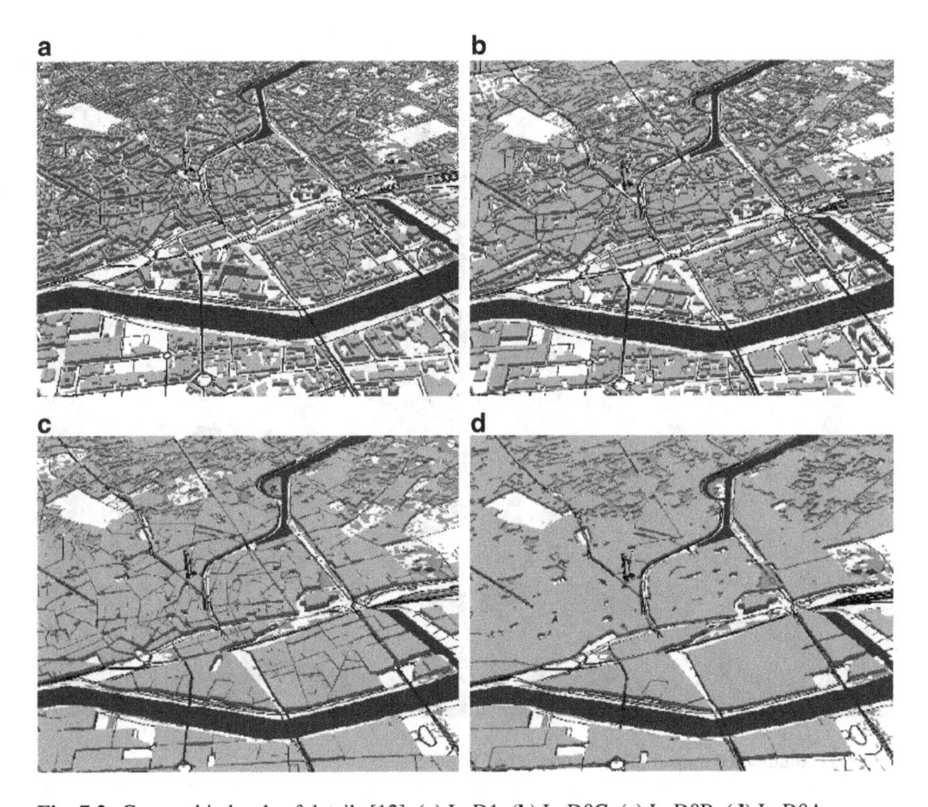

Fig. 7.2 Geographic levels of details [12]. (**a**) LoD1. (**b**) LoD0C. (**c**) LoD0B. (**d**) LoD0A

In order to display the information where it should be onto the physical map, it is necessary to film the physical map via a camera and to compute the pose (i.e. 3D position and 3D orientation of the camera with respect to the physical paper map and thus with respect to the GIS. This process has been achieved by Uchiyama et al. [35] for prepared physical maps where for example the road intersections where manually marked before being detected by the camera and mapped onto the GIS information. The authors are currently carrying ongoing work to adapt this process to any unprepared physical paper map.

These so-called Augmented Maps allow for a physical interaction with the paper maps while augmenting them in real-time with meaningful 2D or 3D information. In our case, we propose to display in real-time the evolution of the GIS information that is updated by the crowd-sourcing mechanisms discussed above.

Apart from the need to visualize real-time information in the command center, we also propose to be able to visualize data on the site of the disaster or crisis. This boils down to being able to use a smartphone or tablet on-site (i.e. outdoors) in order to visualize additional information through the device's screen at the correct position on screen and in real-time. Being able to visualize GIS data on site can help assessing correctness of data collected by the crowd-sourcing mechanisms.

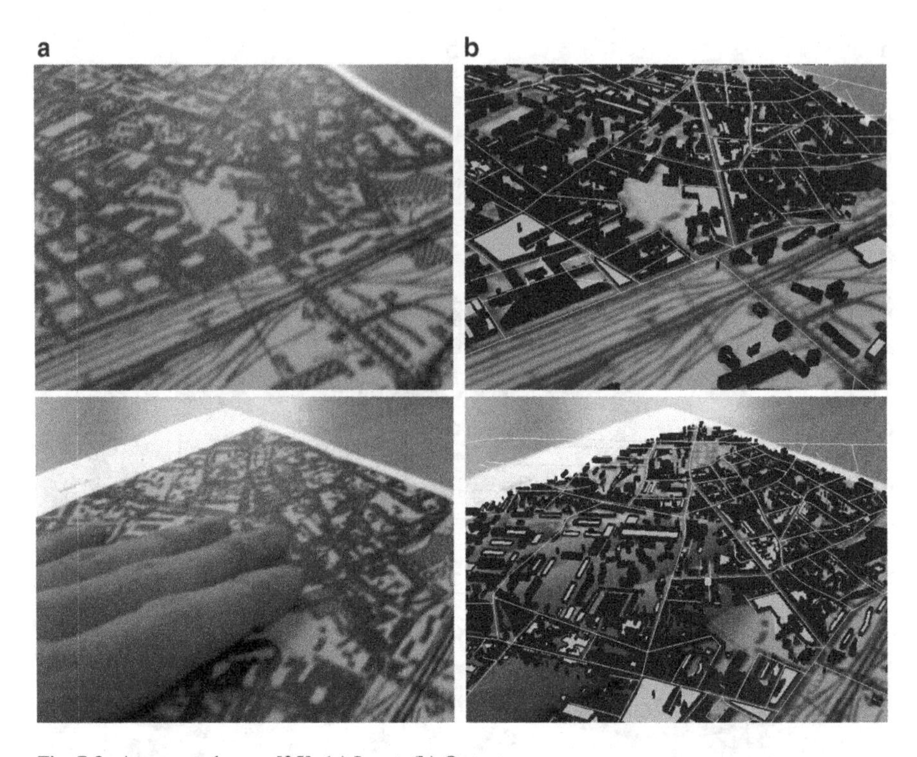

Fig. 7.3 Augmented maps [35]. (**a**) Input. (**b**) Output

Insitu Augmented Reality (AR) techniques can be used to achieve such a goal, but two main scientific locks still need to be addressed:

- Pose computation problem: computing in real-time (at frequencies between 15 and 50 Hz) the pose of the device with respect to the world.
- Data visualization: smartphones or tablets have (relatively) small screens implying to carefully select and layout information in order to avoid overlap or clutter.

Recent democratization of mobile devices with embedded geolocalization, gyroscopic, accelerometer and imaging capabilities allowed for the development of mobile AR applications. Nevertheless, GPS information is relatively unprecise with common embedded systems, especially in dense urban areas. Although there exist professional positioning systems but they remain expensive and cumbersome. Applications such as CityViewAR [17] allow for visualizing geolocalized 3D models in augmented reality based only on GPS information, but suffer from constrained initialization steps and do not allow for real-time tracking of the displayed information.

As mentioned in Sect. 7.4.1, many approaches proposed to combine GPS and gyroscopes or accelerometers data with computer vision techniques based on images obtained by the devices embedded cameras but also suffer from limitations such as the need for a preliminary 3D acquisitions of the surroundings [1], or are based on panoramic images [2, 39]. Nevertheless, such data fusion techniques lay the groundwork for real-time AR 3D information display on mobile devices [25] or even for real-time video/GIS registration [45].

7.4.4 Ethics and Legal Issues

Crowd-sourcing raises major challenges w.r.t. the respect of legislation. As described by [42], many fields are impacted, e.g. inventorship, copyright and data security. As we consider here an approach directly provided by researchers and engineers, the patent issue is not as critical as in the case where external workers participate in the design of the solution. But considering our methodology, the most immediate problem concerns the copyright of the processed images. Recent trials remind us that it is not because an image is shared publicly on a social network that it can be used by a third party, even for the benefits of illustration of an article: in November 2013, it was judged that Agence France-Presse and Getty Images wilfully violated the Copyright Act when they used photos taken by Daniel Morel in Haiti after the 2010 earthquake [37]. The companies were condemned to a fine of $1.2 million. In addition, the copyright issue is a worldwide challenge, as most countries have adopted similar regulations thanks to the Berne convention.

Another major issue relates to personal data processing, meaning: how images that may be filled with personal information (for example faces but also GPS coordinates) can be inserted and managed in a giant database without the consent of the related persons? Especially in Europe, personal data processing is a very hot topic and requires extreme caution not only when gathering data, but also when ensuring the database's security. This is enforced by national commissions which role is to supervise every personal data processing. If the disaster to be monitored takes place in Europe, the United States or Japan, the legislative framework w.r.t. personal data processing to consider is very different [32]. How can we face such differences? These challenges can be addressed in two different ways. The first – and most idealistic one – would be to imagine that countries agree to enforce a law considering a very special status to every document shared publicly on the web in case of a major disaster event. As long as data are used for rescue purposes only and are stored and processed only during a short amount of time, there may be an exception regime, allowing to circumvent every national regulations towards copyright, personal data processing, etc. This may be the same approach as the one adopted in the case of war [11] or health issues, but even exceptions are generally quite hard to handle [24]. In addition, even if large-scale disasters (like the 3.11 Earthquake in Japan) may help to build a social consensus to rely on exceptional measures for managing data in case of a large event, it is still very unlikely.

That is why a more pragmatic project should build a social platform dedicated to sharing information in case of a disaster. This platform would be identified as a service per se, answering a very specific – yet important – target, which is the reaction to disasters. In the General Condition of Use, it should be stated that every uploaded resource would be shared under a license (for example Creative Commons), allowing the processing of images for scientific motivations and that every third-party personal data that may be processed (for example the information related to other people appearing in an image or a video) will not be shared with others. This is connected to crucial ethics considerations, especially about the respect of human dignity (avoid the spread of images of recognizable dead corpses or badly injured bodies).

7.5 Conclusions

In this chapter, we have presented what we believe could be an interesting system for gaining a better situation awareness in disaster situations. It is based both on digital images and crowd-sourcing. Indeed, we believe that this image-based crowd-sourcing would be of interest because crowd-sourcing has shown to be able to fill large databases and to be able to provide coverage of large disasters through social networks. Moreover images may be superior to speech(text)-based systems because they convey more information that can be further analyzed by skilled professionals rather than the general public and may be less subjective content. This proposal implied a minimum level of network connectivity to have a chance of working.

Yet, such a system raises a number of challenges in terms of image geo-referencing, spatio-temporal data model and visualization. We have shown the state-of-the-art in the respective areas of those challenges and proposed some of our solutions. The complete system remains a proposal, we have only tried to show that the challenges raised by our proposal could be addressed.

References

1. Arth C, Wagner D, Klopschitz M, Irschara A, Schmalstieg D (2009) Wide area localization on mobile phones. In: ISMAR, Orlando, pp 73–82
2. Arth C, Klopschitz M, Reitmayr G, Schmalstieg D (2011) Real-time self-localization from panoramic images on mobile devices. In: ISMAR, Basel, pp 37–46
3. Assilzadeh H, Levy JK, Wang X (2010) Landslide catastrophes and disaster risk reduction: a GIS framework for landslide prevention and management. Remote Sens 2:2259–2273. doi:10.3390/rs2092259
4. Beard K (2006) Modeling change in space and time: an event based approach. In: Innovations in GIS: dynamic and mobile GIS: investigating change in space and time. Taylor & Francis, Hoboken, pp 55–74
5. Bioret N (2010) Image geolocazation using a GIS. PhD thesis (in french), École Centrale de Nantes
6. De Runz C (2008) Imperfection, temps et espace: modélisation, analyse et visualisation dans un SIG archéologique. PhD thesis, université de Reims
7. Desjardin E, de Runz C, Pargny D, Nocent O (2012) Modélisation d'un SIG archéologique et développement d'outils d'analyse prenant en compte l'imperfection de l'information. RIG 22(3):367–387. doi:10.3166/rig.22.367-387

8. Duda KA, Abrams M (2012) Aster satellite observations for international disaster management. Proc IEEE 100(10):2798–2811
9. Friis-Christensen A, Tryfona N, Jensen C (2001) Requirements and research issues in geographic data modeling. In: Proceedings of the 9th ACM international symposium on advances in geographic information systems, Atlanta, pp 2–8
10. Gupta A, Lamba H, Kumaraguru P, Joshi A (2013) Faking sandy: characterizing and identifying fake images on twitter during hurricane sandy. In: International conference on world wide web companion, WWW'13 Companion, Rio de Janeiro, pp 729–736
11. Hayes C, Kesan J (2012) At war over CISPA: towards a reasonable balance between privacy and security. Illinois public law research paper (13-03)
12. He S, Moreau G, Martin J (2012) Footprint-based generalization of 3D building groups at medium level of detail for multi-scale urban visualization. Int J Adv Softw 5(3&4):377–387
13. Jha MN, Levy J, Gao Y (2008) Advances in remote sensing for oil spill disaster management: state-of-the-art sensors technology for oil spill surveillance. Sensors 8(1):236–255
14. Kato H, Billinghurst M (1999) Marker tracking and HMD calibration for a video-based augmented reality conferencing system. In: Proceedings of IEEE and ACM international workshop on augmented reality, San Francisco, pp 85–94. doi:10.1109/IWAR.1999.803809
15. Klein G, Drummond T (2004) Tightly integrated sensor fusion for robust visual tracking. Image Vis Comput 22(10):769–776
16. Laplanche F (2002) Conception de projet SIG avec UML. Liège: bulletin de la Société géographique de Liège 42:19–25
17. Lee GA, Dünser A, Kim S, Billinghurst M (2012) CityViewAR: a mobile outdoor AR application for city visualization. In: ISMAR AHM, Atlanta, pp 57–64
18. Minout M (2007) Modélisation des Aspects Temporels dans les Bases de Données Spatiales. PhD thesis, université libre de Bruxelles
19. Miralles A (2005) Ingénierie des modèles pour les applications environnementales. PhD thesis, Montpellier university
20. OpenStreetMap. http://www.openstreetmap.org
21. Parent C, Spaccapietra S, Zimanyi E (2006) Conceptual modeling for traditional and spatio-temporal applications: the MADS approach. Springer, Berlin
22. Parent C, Spaccapietra S, Zimanyi E (2009) Semantic modeling for geographic information systems. In: Encyclopedia of database systems. Springer, New York, pp 2571–2576
23. Pelekis N, Theodoulidis B, Kopanakis I, Theodoridis Y (2005) Literature review of spatio-temporal database models. Knowl Eng Rev 19:235–274
24. Regidor E (2004) The use of personal data from medical records and biological materials: ethical perspectives and the basis for legal restrictions in health research. Soc Sci Med 59(9):1975–1984
25. Reitmayr G, Drummond T (2006) Going out: robust model-based tracking for outdoor augmented reality. In: Proceedings of IEEE ISMAR'06, Santa Barbara, pp 109–118. doi:10.1109/ISMAR.2006.297801
26. Richter S, Hammitzsch M (2013) Development of an android app for notification and reporting of natural disaster such as earthquakes and tsunamis. General Assembly European Geosciences Union
27. Roche S, Propeck-Zimmermann E, Mericskay B (2013) GeoWeb and crisis management: issues and perspectives of volunteered geographic information. GeoJournal 78:21–40. doi:10.1007/s10708-011-9423-9
28. Seo B-K, Park J-I, Park H (2011) Camera tracking using partially modeled 3-D objects with scene textures. In: ISVRI, Singapore, pp 293–298
29. Simon G (2011) Tracking-by-synthesis using point features and pyramidal blurring. In: ISMAR, Basel, pp 85–92. doi:10.1109/ISMAR.2011.6162875
30. Simon G, Fitzgibbon A, Zisserman A (2000) Markerless tracking using planar structures in the scene. In: Proceedings of international symposium on augmented reality, Darmstadt, pp 120–128

31. Steinhoff U, Omercevic D, Perko R, Schiele B, Leonardis A (2007) How computer vision can help in outdoor positioning. Ambient Intell Lect Notes Comput Sci 4794:124–141. doi:10.1007/978-3-540-76652-0_8
32. Tan DR (1999) Personal privacy in the information age: comparison of internet data protection regulations in the United Stats and European union. Loy LA Int'l & Comp LJ 21:661
33. Tronin AA (2009) Satellite remote sensing in seismology. A review. Remote Sens 2(1):124–150
34. Tuytelaars T, Mikolajczyk K (2008) Local invariant feature detectors: a survey. Found Trends Comput Graph Vis 3(3):177–280
35. Uchiyama H, Saito H, Nivesse V, Servières M, Moreau G (2008) AR representation system for 3D GIS based on camera pose estimation using distribution of intersections. In: International Conference on Artificial Reality and Telexistence (ICAT), Yokohama, pp 218–225
36. Uddin K, Gurung DR, Giriraj A, Shrestha B (2013) Application of remote sensing and GIS for flood hazard management: a case study from Sindh Province, Pakistan. Am J Geogr Inf Syst 2(1):1–5. doi:10.5923/j.ajgis.20130201.01
37. U.S. District Court for the Southern District of New York, Case agence france-presse v. morel, No. 10-02730, 11 (2013)
38. Vacchetti L, Lepetit V, Fua P (2003) Fusing online and offline information for stable 3D tracking in real-time. In: CVPR, II, Madison, vol 2, pp 241–248. doi:10.1109/CVPR.2003.1211476
39. Ventura J, Hollerer T (2012) Wide-area scene mapping for mobile visual tracking. In: ISMAR, Atlanta, pp 3–12
40. Voigt S, Kemper T, Riedlinger T, Kiefl R, Scholte K, Mehl H (2007) Satellite image analysis for disaster and crisis-management support. Geosci Remote Sens 45(6):1520–1528
41. Wagner D, Reitmayr G, Mulloni A, Drummond T, Schmalstieg D (2010) Real-time detection and tracking for augmented reality on mobile phones. IEEE Trans Vis Comput Graph 16(3):355–368
42. Wolfson S, Lease M (2011) Look before you leap: legal pitfalls of crowdsourcing. Am Soc Inf Sci Technol 48(1):1–10
43. Zaki C (2011) Modélisation Spatio-temporelle multi-échelle des données dans un SIG urbain. PhD thesis, École Centrale Nantes
44. Zhang F, Tourre V, Moreau G (2013) A general strategy for semantic levels of details in 3D urban visualization. In: Eurographics workshop on urban data modelling and visualisation, Girona, pp 33–36
45. Zhu S, Morin L, Pressigout M, Moreau G, Servières M (2013) Video/GIS registration system based on skyline matching method. In: IEEE ICIP, Melbourne, pp 3632–3636

Chapter 8
Foresight and Forecast for Prevention, Mitigation and Recovering after Social, Technical and Environmental Disasters

Natalya D. Pankratova, Peter I. Bidyuk, Yurij M. Selin, Illia O. Savchenko, Lyudmila Y. Malafeeva, Mykhailo P. Makukha, and Volodymyr V. Savastiyanov

Abstract The methodology for foresight and forecasting of social, technological and environmental disasters for prevention, mitigation and recovering after disaster situations is considered. The proposed approaches can be effectively applied both for a preliminary forecast of possible disasters and for modelling, preparation and evaluation of measures to prevent and control potential disasters. Methods also allow obtaining and improving of the control and managing in disasters, and provide decision-making support for disaster situations. Developed approaches are very flexible to use in systems of various nature, such as technical, environmental, social, human, economic and others. Developed tools allow reducing the time, financial and human resources for recovering from the effects of disasters in conditions under inaccuracy, incompleteness, fuzziness, untimeliness, noncredibility, and contradictoriness of information.

8.1 Introduction

Throughout its life, humanity has been facing natural disasters. Storms, volcanic eruptions, typhoons, rockfalls are all the processes of nature. With the development of civilization the humanity starts to influence on nature. Phenomena such as, for example, increased levels of CO_2, which changes the landscape, reducing the level of ozone are a consequence of human activity.

After humanity reached industrial era with a certain level of technology, it had faced to another type of disasters – technological disasters. Thereby accidents at

N.D. Pankratova (✉) • P.I. Bidyuk • Y.M. Selin • I.O. Savchenko • L.Y. Malafeeva
M.P. Makukha • V.V. Savastiyanov
Institute for Applied System Analysis NTUU "KPI" 37, 35 building,
Peremogy Avenue 37, Kyiv 03056, Ukraine
e-mail: natalidmp@gmail.com

H.-N. Teodorescu et al. (eds.), *Improving Disaster Resilience and Mitigation - IT Means and Tools*, NATO Science for Peace and Security Series C: Environmental Security,
DOI 10.1007/978-94-017-9136-6_8,

nuclear power plants, chemical plants, train wrecks, chemical products evolved in the air also became a part of our lives.

In recent years, people are concerned with other type of disaster – social. For example, that could be humanitarian catastrophes as a result of military action. Riots in many countries are caused by unreasoned actions of governments, and list could be continued.

All these disasters differ by nature, they all have various reasons and all have a different course. All they could manifest themselves as separate from each other or in a chain with other disasters. In that case we can say that they are closely interrelated. All these disasters as processes have the same features:

- diversity and diverse nature of the causes and factors, as well as actions that lead to their appearance;
- spatial distribution of the conditions of uncertainty in time and space dynamics of the regions and their impact;
- unsteady properties and the uncertainty of their characteristics.

The close relationship of disasters can be uncovered by the following examples. Technogenic accident at the Chernobyl nuclear power plant in 1986 led to the environmental disaster – the contamination of area. That, in turn, led to the social catastrophe – forced evacuation of a large number of people. Abnormal natural phenomenon – the strongest in the history of Japan earthquake and ensuing tsunami led to a man-made disaster – the accident at Fukushima in 2011. And the result was the same – a social disaster, evacuation, etc.

The strongest typhoon "Khayan" in 2013 led to the social catastrophe in the Philippines – a massive destruction of towns and villages, the death of thousands of people. Social catastrophe could be also the result of "unreasonable" actions of government. For example, the actions of the governments of some countries had led to the "Arab Spring." Technological disaster on an oil platform in the Gulf of Mexico in 2010 led to an environmental disaster – coastal oil pollution and changes in the ecosystem of the region.

The sequence of disasters can be arbitrary. Social unrest could lead to the industrial accidents, and then as a consequence, to the ecological catastrophe.

These properties and characteristics determine the practical need to study a variety of properties, relationships, interactions, interdependencies of diverse factors and causes of dangerous processes based on a single system approach to achieve an unified position, defining objectives for the management and control of the situation – the timely prevention and/or minimization of the undesirable consequences. However, analysis shows that at the present time various types of technological, environmental and social processes, their causes, occurrence, effects and scope are investigated separately, excluding relationships, interdependencies, interaction. These approaches do not take into account some important factors that influence these processes, the severity of undesirable effects, the ability and effectiveness of prevention.

Consider the mathematical formalization for modelling processes in order to predict or foresee the possible prevention or mitigation of different natural disasters.

8.2 Application of Modern Forecasting Techniques

To increase quality of managerial decisions, risk management, quality of automatic control for engineering systems and technology it is necessary to develop new forecasting techniques directed towards further improvement of short- and medium term forecasting. Existing forecasting methods that are based on various analytical procedures and logical rules in many cases cannot provide desirable quality of forecasting results. That situation requires from researchers to look for new efforts in enhancing quality of forecasts estimates.

8.2.1 Adaptive Regression Analysis Approach

Correct application of modern modelling and adaptive estimation techniques, probabilistic and statistical data analysis provide a possibility for organizing computing process in a way to get higher quality of forecast estimates in situation of structural, parametric and statistical uncertainties. Such uncertainties arise due to availability of nonstationary and nonlinear process under study, missed data records, noise in the measurements, extreme values and short samples. One of the possibilities for adaptation provide Kalman filtering techniques that generate optimal state estimates together with short term forecasts in conditions of influence of external stochastic disturbances and measurement noise [1, 2].

We propose a concept of a model adaptation for dynamic processes forecasting based on modern system analysis ideas that supposes hierarchical approach to modelling and forecasting procedures, taking into consideration possible structural, parametric and statistical uncertainties, adaptation of mathematical models to possible changes in the processes under study and the use of alternative parameter estimation techniques aiming to model and forecast estimates improvement.

We propose the new adaptive scheme that is distinguished with several possibilities for adaptation using a complex quality criterion. The model structure estimation is a key element for reaching necessary quality of forecasts. It is proposed to define a model structure as follows: $S = \{ r, p, m, n, d, z, l \}$, where r is model dimensionality (number of equations); p is model order (maximum order of differential or difference equation in a model); m is a number of independent variables in the right hand side; n is a nonlinearity and its type; d is a lag or output reaction delay time; z is external disturbance and its type; l are possible restrictions for variables. For automatic search of the "best" model it is proposed to use the following criteria:

$$V_N(\theta, D_N) = e^{|1-R^2|} + \ln\left(1 + \frac{SSE}{N}\right) + \\ + e^{|2-DW|} + \ln(1 + MSE) + \ln(MAPE) + e^U,$$

where θ is a vector of model parameters; N is a power of time series used; R^2 is a determination coefficient; DW is Durbin-Watson statistic; MSE is mean square error; $MAPE$ is mean absolute percentage error; U is Theil coefficient. The power of the criterion was tested experimentally with a wide set of models.

There are several possibilities for adaptive model structure estimation: (1) automatic analysis of partial autocorrelation for determining autoregression order; (2) automatic search for the exogenous variables lag estimates (detection of leading indicators); (3) automatic analysis of residual properties; (4) analysis of data distribution type and its use for selecting correct model estimation method; (5) adaptive model parameter estimation with hiring extra data; (6) the use of adaptive approach to model type selection. These models as well as classification trees and Bayesian networks have been used to forecast the direction of stock price movement. Application of the concept described provides the following advantages: (1) automatic search for the "best" model reduces the search time several times; (2) it is possible to analyze much wider set of candidate models than manually; (3) the search is optimized thanks to the use of complex quality criterion. Testing of the system with stock price data showed that it is possible easy to reach a value of absolute percentage error of about 3–4 % for short term forecasting.

8.2.2 Kalman Filtering

Kalman filtering algorithms could be easily hired for solving short term forecasting problems in the frames of adaptation procedure given above. The models constructed according to the adaptation scheme considered should be transformed into state space representation form that makes it possible further application of the Kalman type optimal filtering algorithm [3]. An advantage of the approach is in the possibility of model adjusting to random external disturbances and taking into consideration possible measurement errors.

8.2.3 Bayesian Networks

Bayesian networks (BN) or Bayesian belief networks are probabilistic models in the form of a directed graph the vertices of which represent selected variables and arcs reflect existing cause and effect relations between the variables [4].

General problem statement touching upon application of Bayesian networks includes the following steps: (1) thorough studying of a process being modelled; (2) data and expert estimates collecting; (3) selection of a known or construction of a new method for model structure estimation (learning); (4) BN parameter learning (conditional probability tables construction); (4) development of a new or selection of known inference method; (6) testing the BN constructed using actual and generated data. In spite of the fact that general theory of BN has been developed quite well, usually many questions arise when a specific practical problem is solved.

8.2.4 Hidden Markov Model

A hidden Markov model (HMM) is a statistical Markov model in which the system being modelled is assumed to be a Markov process with unobserved (hidden) states. A HMM can be considered the simplest dynamic Bayesian network [5]. In a hidden Markov model the state is not directly visible, but output, dependent on the state, is visible. Each state has a probability distribution over the possible output tokens. Therefore the sequence of tokens generated by an HMM gives some information about the sequence of states.

8.2.5 Group Method for Data Handling

The group method for data handling (GMDH) is a powerful modern instrument for process modelling and forecasting developed at the Ukrainian National Academy of Sciences in the second half of last century by Ivakhnenko [6]. It generates the forecasting model in the form of the Kolmogorov-Gabor polynomial that could be used for describing linear and nonlinear systems. The problem statement for application of the technique should include the following elements: (1) selection of partial descriptions that create a basis for the possible final model; (2) selection and adaptation of the model parameters membership functions for a particular application; (3) development of a new or application of known model parameter estimation technique; (4) selection of a model quality criteria for the use at intermediate computation steps and for the final model selection.

8.2.6 Generalized Linear Models

Generalized linear models (GLM) is a class of models that extend the idea of linear modelling and forecasting to the cases when pure linear approach to establishing relations between process variables cannot be applied [7]. GLM constructing can be considered from classical statistics or a Bayesian perspective. The problem statement regarding such type of model construction is touching upon the following elements: type of prior distribution for model parameters; a method for parameters estimation using appropriate simulations techniques; posterior simulation etc. GLM is successfully applied to solving the problems of classification and nonlinear process prediction.

8.2.7 *Combination of Forecasts*

The problem of forecasts combination arises in the cases when one selected technique is not enough for achieving desirable quality of forecasting. In such cases it is necessary to select two or more ideologically different forecasting techniques and to compute combined estimate using appropriately selected weights [8]. In a simple case equal weights are applied to the individual forecasts. Other approaches to computing these weights are based on previously found prediction errors for each method.

8.3 The Foresight System Methodology

8.3.1 *The Foresight Importance and Goals*

Challenges and threats, the potential level of disasters consequences of social, technical or environmental nature, being aware by the modern society, make it necessary to foresee objectively the future scenarios at least approximately. Foresight is the process of decision making for complex systems with the human element concerning their possible future behaviour. This process applies individual methods in a certain sequence while establishing certain interrelations between them. The process is formed with the help of the more universal methodology known as the ***scenario analysis***.

The foresight is required to form a rational and fault-free development strategy of any organized society (nation, country, organization or company) in the world, which is being constantly tested by nature where tough competition and the risks of social, technical and environmental disasters exist.

For complex systems with the human factor the essential uncertainties of data, information and multifactor risks are inherent to various sorts of their behaviour. The expert judgments concerning their qualitative characteristics always have subjective character. Thus, in view of all specified features of these systems concerning their behaviour in the future, the quite certain decisions in form of the scenarios and strategies of their development should be accepted.

The scenario construction can be realized with the help of the universal set of means and approaches, named scenario analysis methodology, which is the complex of mathematical, software, logical, and organizational means and tools for determination of the sequence of method application, their interconnections, and the formation of the foresight process, in general.

In this methodology several qualitative and quantitative analysis methods are used for solving the foresight problem. These methods are used at four stages of the foresight.

The first stage – is the preliminary study of the foresight problem. At the first stage for the preliminary study of the problem it is necessary to analyze, in more detail, its characteristics, to define research directions (focuses or platforms) and form the most important criteria and goals for the given task. Methods used here in the essence and under the organizational forms are very simple. A correct application of the methods has an essential importance, since the loss of the input information at this stage will result in significant mistakes and unjustified expenses at performance of the complex of works on the technology foresight.

The second stage – is the qualitative analysis of the problem. At the second stage several methods are used to perform the qualitative analysis of the objects, events and phenomena in research. Many of these methods were thoroughly researched and improved by IASA and some are shortly described below.

The third stage – is the scenario construction. At the third stage, an empirical nine-step procedure of scenario construction using STEEPPV program can be applied. In literature this procedure is known as "The Method of Scenario Writing".

The fourth stage – is the analysis and selection of scenarios. At this stage the scenarios are presented to a group of people who are to take strategic decisions; and the comprehensive analysis of these scenarios is carried out considering the reality and feasibility level determination for each scenario; probability estimation of events on which scenarios are based; risks connected with each of the scenarios.

When solving the foresight problems it is important to find experts from the most skilled specialists in the subject domain, and also to use the newest mathematical support and powerful information technologies. This should guarantee the most accuracy and adequacy of possible scenarios of the future.

The objective knowledge and creative assumptions approach in an interactive man-machine procedure allows one to increase the accuracy of the development scenarios for processes, phenomena, and events under investigation. The above-mentioned process is realized on the basis of creation and usage of the universal set of means and approaches, named the Scenario Analysis Information Platform (SAIP) [9]. This platform is a complex of mathematical, software, logical, organizational and technological means and tools for the foresight process realization on the basis of person's interaction with specially created for this program-technical environment. It should be noted that in solving practical problems of the foresight on the basis of SAIP, as a rule, the estimates are received in the on-line mode via Internet.

The software used in this platform is the network-based information system of decision making for future scenarios construction that includes the special mathematical apparatus and the convenient and flexible user interface. With the help of Internet, this interface allows one to involve experts from any part of the world; quickly obtain and process their decisions; and organize communication, interaction, and information exchange among them. All of that, together with the employed analysis and expert information processing methods, allows us to obtain the well-grounded high quality solutions for problems related to every specific expertise.

8.3.2 *Expert Estimations Based on Modified Delphi Method*

In the process of solving the foresight problems experts form possible events, conditions, solutions, evaluate the accuracy of the assumptions and hypotheses, the importance of the objectives on the basis of information given to them about the subject area, considered objects and relationships, characteristics, and performance [9]. The proposed method of construction of the survey forms for expert evaluation method based on the Delphi method helps to solve problems related to:

- automated generation of the survey forms;
- correctness of the wording and presentation of the issue in order to avoid ambiguity interpretation of its meaning;
- complete description of the context of the situation with its consequences, in which the question is asked, as well as the elimination of contextual redundancy;
- ergonomic structure of the survey forms with a focus on the most important experts of key issues;
- obtaining the results of the survey in the form of kid-machine interface and decision-makers;
- inevitability of errors due to manual procedures for processing large amounts of information.

To achieve the objectives, a formalization of creating profiles stages was carried out and a module of automated generation of the survey forms, which consists of an issue generation procedure and directly questionnaire form generation procedure was developed.

A universal approach to the construction of the survey forms is to combine different applications of the method of expert assessment based on the Delphi method in the analysis phase of the process of technological foresight that has been realized in complex algorithmic procedures.

8.3.3 *Scenario Analysis Using Morphological Models*

One of the highly productive ways of dealing with complex, unstructured problems is applying the morphological analysis method. The essence of this method is in analyzing objects, processes or phenomena by picking several their characteristics, which can have alternate values. By combining different values for each characteristic, a very large multitude of object configurations is produced and can be easily analyzed using expert estimations for the constituent parts [9].

This technique is especially useful when considering scenarios for a problem. This model identifies the key factors and parameters of the scenario and provides the means to evaluate their probabilities (for non-controlled parameters) or efficiencies (for controlled parameters), taking into account the connections and interdependencies between them. These values can be used to determine the most important or critical states of the scenario parameters, to rank these states by the probability or efficiency of their implementation, to select the most probable scenario configurations, and also as the input data for other methods. Thus a great multitude of scenarios and response strategies is analyzed at once without burdening the experts with too much questions.

A morphological table can be used to assess different scenarios of disastrous situations and prepare response strategies for these scenarios. This can be done both for planning multi-disaster management measures and for preparation of strategies against a specific type of disaster.

Several other add-ons for the method are implemented, such as classification of scenarios by degree and level of risks and supporting the morphological model on time intervals, considering changes to probabilities of the alternatives with the flow of time.

8.3.4 Hierarchy-Based Qualitative Analysis

This method originates from the classic analytical hierarchy process, designed by the American scientist T. Saaty as a tool for decision making [10]. The method is based on using the so-called hierarchical networks to construct a model aimed at calculating the estimates of the potential future scenarios. The levels of the hierarchy contain alternate factors that directly or indirectly contribute to the main goal of the problem. The method employs pair-wise comparisons between these alternatives to assess their importance, and different synthesis methods to evaluate their influence on the top of the hierarchy that constitutes the main goal of the problem. The hierarchies can be quite large, providing the means to construct close to real world models and support decision making in problems with many contradictory goals, criteria and actors, which is the usual situation in the foresight problems.

Contrary to the other similar methods, the idea of this method is to achieve focus, or convergence to a single concept from the expert conclusions and actions of the numerous participants in the foresight process. In this case the method is based on a causal view at the processes that provide the foundation for the construction of the future scenarios.

A system approach was introduced in IASA to get a reliable solution based on a modified analytical hierarchy method [11]. This approach consists of the modified analytical hierarchy method for treating fuzzy expert estimations, the method of assessing the solution alternatives based on their benefits, opportunities, costs, risks (BOCR), the approaches to evaluate the rank reversal risks in the synthesis method and a complex sensitivity analysis procedure.

The application of the proposed above methodologies and techniques to solve problems associated with environmental, technical and social disasters as a software tool is developed [9].

8.4 Modelling and Technologies for Restoration of Oil Polluted Soils and Water Bodies

On the basis of system methodology using the forecast and foresight methods the project in direction of environmental disasters was realized. The problem of cleaning soils and water bodies from oil pollution, development of new technologies of system restoration of oil-polluted lands is a high priority problem. Oil discharges and accidental oil spills are a particularly important problem, because it often causes contamination of the underground water and drinking water sources. Oil pollution is the essence of environmental threats, which every country risks to face every day and hour with.

The absence of a modelling tool set of natural and manmade disasters does not allow timely and systematically implementation of their prevention, foresight and forecast. In this project the system restoration of oil polluted soils and water bodies on the basis of integrated tools for modelling and microbiological technologies is proposed.

8.4.1 Integrated Tools Set of Modelling and Microbiological Technologies for Restoration of Oil-Polluted Soils and Water Bodies

The proposed integrated tools are based on the foresight methodology, forecasting processes and mathematical models, methods of system analysis, as well as the technological principles of obtaining carbon bioactive materials of bio-destructive type [12].

The system methodology of foresight makes it possible to build and estimate alternative variants of application of microbiological technologies in the given region using objective quantitative information and experts' judgements about the status of the region [9]. These variants are compared in terms of various economic, ecological, political, social and other criteria, situational and other risks. When the system methodology of foresight is used, experts at different fields, for instance, environmentalists, government, public organizations and other are involved in the process of alternatives estimation. In this case alternative variants of application of microbiological technologies in the region are compared in terms of goals of the involved actors. As a result one or several first-priority alternatives are obtained [13].

The adaptation of mathematical models, according to realistically investigated contaminated soil and water bodies, is realised in Siberia, Russia and in the Crimea, (former) Ukraine [14, 15]. These models are based on equations of mathematical physics, allowing to estimate the boundaries of latent contamination, the parameters of the distribution pollution, the concentration sphere of the pollutant, the points and areas to maximize its content, direction and rate of its drifting in the environment [16]. The modified methods of hidden Markov chains are used to determine the probability of generating the parameters of the pollutants spread, improve the quality of the whole modelling process and evaluation of foresights that provide a distribution sphere of harmful substances in water-saturated soil layer and solid. The adequacy of the constructed mathematical models and their compliance with the basic dynamic processes, which occurred in biological systems subjected to anthropogenic impacts and the effectiveness of the proposed approaches to the control of natural restoration processes and technology systems are shown. The new and modified methods of system analysis as the most universal and fitting the current requirements are applied. In its turn, this kind of analysis is the application of the methodological, mathematical and organizational tools, which are designed to identify internal and external relationships and interactions between the processes of object, the means of description, the estimation of parameters, modelling, and as a consequence – forecasting and scientific foresight [9]. This allows building and evaluating alternative options for the application of bioactive materials in the region, taking into account objective information and expert opinions on the state of the region. In its turn, it allows the decision maker to get the most complete, accurate and most important, timely information about possible or unavoidable adverse impacts of hazardous processes of eco environment.

Modelling of microbiological technologies of petropolluted grounds and reservoirs is realized on the basis of available aprioristic data of monitoring, expert estimations and results of prediction technology. The production of biosorbents is based on adsorptive material with immobilized oil-oxidizing microorganisms, which have the ability to localize and destroy oil and oil products in their localized state [12]. Destruction of oil can be conducted up to the last stages, when only products of oil are left, which are carbon dioxide, water and other ecologically inert components. The residual part of the biosorbent is the initial base of the sorptive material. The process of oil biodestruction is conducted both on the surface and in the deeper layers of ground. Biosorbents often are the only means of dealing with oil pollution. Application of sorbent of this type allows quickly and effectively localizing accident-related oil spills and providing the destruction of oil and oil products.

For correct restoration of polluted soils it is necessary to investigate the distribution of contamination. The migration of pollution in ground water is determined by geological conditions of the surface layers of soil. Analysis of pollution distribution is made by numerical simulations using algorithms, which are based on the finite element method.

Stage of Modelling for Restoration Stage of modelling is a scientific basis for the calculation of field distribution of oil pollution. This allows one to determine the geometrical parameters of pollution, its density, concentration, rate of distribution, study the prognosis and foresight of the spread of contamination and then efficiently solve the problem of calculating the required volume and concentration of the sorbent. Traditionally, this stage is absent when the clean-up of soil and water from pollution with oil and evaluation of these parameters is not performed. Such a strategy without justification forecast and foresight often leads to significant long-term process of restoring soil and water pollution and is costly.

In order to combat oil pollution it is important to know the main characteristics of the oil slick, which could be most easily found from forecasting mathematical models, which include:

- forecasting based on different time series models;
- exponential smoothing;
- model of linear growth;
- Bayesian networks technique.

All of the named methods and techniques were scientifically proved and tested in a number of projects. Many of these methods were significantly improved compared to the analogues that are used in the world. On the basis of proposed models the following problems are solved:

- identification of boundaries and depth of the affected environment;
- identification of the physical characteristics of the environment;
- purification velocity of the used sorbent for cleaning the contaminant – oil inclusions;
- short-and medium-term forecast of the processes of spreading and dissipation of pollutants;
- minimization and optimization of the places where sorbent is used;
- minimization and optimization of the placement and operation modes of the gaging equipment;
- computer graphics implementation of the simulation results in a given time scale.

With the help of foresight methodology and methods of forecast the following problems are solved:

- scientific ground of application expedience of bioactive materials in specific region on the basis of the technological foresight;
- identify the causes of the pollution;
- definition of the geometric parameters (3-D) of the border pollution;
- calculation of the concentration of the field of pollution;
- calculation of the trend of the pollution;
- determination of risks that have emerged due to the presence of pollution;
- calculating the required volume (and quality) of the sorbent.

Process of water bodies purification is shown in Fig. 8.1. After cleaning water bodies the biosorbents, saturated by oil products, gradually sink to the bottom and

Fig. 8.1 Process of water bodies purification

utilize themselves as a result of oil product biodestruction. After cleaning solid surfaces the biosorbents, saturated by oil products, are moved to a sump, where they utilize themselves as a result of oil product biodestruction.

The main goal of the proposed strategy is to guarantee a rationally justified reserve of survivability of a complex system in real conditions of fundamentally irremovable information and time restrictions.

The main idea of the strategy is to ensure the timely and credible detection, recognition, and estimation of risk factors, forecasting their development during the definite period of operation in real conditions of complex objects operation, and on this basis ensuring timely elimination of risk causes before the occurrence of failures and other undesirable consequences.

The benefits of successful implementation of the integrated tools are as follows:

- return of the restored areas into economic use;
- recultivation and fertility increase of the land;
- preventing the possibility of oil penetration into underground water layers;
- increase in quality of life for people in the region;
- strengthening of NATO positive image and improvement of public mood.

The cost of land restoration includes:

- data collection, monitoring, gathering samples of soils for the area;
- research and detailed analysis of the problem;
- manufacturing, transfer and spread of sorptive materials;
- organizational activity and popularization.

8.5 Foresight and Early Prediction of Social Negative Trends, Abnormal Situations and Disasters Based on Social Media Analysis

An example of situations relating to social disaster and the approaches to uncover the weak signals of future social disaster are given below.

Today Twitter, Instagram and local bloggers are in some cases much more faster or accurate in displaying of picture in disaster area. In other cases the social media are the information weapon of different groups of interest (political, financial, etc.),

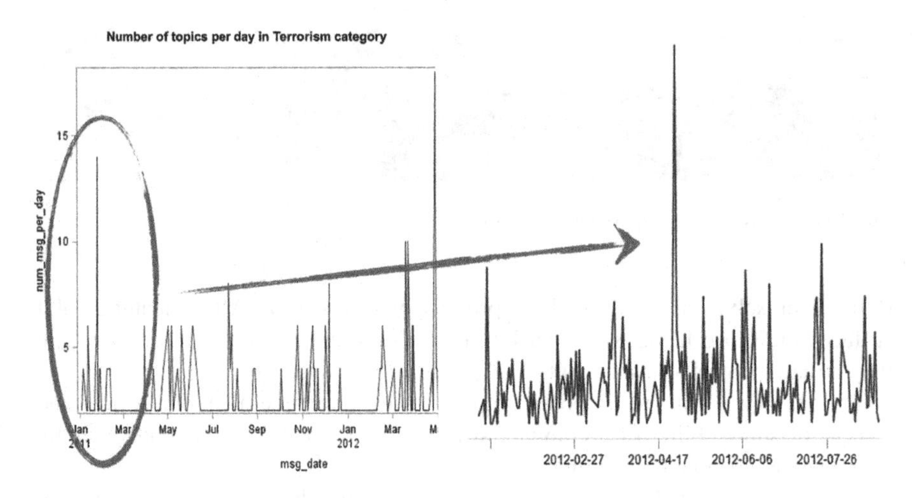

Fig. 8.2 Information peaks on terrorism messages in social media

which are using facts beyond the context to achieve their goals. Analysing the threats and social reaction in past, i.e. not only a year ago, also yesterday, especially in case of disaster, we study how people deal with this disaster and what could be helpful to overcome this disaster.

Internet and global interconnection, smartphones and social networks made us to find new way how to discover the information. Any significant situation from social point of view immediately rises up traffic and count of messages. Of course, if only you have an instrument to categorize it and do "measurement".

For example, the tragical event in Dniepropetrovsk (Ukraine) at the 27th April 2012 – terrorist attack immediately had a peak in social media (Fig. 8.2). But not all persons or organizations were able to display the clear information.

Even the simple analyses showed how unclear and unsure information was published, not synchronized in this day even among government or social forces. What about other sources – the bloggers tried to make a sensation, writing by the principle: the more panic – the better shock for audience and more "clicks".

Also political forces tried to make a political game, making sometimes really strange or irrational statements. Among other situations some really bad consequences of panic under terrorist attack were discovered, when, for example, the panic was the background for bank robbery. Therefore, some social forces will definitely use disaster as a good reason to make a business or crime. There are large numbers of such examples, marauding in Russia in flood areas, and so on ...

So the way is to make text analytics to uncover situation before, under and after critical circumstances through the social media analysis to overcome social crisis and shock. The goal is to develop a system methodology of foresight and prediction [9] of negative social trends and disasters based on early indicators from social media and semistructured data. The project consists of the following stages: (1) Identify examples of past negative social trends, abnormal situations, and disasters;

describe those cases in a formalized way; (2) Use foresight methodology for identifying key factors, risks, and possible consequences of investigated processes; (3) Identify relevant data sources: structured (statistical) and semistructured [17] (social media, news, blogs); (4) Collect data from the sources selected, transform and prepare data for the analysis; (5) Build and assess descriptive and predictive models of social processes under investigation; (6) Develop a system methodology of foresight and prediction of negative social trends and disasters based on early indicators from social media.

One of the hardest stages is the linguistic modelling of problem domain, especially if the domain is social, like politic or law making. If the classification and clustering of the subject areas is more convenient, the extraction of social reaction and sentiments are more complicated because of structure of language. The model in politic domain was developed on basis of analyzing sentiments in articles and messages on Eastern Partnership summit in Vilnius in November 28, 2013 and the European integration sentiments in society.

The model consists of number of situations, described by a set of rules on basis of which the analysis was done in aspect-level sentiment mining [18]. The model was built using SAS(R) Sentiment Analysis (SA) Studio syntax [19] adopting with N-gram situation based lemma rule set for the flexible Russian and Ukrainian languages on the first level, combining with aspect based rules on the second level.

The results were obtained on the Eastern Partnership summit messages in terms of "political crisis and social disaster in case of armed confrontation" topic. The quality of results was checked manually using SAS SA Workbench software for $\sim$1,000 preselected messages and the rule set of 800 rules does cover about 76 % of identified aspects.

The results were processed with several statistical methods to uncover the trends in social reaction, the time gap between reaction of political forces inside and outside the country, identifying the related subject (i.e. criminal, economical and financial activities) until social disaster situation. Also the groups of collaborating and confronting media sources were detected based on sentiment per aspect profile combined with subject correlation profile in case of poor data. The trends and results detected in the research could be used in the actions to overcome the consequences of social crisis in political, social, legal and other areas and uncover and measure the sustainability of democracy system in the country before, until and after social crisis avoiding the information (i.e. knowledge) noise from corrupted media sources.

References

1. Balakrishnan AV (1984) Kalman filtering theory. Optimization Software, Inc., New York, 222 p
2. Korbicz J, Bidyuk PI (1993) State and parameter estimation. Technical University of Zielona Gora, Zielona Gora, 303 p
3. Zgurovsky MZ, Podladchikov VM (1995) Analytic methods of Kalman filtering. Naukova Dumka, Kyiv, 285 p

4. Cowell RG, Dawid AP, Lauritzen SL, Spiegelhalter DJ (1999) Probabilistic networks and expert systems. Springer, New York, 321 p
5. Baum LE, Petrie T (1966) Statistical inference for probabilistic functions of finite state Markov chains. Ann Math Stat 37(6):1554–1563
6. Ivakhnenko AG (1982) An inductive self-organization method for complex systems models. Naukova Dumka, Kyiv, 296 p
7. Dobson A (2002) An introduction to generalized linear models. CRC Press Company, New York, 221 p
8. Bidyuk PI, Romanenko VD, Timoshchuk OL (2013) Time series analysis. Polytechnika, NTUU "KPI", Kyiv, 607 p
9. Zgurovsky MZ, Pankratova ND (2007) System analysis: theory and applications. Springer, Berlin/London, 475 p
10. Saaty TL (2008) Decision making with the analytic hierarchy process. Int J Serv Sci 1(1):83–98
11. Pankratova N, Nedashkovskaya N (2013) Estimation of sensitivity of the DS. AHP method while solving foresight problems with incomplete data. Intell Control Autom 4(1):80–86. doi:10.4236/ica.2013.41011
12. Pankratova N, Khokhlova L (2012) Integrated tools for restoration of oil polluted soils and water bodies. Int J Inf Theories Appl ITHEA SOFIA 12(1):39–49
13. Pankratova ND, Savastiyanov VV (2009) Modeling of alternatives in the technology foresight scenarios. Syst Res Inf Technol 1:22–35
14. Khohlova L, Shvets D, Khohlov A (2004) Patent of Ukraine № 43974. CO2F3/34, BO1J20/20. A sorption bio-destructive material for cleaning up the soil and water subsurface from oil and petroleum products. Published on 15.03.2004, Bulletin №3
15. The application of bioactive sorbent in oil spill in the fields of OGDP Barsukovneft (Rosneft) (2001–2003) Russia № 00–2 ПН-01, Joint research project UKRAINE-Russia, Area of investigation: ecology
16. Pankratova ND, Zavodnik VV (2004) System analysis and evaluation of dynamics for ecological processes. Syst Anal Inf Technol 2:70–86
17. Buneman P (1997) Semistructured data. In: Proceedings of the ACM symposium on principles of database systems. Abstract of invited tutorial, Tucson, pp 117–121
18. Hu M, Liu B (2004) Mining opinion features in customer reviews. In: Proceedings of nineteenth national conference on artificial intelligence (AAAI-2004), San Jose, July 2004
19. Reckman H, Baird C, Crawford J, Crowell R, Micciulla L, Sethi S, Veress F (2013) Rule-based detection of sentiment phrases using SAS Sentiment Analysis. http://www.cs.york.ac.uk/semeval-2013/accepted/79_Paper.pdf

Chapter 9
Disaster Early Warning and Relief: A Human-Centered Approach

Ernesto Damiani, Mustafa Asim Kazancigil, Fulvio Frati, Babiga Birregah, Rainer Knauf, and Setsuo Tsuruta

Abstract Effective need collection is a major part of post-disaster assessment and recovery. A system for matching needs with available offers is an essential component for the recovery of communities in the aftermath of natural disasters. In the classic approach, the needs of disaster-stricken communities are collected by rescue personnel sampling a geographic grid and using emergency communication facilities. Subsequently, needs are matched to offers available at some central server and relief interventions are planned to deliver them. In this chapter, after an introduction to this approach, we explore the new notion of *peer-to-peer community empowerment* for some key activities of disaster management, including *early warning* systems, *needs* collection, and *need-to-offer* matching. Then, we present the design of a framework that leverages on the capacity for the communities to self-organize during crisis management, proposing advanced algorithms and techniques for need-offer matching. Our framework supports pre- and post-disaster use of social networks information and connectivity via an evolvable vocabulary, and supports metrics for resilience assessments and improvement.

E. Damiani (✉) • M.A. Kazancigil • F. Frati
Computer Science Department, Università degli Studi di Milano, via Bramante, 65, I-26013 Crema (CR), Italy
e-mail: ernesto.damiani@unimi.it

B. Birregah
CNRS Joint Research Unit in Sciences and Technologies for Risk Management, University of Technology of Troyes, Troyes, France

R. Knauf
Faculty of Computer Science and Automation, Technische Universität Ilmenau, Ilmenau, Germany

S. Tsuruta
School of Information Environment, Tokyo Denki University, Tokyo, Japan

H.-N. Teodorescu et al. (eds.), *Improving Disaster Resilience and Mitigation - IT Means and Tools*, NATO Science for Peace and Security Series C: Environmental Security, DOI 10.1007/978-94-017-9136-6_9,

9.1 Introduction

In this chapter we present the design and implementation of a framework that leverages on the self-organization capacity of disaster-stricken communities during crisis management. Our framework deals with some crucial activities to be performed before an emergency situation, like *Early Warning* (EW), and with other post-emergency activities like *Need/Offer Matching* (N/OM) as essential components of efficient relief operations. Experience gained in the field shows that EW may considerably alleviate the impact of disasters, while N/OM plays a crucial role in improving and facilitating community and critical infrastructures resilience.

We start from the basic, informal notion of *process model* as a "recipe for achieving a goal" [1] to provide a simple process model for disaster management that will be used throughout the chapter. Like most process models, our model is composed of *activities* to be executed (conditionally, serially or in parallel) by *actors* playing pre-defined *roles*. Furthermore, our process model is expected to be monitorable and measurable (e.g. via Key Performance Indicators – KPIs). The traditional KPIs include *business continuity*, *time-to-recovery*, etc.

Case-study analysis has shown that major activities in the disaster management process model include *Early Warning*, *Needs Collection*, *Need/Offer Matching*, and *Offer Delivery*. Each of these activities will be described in detail in the next sections; they can be all modeled as *sub-processes*, i.e. composite activities composed of finer-granularity steps. For instance, the N/OM activity consists of two sub-activities: *Rapid Needs Assessment* and *Offer Matching*. Rapid Needs Assessment aims to quickly provide accurate information to match a community's emergency needs with available resources in the aftermath of a disaster. Early assessments combined with rapid mobilization of resources can significantly reduce the consequences of a disaster [2]. Rapid needs assessments can establish the extent of the emergency; measure the emergency's impact; identify existing response capacity; and inform priority response actions [3].

Usually, activities in the disaster process model rely on different IT platforms: early warning systems (Sect. 9.3) use existing cellular networks and digital television channels, while need/offer matching (Sect. 9.4) uses the Internet or relies on ad-hoc telecommunication networks established as part of relief operations.

Actors performing our process model's activities play roles recognized by national and international regulations and treaties dealing with rescue and relief operations, like Governmental Relief Agencies and NGOs (Non-Governmental Organizations.)

In this chapter we tackle (the community's correct response to) Early Warning and Need/Offer messages from a novel perspective, i.e., as emergent collective behaviors to be facilitated. Our vision is backed by the experience of recent earthquakes in Italy, the post-tsunami nuclear emergency in Japan, or the hurricane Sandy emergency on the East Coast of the United States; which all demonstrated that communities should be empowered to establish and run their own need/offer marketplace, for a quicker and more efficient recovery process. The assessment of needs and efficient pooling of common resources emerge as people and organizations use social networks to express their needs and submit offers to an intelligent marketplace capable of recommending suitable matches.

A key aspect when enabling peer-to-peer need/offer matching is checking the trustworthiness of the offers. Lack of trustworthiness is sometimes due to user maliciousness, but more often to lack of information: experience has shown that only 8–20 % of tweets generated during a crisis include enough situational data (time, location, coordinates, etc.) to be practically useful [4]. Also, needs expression may be impaired by users' lack of information on the vocabulary to be used in their requests. We start from the idea that communities need to express emergency and post-emergency needs according to their own conceptualization. This way, it is hoped that pooling of common resources will spontaneously emerge as people use their own vocabulary to submit offers and needs to a marketplace capable of recommending suitable matches.

The five major features of our approach are: *(i)* modeling disaster-stricken communities' needs/offers using dynamically negotiable vocabularies, allowing communities to express their emergency and post-emergency needs according to their own conceptualization (rather than the one from a regulatory body); *(ii)* use of existing social networks like Twitter for geo-located needs/offers submissions with the help of GIS technology; *(iii)* trusted checking of user location and other context information to assess the credibility of needs/offers; *(iv)* innovative artificial intelligence models able to evolve and adapt matching algorithms based on context changes and feedback; *(v)* providing advanced spatio-temporal analytics modules to assess the recovery process.

9.2 An Evolvable Vocabulary for Warnings, Needs and Services

Today, most disaster warning and assessment systems are *interview-based*. Interviews rely on geographic grids in order to ensure adequate coverage. Once the interviews have been carried out, matchmaker systems are used as focal points to receive donors' offers (in terms of goods, services, and human resources) and match them to the needs of a disaster-stricken community, identifying gaps in assistance and finding ways to fill them. Some matchmakers use operation research algorithms to compute the "best" N/OM towards reaching a preset Recovery Point Objective (RPO) or Recovery Time Objective (RTO). In most of open crisis management platforms, the matching process (when it exists) is not explicitly highlighted. It is up to the user to make his/her own matching between needs and offers that are recorded or displayed by the platform. This process can be tedious in several cases, especially when one must deal with huge amounts of heterogeneous needs and offers. Though it is possible to use crowdsourcing-based approaches in many situations, disaster management platforms need semi-automatic approaches to foster the efficiency of the matching process [4].

Table 9.1 Example of needs and offers classes

Health care		**Energy**
Emergency unit		Power plant
Hospitals		Filling-station (diesel, gas, ...)
Doctors		Electricity distribution network
Transportation		**Telecommunication**
Main Highway	Airport	Telephone network (landline and mobile)
Traffic condition	Railway	Internet access
Train station	Heliport	
Accommodation		**Supply warehouse**
Temporary accommodation centre		Food Stocks
Shelter		Water

Outside the disaster management community, N/OM has become an increasingly popular feature of general-purpose social networks. Tools and platforms like *Google Persons Finder, Ushahidi, Sahana*, etc., have introduced the notion of N/OM to the general public, and have reached a good level of user acceptance.

Recently, some specific platforms that attempt to connect the needs and services offering in the wake of disasters have emerged. During the last hurricane Sandy (on October 29th, 2012, along the northeast coast of the United States), one witnessed the launch of several platforms with the aim of helping to establish a match between the needs of the communities and the efforts of individuals, organizations and groups willing to assist in the recovery process. For example, the CUNY Graduate Centre initiated a map of community services and needs called *Rockawayrecovery*. NY Tech Responds mobilized New York's tech community to support hurricane Sandy recovery. More recently, the *@wehaveweneed* initiative proposes a platform to supply sharing for disaster relief. Recently, Microsoft has launched its *HelpBridge* [5] application for iOS, Windows Phone and Android, which can be used to alert friends and relatives when disaster strikes. The users of such applications can offer time, money and other resources to support the recovery effort.

The emergence of these platforms for displaying needs and offers shows a new dimension of the concept of neighbourhood resilience.

Our research uses a different though related approach to bring this "social network" vision of N/OM to disaster management activities. It is based on vocabulary negotiation involving all stakeholders. The warning (Sect. 9.3) and matchmaking (Sect. 9.4) techniques described in this chapter rely on an evolvable vocabulary listing entities and attributes used when expressing warnings, needs and offers.

Table 9.1 gives a skeleton of our evolvable vocabulary of needs and offers for each class. Based on this classification, the community can add other terms or modify the relation between terms via explicit or implicit voting [6].

Explicit votes are suggestions for new terms in the vocabulary. Suggestions are accepted once they reach a threshold based on the credibility and/or the number of their proposers. In turn, *implicit votes* can be inferred by the use of terms in previous social network (e.g. Twitter) communications about disaster-related hashtags. In a previous research [7], we described our own software tool for consensus design, the *Innovation Factory*, which is used here for supporting vocabulary evolution.

9.3 Earthquake Early Warning System

In this section, we focus on an *Early Warning* (EW) system aimed at conveying to the general public timely and, above all, useful information regarding seismic warning data, fostering the emergence of collaborative behavior.

Current *Earthquake Early Warning* (EEW) systems follow two main approaches: *Regional* and *On-site* warning [8, 9]. In the regional warning approach, the traditional seismological method is used to locate an earthquake, and determine the magnitude from stations at close epicentral distances, and estimate the ground motion at other distant sites. In the on-site warning approach, the beginning of the ground motion (mainly by measuring the Primary waves, often shortened as *P-waves*) observed at a site is used to predict the ensuing ground motion at the same site by measuring the Secondary waves, often shortened as *S-waves*; and the related but distinct *Surface waves* [8, 9].

An EEW system based on the detection of the initial *P-waves* can save a significant amount of time, especially in short distances from the fault lines [8]. *On-site warning* is usually based on individual sensors, while *regional warning* requires seismic networks. The *regional warning* approach is more reliable, but requires more time, and therefore can not be used for EEW in cities and sites which are located very close to a seismic fault line. In contrast, the *on-site warning* approach is less reliable, but is comparatively much faster and can provide EEW for cities and sites that are located even in close proximity to a seismic fault line [8, 9].

Our approach to EW is based on the analysis of the aftermath of the devastating Izmit earthquake of 17 August 1999, when Kandilli Observatory and Earthquake Research Institute (KOERI) of Boğaziçi University in Istanbul was assigned with the task of establishing the Istanbul Earthquake Early Warning and Emergency Rapid Response System [10]. Previous research in this domain include seminal work by Rascioni et al. who addressed emergency and alert dissemination via digital terrestrial television (DVB-T) environment, with special emphasis on the Civil Protection Operational Centre (CPOC) network in the Marche region of Italy [9]. Another seminal research project by Pau and Simonsen focused on the use of Cellular Broadcast messaging services to reach wide audience [11]. In turn, Hester et al. have based their research on the integration of information systems for post-earthquake research response [12]; while Fortier and Dokas have developed an Early Warning System based on advanced information modeling [13] which provided some of the background for our own work.

Japan has put in place an EW system that reaches out to the general public. In January 2007, the Japanese national TV channel NHK began broadcasting earthquake and tsunami warning information through datacasts, using DVB-S

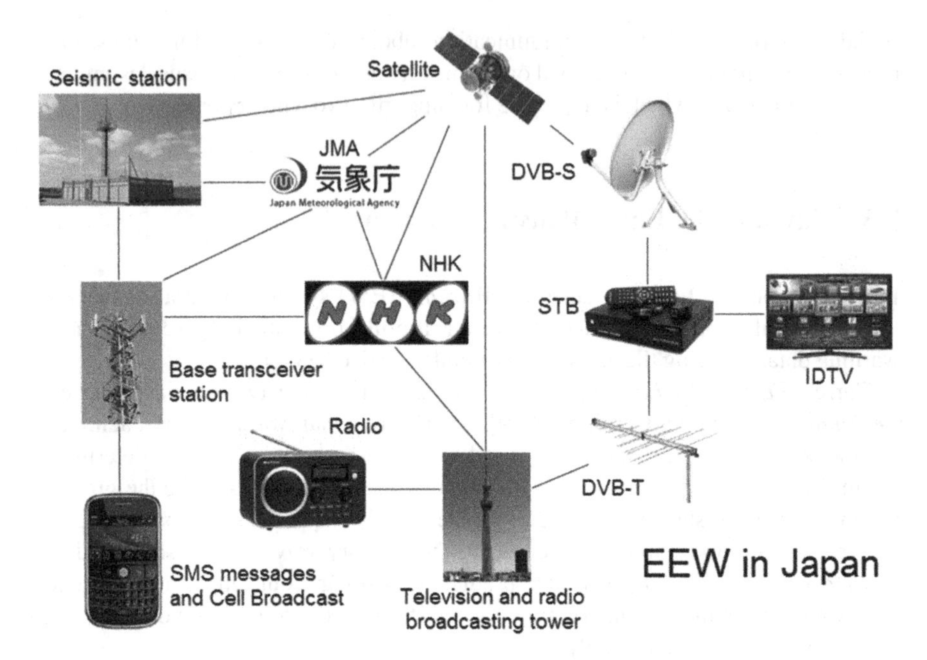

Fig. 9.1 EEW communication services in Japan

(Broadcasting Satellite) digital broadcasts, terrestrial digital broadcasts (ISDB-T), and terrestrial digital broadcasts for mobile receivers [8]. Today, Japan is the only country in the world with a fully active earthquake and tsunami early warning system, which is integrated with all available types of mass communication systems and networks in the country [8].

When tsunami warnings are issued, the *Emergency Broadcast System* breaks into the signal and automatically tunes all radios and televisions in the warned areas to NHK, so that the warnings are received properly by the public [8].

The EW activity in Japan can be modeled according to the following sequence of elementary steps: *(i)* seismometers detect the first shockwave; *(ii)* computers analyze the *P-wave* and estimate how powerful the *S-wave* will be; *(iii)* if the *S-wave* is estimated to be more powerful than a certain threshold (*Lower 5* on the Japanese national scale) an alert is issued [8].

The alert, issued by Japan Meteorological Agency (JMA), is disseminated to the public by radio and television broadcasters, mobile phone operators, and internet websites. The success rate of the *on-site warning* approach in Japan is 87.4 %, based on the detection of *P-waves* to predict the ensuing ground motion (*S-waves* and *surface waves*) at the same site, according to JMA data [8].

Figure 9.1 displays the current status of the earthquake early warning (EEW) communication systems in Japan. Still, the messages conveyed to the general public by the Japanese system use a fixed, pre-set vocabulary.

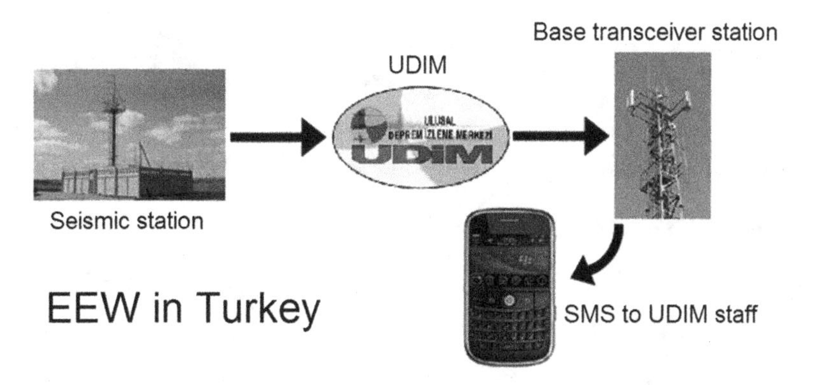

Fig. 9.2 Current EEW communication service in Turkey. Information on earthquakes larger than 3.0 on the Richter scale are sent to UDIM; those larger than 4.0 are sent to the Turkish government institutions like AFAD; those larger than 5.0 are sent to rescue organizations like AKUT. SMS messaging is provided by cell phone operator Turkcell

Following the Izmit earthquake of 17 August 1999, Kandilli Observatory and Earthquake Research Institute (KOERI) of Boğaziçi University in Istanbul was assigned with the task of establishing the Istanbul Earthquake Early Warning and Emergency Rapid Response System [10]. The construction of the system is realized by the GeoSIG and EWE consortium. Communications, related only to the Rapid Response System (as of early 2013) are provided by the Turkish GSM service provider Turkcell [10, 14].

UDIM (National Earthquake Monitoring Center) bound to KOERI sends SMS text messages, e-mails, electronic fax, radio text signals and Twitter messages to the staff of KOERI; to the Turkish government (Prime Ministry, AFAD (Disaster and Emergency Administration Presidency), Governorate of Istanbul, etc.); and to the AKUT search and rescue team, once initial estimation results regarding the epicenter and magnitude of an earthquake arrive [10, 14]. As of early 2013, SMS messages regarding seismic warning data are not broadcasted to the common citizens in Turkey, also due to the lack of Cell Broadcast services in the country [14].

Figure 9.2 displays the current status (as of early 2013) of the available EEW communication systems in Turkey. Earthquake data comes from sensor stations to KOERI as packages of 24-bit information, with an arrival time of 500 milliseconds [14]. The seismic measuring and recording stations in Turkey utilize data communication types including Satellite, GSM-GPRS, Turkcell 3G, Fiber+Satellite, Network, and the Internet [14].

9.3.1 Interactive Early Warning

The main goal of our EW system is to use digital terrestrial television (iDTV) as an efficient medium to convey timely and pervasive information regarding

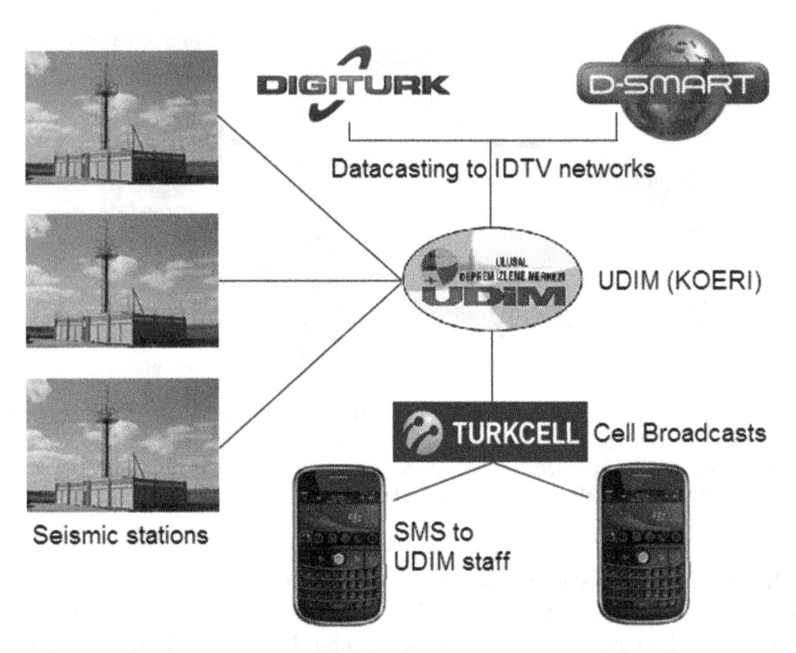

Fig. 9.3 EEW communication services according to our approach currently under consideration in Turkey

seismic warning data to disaster-stricken communities. The overall processing and communication architecture proposed for Turkey, shown in Fig. 9.3, is inspired by the Japanese model.

We will now focus our interactive application for Early Warning, developed via the Java Xlet development kit. The application relies on a simple EEW algorithm based on quick *P-wave* estimation. It is based on a standard middleware layer that includes a Java virtual machine; APIs; an application manager and application specific libraries [15–17] including class libraries that comprise the DVB-J platform for accessing digital television broadcast [17]. A real-time operating system (RTOS) and related device-specific libraries control the hardware via a collection of device drivers [15–17].

A major feature of our application is its openness to an external evolvable vocabulary for messages. Specifically, EEW modules aimed at the general public (Sects. 9.3.1.1 and 9.3.1.2) can dynamically access external information (including the vocabulary) via the standard Xlet interface for online data retrieval [14]. Also, the EW application can monitor submission of feedback information through the return channels for user interactivity (Sect. 9.3.1.3). Our application has a simple user interface (interaction takes place via a television remote control) making it easy to use for all age groups, overcoming the traditional hurdle of digital divide. The application libraries used by the resident applications are also stored on the

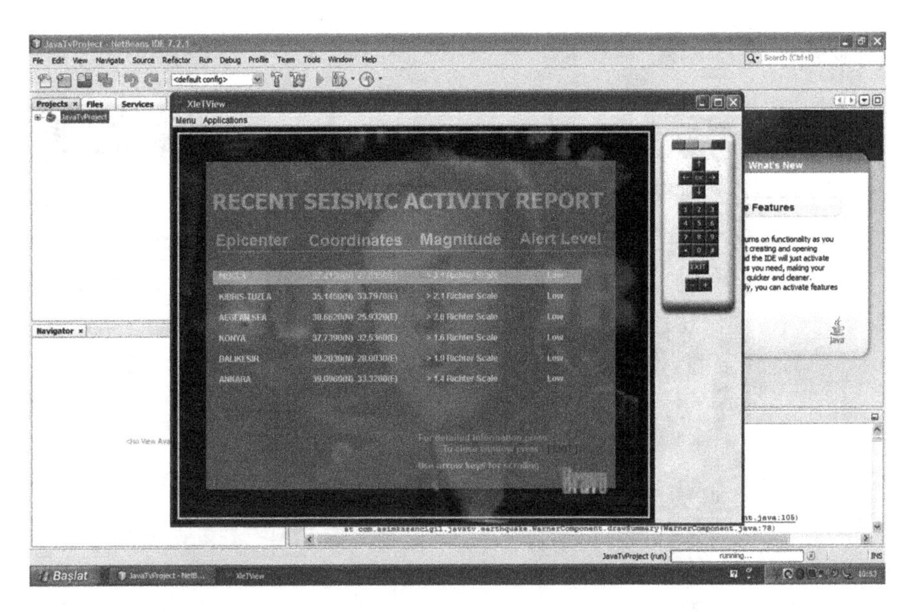

Fig. 9.4 The "Up Arrow" and "Down Arrow" buttons on the remote control device are used for selecting a recent earthquake from the list, and the "OK" button is dialed to get detailed information

television set-top box. One of the basic requirements for middleware is a small footprint and, in addition, the middleware can be stored in Flash ROM to improve performance speed in a set-top box [15–17].

While user interface design is not the focus of this chapter, we remark that many constraints exist for user interface development in our setting, including TV remote control type buttons with limited functionality [15–17]. It is essential to use a simple remote control event model and lightweight drawing components. In the next subsections, we briefly describe the different modules that compose our EW applications.

9.3.1.1 Recent Seismic Activity Report Module

The "Recent Seismic Activity Report" module provides information on the latest seismic activities within the last 24 h. The earthquakes are classified as *Low risk* (for up to 3.5 on the Richter scale, highlighted in yellow); *Medium risk* (between 3.5 and 5.0 on the Richter scale, highlighted in orange); and *High risk* (for 5.0 and above on the Richter scale, highlighted in red).

In the current version, the earthquakes are listed in the application window according to the size of their magnitude on the Richter scale (from the largest to the smallest) as seen in Figs. 9.4 and 9.5. The Earthquake Early Warning module accesses live GeoSIG and GeoDAS data, which can also be retrieved through an

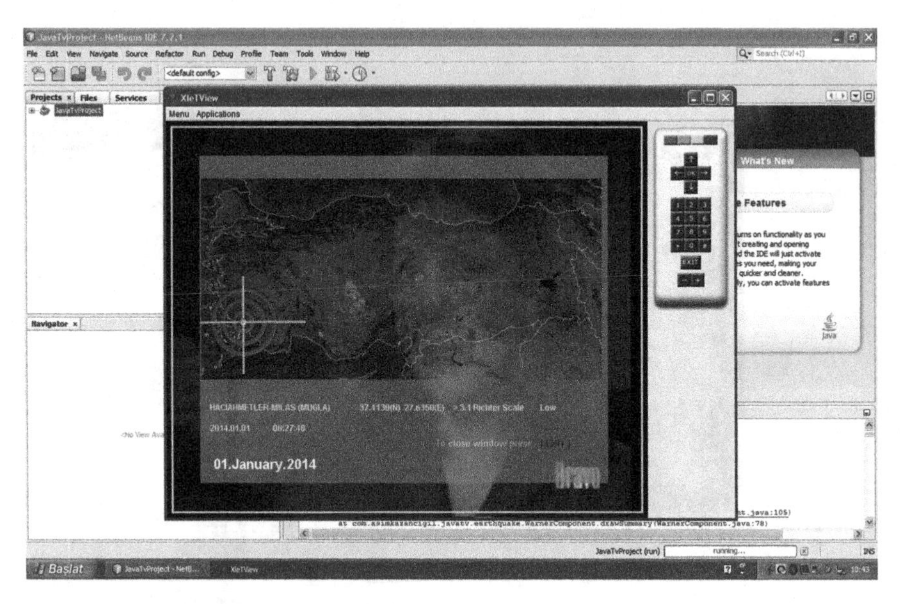

Fig. 9.5 Once the "OK" button is pressed, detailed information given. In this example, the earthquake has a magnitude of 3.1 and is therefore classified as Low Risk (*highlighted in yellow*) according to the vocabulary

automatically updated text file on the KOERI website (located in the URL link http://www.koeri.boun.edu.tr/sismo/eng3.txt), which provides live data on the latest seismic activities in and around Turkey.

9.3.1.2 Earthquake and Tsunami Early Warning

The "Earthquake and Tsunami Early Warning" module enables observatories to send live information regarding seismic activities (such as earthquakes, volcanic eruptions and tsunamis) to the digital video broadcasting networks. This application aims to provide a live information feed between the Kandilli Observatory and Earthquake Research Institute (KOERI) at the Boğaziçi University in Istanbul, Turkey, and the IDTV networks in the country such as Digiturk and D-SMART. KOERI is currently working on methods to provide live seismic early warning alerts and quick seismic information reports to governmental institutions, schools, transportation networks and energy networks through the use of cell phone text messages and the internet, in cases of increased seismic activity detection [14].

Our aim was to develop an Xlet-based IDTV application that automatically detects any signal of abnormally high *P-wave* activity and promptly warns television viewers via automated pop-up applications that instantly appear on their television screens, enabling them to receive near-real-time online information. Local or global broadcasting can be performed selecting the appropriate channels and multiplexers (MUXs) of the digital terrestrial television system.

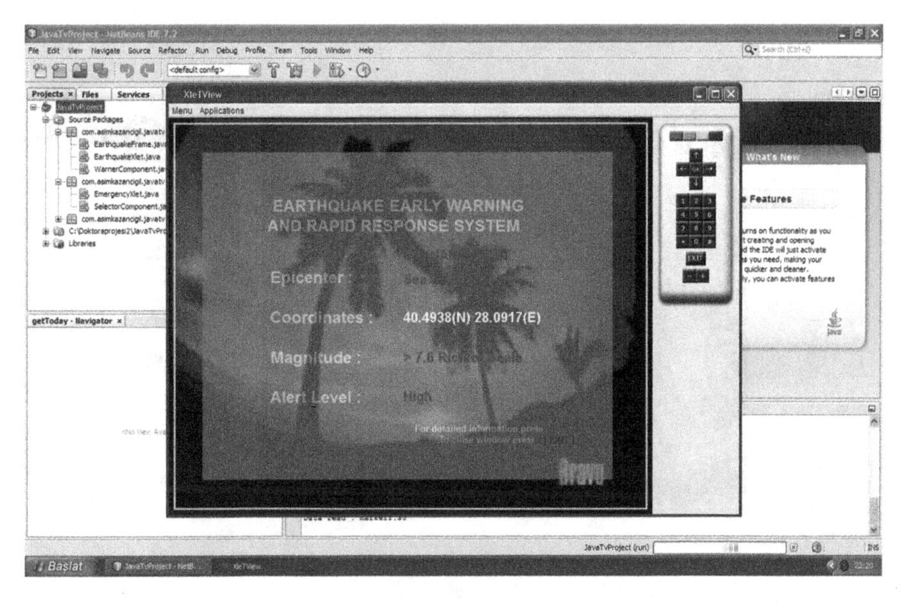

Fig. 9.6 After calculations lasting 1–5 min, the earthquake's epicenter and magnitude appear in the module's pop-up window, together with a "Tsunami Alert". It is estimated that tsunami waves would reach the Sea of Marmara coastline of Istanbul in about 5–10 min

Once the module will be deployed, the live seismic activity information feed will be received from the KOERI servers of Boğaziçi University (http://www.koeri.boun.edu.tr). For testing, we used a "simulated channel" that provides earthquake warning information to our Xlet application through the internet. Figure 9.6 shows a sample broadcast for a *High Risk* level earthquake warning (7.6 on the Richter Scale, with its epicenter at coordinates 40.4938(N) 28.0917(E) in the Sea of Marmara), the data of which are received by the Xlet via the internet.

Epicenter and magnitude calculation takes around 1 min when fully automated [14], or between 3 and 5 min when calculation is done by manually feeding into the software additional data arriving from at least three different measurement stations in the area, thus forming a triangle around the epicenter [14]. The more data from different stations are added into the calculation, the more accurate the epicenter coordinates and magnitude figures become [14]. As of October 2012, there were 205 seismic activity measuring stations across Turkey [14].

9.3.1.3 Emergency Services Module

The main window of the "Emergency Services" module, seen in Fig. 9.7, consists of three service options (police, emergency healthcare, fire brigade).

The detailed information windows of the services have similar content, which consist of a brief description of the service (using the current version of the

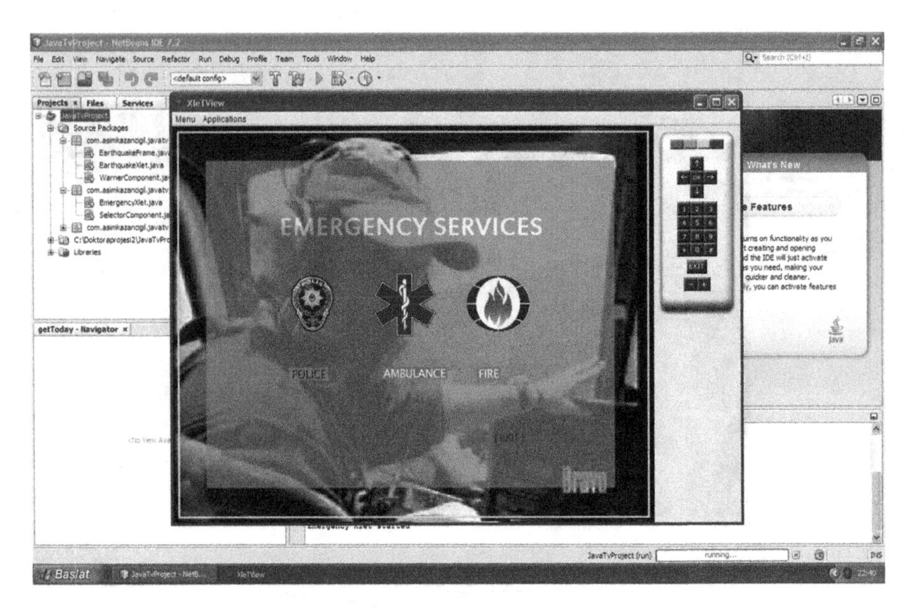

Fig. 9.7 Main window of the "Emergency Services" module

vocabulary) and a request for the user (in this case, a digital television network subscriber) to verify and confirm his/her address. If the address is correct, the user can press on the "OK" button for expressing a need, requesting assistance at his/her address.

Our application allows digital TV viewers to quickly express their need for emergency services via the TV remote control. Digital television network subscribers who agree to share their identity and home address will be able to take advantage of such a rapid service. In the future, specifically designed remote control buttons may also improve the speed and efficiency of such applications, especially for the senior citizens and people who have health problems.

9.4 Need/Offer Matching

In large-scale disasters involving interconnected critical infrastructures and a large population, data collection about offers and needs can turn into big data problem. The multiplicity and the heterogeneity of needs and offers, coupled with the urgent nature of the situations, require semi-automatic pre-processing algorithms to help players. In this section we describe our framework's algorithms and components for N/OM.

Table 9.2 Weight functions

Weight function	Formula
Quantity weight function	$f_{quantity} = \begin{cases} 1\, if\ Q_{need} \leq Q_{offer} \\ 0\, if\ Q_{need} > Q_{offer} \end{cases}$
	$Q_{need} = needed\ quantity$
	$Q_{offer} = offered\ quantity$
Distance weight function	$f_{distance} = \begin{cases} 1\, if\ D \leq x \\ 0\, if\ D > x \end{cases}$
	$D = physical\ distance\ between\ offer\ and\ need$
	$x = maximal\ authorized\ distance$

9.4.1 The Matcher

The Matcher component is based on our evolvable vocabulary (Sect. 9.2) in which the postings (needs and offers) can be classified as well as initial relevant properties of each vocabulary item. The Matcher's algorithm identifies for each offer its class of needs, then uses a global correspondence function to match to a need the best offer. The algorithm follows these steps:

1. Input: needs and offers list.
2. For each need N, get its class C.
3. Select set of offers O with same C.
4. Call weight functions for each attribute considered in the calculation of the global correspondence function (see Table 9.2).
5. Call the global correspondence function to calculate the overall weight for each offer O.
6. Sort all global functions of matching in ascending order.
7. Select the offer with best weight.
8. Output: pairs offer/need.

After the generation of the data, the matching algorithm is launched. The output of the algorithm is a matching-list consisting in a set of couples (Offer, Need). One can visualize the matching-list as an oriented graph, where the set of edges is the matching-list. The size of the nodes representing a given offer depends on the number of needs that it was attributed by our algorithm.

Beyond the simple matching, the proposed approach can help to identify critical resources or services during a disaster.

The next step in the design of the N/OM matcher will be to include ontology-based approaches to improve the process of classification of relevant information in the posts. The component also deals with ontology negotiation by the stakeholders, including access rights, consistency checks, and synchronization issues. A major feature of the Matcher is adapting property weights according to users' feedbacks about proposed matches.

9.4.2 Spatio-temporal Analytics Module (SAM)

The SAM component implements relevant metrics to assess spatiotemporal trustworthiness and reliability of needs and offers. Moreover, these metrics help to measure the rapidity of the matching process and geographical distribution of the active communities and agencies. SAM integrates:

(i) a timeline of the process of the matching process which will help to monitor the rapidity,
(ii) a clustering engine to identify spatiotemporal clusters that emerge in the community, and
(iii) an algorithm for trustworthiness estimation for each set of need/offer. Using anomaly detection algorithm on temporal collocation graphs, one can detect critical zones that need additional intervention.

9.5 Trustworthiness and Reputation Manager (T&R Manager)

The T&R manager module deals with authentication, reputation and privacy mechanisms. Authentication relies on standard service interfaces for token distribution. Such interfaces will be acting as intermediaries between our framework and information available to mobile phone operators after SIM-based authentication of potential requestors. Each user, be it an individual or an organization, a donor or a requestor, is the sole master of how much information will pass through these interfaces, by means of the attributes included in each token. Data supporting the system reputation infrastructure can be joined with geographic information and other geographical open data. The resulting reputation can be used to: (i) compute credibility in vocabulary negotiation; (ii) filter out phony or non-reliable postings in N/OM.

The core module of our T&R Manager has been implemented using a Social Network Analysis library. The library relies on the following tools:

(i) The Real-time Business Process Monitoring platform ZEUS [18], a cloud-enabled, grid-based triple-store applied to real time process monitoring;
(ii) The open source platform Jena [19], used to build semantic network-based application and designed to deal with linked data web contents.

The library implements a set of core metrics that are applied to the social network graphs, extracted by the above tools, describing the connections between process actors. Data coming from the metrics are then used to evaluate users' trustworthiness and reputation. In particular, the metrics listed in Table 9.3 have been implemented, aimed at evaluating the centrality of a node within the whole network. In our scenario, nodes are both process actors, namely citizens, governmental bodies,

Table 9.3 First subset of Social Network metrics

Name	Description
No. of relations (X)	Indicates the number of nodes that are connected to X. Indicates how many actors and/or resources are connected to a node
NodeBetweenness (X)	Indicates how "far" a node is w.r.t. the other nodes. The measure is useful to identify possible bottlenecks in terms of actors or resource distribution, showing the need for a better distribution
NodeIncloseness (X)	Calculated as the inverse of sum of the distances of the node w.r.t. the other nodes, normalized to the number minus 1. Nodes with low values are used only for the initialization activities, and lose importance in the enactment of the process. Indicates those resources and actors that must be preserved in the first stage of the disaster
NodeOutcloseness (X)	Calculated as the inverse, normalized to 1, of the sum of the distances of the node w.r.t. the other nodes. Nodes with low values indicate that the node is far from the center and the resources/actors will be important in the later phase of the recovery process

humanitarian organizations, and resources that are involved in the N/OM matching. Values resulting from the metrics are then used as initial parameters for the calculation of trustworthiness and reputation functions. In fact, knowing the position and the importance of a specific actor/resource in the network is important for the evaluation of the overall importance of such a node.

9.6 Conclusions

We have presented a framework supporting some key activities of the disaster management process. Although the whole framework is currently at the design stage, partial implementations of several components (Sects. 9.3 and 9.4) are already available. We are fully aware that our approach assumes that the ICT infrastructure (e.g. Internet and the social networks) will stay available in some form even after a major disaster. This assumption however is less far-fetched than it could appear at first sight, as recent experience has shown that the global network stays available around the crater of disasters, and wireless technology allows it to operate even in case of a breakdown. Indeed, many examples suggest [20] that the ICT infrastructure is more resilient than other infrastructures such as power and water supply. The prototyping of Xlet-based interactive applications for Seismic Early Warning and Emergency Services provided additional methods to communicate information about natural disasters such as earthquakes and tsunamis in a timely and efficient way, through digital television networks.

Acknowledgments The authors wish to thank Prof. Mustafa Erdik, Dr. Doğan Kalafat and Dr. Mehmet Yılmazer from KOERI for the valuable information they kindly provided. This work was co-funded by the European Commission under the Information and Communication Technologies theme of the 7th Framework Programme, Integrated Project ARISTOTELE (contract no. FP7-257886).

References

1. van der Aalst W et al (2012) Process Mining Manifesto. In: Lecture Notes in Business Information Processing, vol. 99, Springer Berlin Heidelberg, doi: 10.1007/978-3-642-28108-2_19
2. Lillibridge SR, Noji EK, Burkle FM (1993) Disaster assessment: the emergency health evaluation of a population affected by a disaster. Ann Emerg Med, Elsevier, Amsterdam, 22:1715–1720
3. World Health Organization (1991) Rapid health assessment protocols for emergencies. World Health Organization, Geneva
4. Birregah B, Top T, Perez C, Châtelet E, Matta N, Lemercier M, Snoussi H (2012) Multi-layer crisis mapping: a social media-based approach. In: Proceedings of the 2012 IEEE international conference on enabling technologies: infrastructures for collaborative enterprises (WETICE 2012), Toulose, France, pp 379–384
5. Microsoft Corporation (2013) HelpBridge mobile application. http://www.microsoft.com/about/corporatecitizenship/en-us/nonprofits/Helpbridge.aspx. Accessed 10 Jan 2014
6. Ceravolo P, Damiani E, Viviani M (2007) Bottom-up extraction and trust-based refinement of ontology metadata. IEEE Trans Knowl Data Eng 19(2):149–163
7. Bellandi V, Ceravolo P, Damiani E, Frati F, Cota GL, Maggesi J (2013) Boosting the innovation process in collaborative environments. In: Proceedings of the 2013 IEEE international conference on systems, man, and cybernetics (SMC), Manchester, UK, pp 1432–1438
8. Nomoto Y (2007) Earthquake and tsunami information services via data broadcasting by NHK Japan. ABU technical committee annual meeting, Tehran, 30 October–1 November 2007. Doc T-7/41-2
9. Rascioni G, Gambi E, Spinsante S, Falcone D (2008) DTT Technology for rural communities alerting. In: Fiedrich F, van de Walle B (eds) Proceedings of the 5th international ISCRAM conference, Washington, DC, USA, May 2008
10. Alcik H, Ozel O, Wu YM, Ozel NM, Erdik M (2010) An alternative approach for the Istanbul Earthquake Early Warning System. Soil Dyn Earthq Eng. doi:10.1016/j.soildyn.2010.03.007
11. Pau LF, Simonsen P (2008) Emergency messaging to general public via public wireless networks. In: Fiedrich F, van de Walle B (eds) Proceedings of the 5th international ISCRAM conference, Washington, DC, USA, May 2008
12. Hester NC, Horton SP, Wilkinson J, Jefferson TI (2008) Integration of information systems for post-earthquake research response. In: Fiedrich F, van de Walle B (eds) Proceedings of the 5th international ISCRAM conference, Washington, DC, USA, May 2008
13. Fortier SC, Dokas IM (2008) Setting the specification framework of an early warning system using IDEF0 and information modeling. In: Fiedrich F, van de Walle B (eds) Proceedings of the 5th international ISCRAM conference, Washington, DC, USA, May 2008
14. Kazancigil MA (2012) Interviews with Prof. Mustafa Erdik, Dr. Doğan Kalafat and Dr. Mehmet Yılmazer on the Istanbul Earthquake Early Warning System. Kandilli Observatory and Earthquake Research Institute (KOERI), Boğaziçi University, Istanbul
15. Peng C (2002) Digital television applications. Doctoral dissertation, Helsinki University of Technology Press, Helsinki. ISBN: 951-22-6171-5
16. ETSI (1997) Digital Video Broadcasting (DVB): framing structure, channel coding and modulation for 11/12 GHz satellite services. European Telecommunications Standards Institute France, August 1997. EN 300 421

17. Calder B, Courtney J, Foote B, Kyrnitszke L, Rivas D, Saito C, Loo JV, Ye T (2000) Java TV API Technical Overview. The Java TV API White Chapter, Version 1.0. Oracle White Papers, Cupertino, CA, 14 November 2000
18. Leida M, Chu A, Colombo M, Majeed B (2012) Extendible data model for real-time business process analysis. Khalifa University Science Editor, Abu Dhabi
19. Apache Foundation (2013) Apache Jena. http://jena.apache.org. Accessed 10 Jan 2014
20. Kessler S (2013) Social media plays vital role in reconnecting Japan quake victims with loved ones. http://mashable.com/2011/03/14/internet-intact-japan. Accessed 10 Jan 2014

Chapter 10
Expert Systems for Assessing Disaster Impact on the Environment

Mykola M. Biliaiev, N. Rostochilo, and M. Kharytonov

Abstract This paper presents expert systems and the numerical models to simulate atmosphere pollution after accidents with toxic substances. The expert systems allow one to assess impact on the person or environment after accidents of different types at chemical enterprises or storages. The process of toxic gas dispersion in the atmosphere is computed using the convective-diffusive equation. To compute the flow field among the buildings, two fluid dynamic models are used. The first fluid dynamic model is based on the 3-D equation of the potential flow. The second fluid dynamic model is based on the 2-D equations of the inviscid separated flows. To solve the convective-diffusive equation of the toxic gas dispersion in the atmosphere, the implicit change-triangle difference scheme is used. To solve the equations of the fluid dynamic models the A.A. Samarski's difference scheme of splitting and change-triangle scheme are used. The developed numerical models have two submodels. The first submodel was developed to simulate the process of the atmosphere pollution after the accident spillages and the evaporation of the toxic substance from the soil. The second submodel allows predicting the air pollution inside the rooms in the case of the outdoor toxic gas infiltration into the room. The developed systems allow assessing the safety of the evacuation route. The developed expert system can be used to solve some specific problems such as the assessment of efficiency of protection measures which are developed to reduce the negative impact on the environment.

M.M. Biliaiev (✉) • N. Rostochilo • M. Kharytonov
Dnipropetrovsk National University of Railway Engineering,
Lazaryana Str. 2, Dnipropetrovsk 49010, Ukraine
e-mail: envteam@ukr.net

H.-N. Teodorescu et al. (eds.), *Improving Disaster Resilience and Mitigation - IT Means and Tools*, NATO Science for Peace and Security Series C: Environmental Security,
DOI 10.1007/978-94-017-9136-6_10,

10.1 Introduction

Ukraine has powerful chemistry industry and it means that there are a lot of chemical enterprises which use or produce different toxic substances. Also there are more than 2,000 storages with toxic substances in the country. As a rule the huge amount of toxic substances is transported by railway. It's obvious that all these objects are the potential sources of threat for people and environment in the case of accidents. Accidences cause emission of toxic substances, as a rule, into the atmosphere and the following their dispersion in this medium. To develop the adequate protection measures for people and estimate the impact to the environment it is necessary to know the level of the danger in the case of accidents. To solve this problem the expert systems are needed. The main tasks of the expert systems are the following:

- prediction of the atmosphere contamination after the accident;
- assessment of the evacuation routes safety;
- assessment of the protection measures efficiency;
- prediction of toxic gas flows into the rooms of the buildings;
- to solve the problems very quickly.

Nowadays in emergency service of Ukraine the empirical model to predict the dimensions of the hitting area after accidents is used. Using this model it is possible to calculate some parameters of the contaminated area, for example, to calculate the width of the contaminated area using the following equation:

$$D = 0.3L^n,$$

where L is the length of polluted area (the L value is determined using the special Table); n is a parameter which depends on the atmosphere condition (stable, neutral, etc.). Worthy of note that this model does not take into account the wind speed and parameters of the atmosphere diffusion. It also does not take into account the influence of buildings (Fig. 10.2) or complex terrain on the process of the toxic gas dispersion. This model is not realistic and widely criticized in scientific papers. To predict the level of danger in the case of accidents at the chemistry enterprises or chemical storages it is necessary to develop the expert systems which are based on the adequate mathematical models.

10.2 Mathematical Model of Toxic Gas Dispersion in the Atmosphere

To simulate the dispersion process of the toxic gases in the atmosphere the advection-diffusion model is used:

$$\frac{\partial C}{\partial t} + \frac{\partial uC}{\partial x} + \frac{\partial vC}{\partial y} + \frac{\partial (w-ws)C}{\partial z} + \sigma C = \frac{\partial}{\partial x}\left(\mu_x \frac{\partial C}{\partial x}\right) + \frac{\partial}{\partial y}\left(\mu_y \frac{\partial C}{\partial y}\right) + \frac{\partial}{\partial z}\left(\mu_z \frac{\partial C}{\partial z}\right) + \\ + \sum Q_i(t)\delta\left(x - x_i\right)\delta\left(y - y_i\right)\delta\left(z - z_i\right), \quad (10.1)$$

where C is the concentration of toxic gas; u, v, w are the wind velocity components; ws is the speed of gravity fallout; σ is the coefficient taking into account the process of pollutant decay or rain washout; $\mu = (\mu_x, \mu_y, \mu_z)$ are the diffusion coefficients; Q_i is the intensity of point source of ejection; $\delta(x - x_i)\delta(y - y_i)\delta(z - z_i)$ are Dirac's delta functions; $r_i = (x_i, y_i, z_i)$ are the coordinates of the ejection source.

In the developed numerical model the following approximations for the wind speed and diffusion coefficients are used [1]:

$$u = u_1 \left(\frac{z}{z_1}\right)^n, \quad \mu_x = \mu_y, \quad \mu_y = k_0 u, \quad \mu_z = 0.11 z,$$

where u_1 is the wind speed at the height $z_1 = 10$ m; $n = 0.15$; k_0 is the empirical parameter.

10.3 Hydrodynamic Model

To take into account the complex terrain or buildings influence on the pollutant dispersion two Fluid Dynamic Models are used.

The model of potential flow In this case the governing equation is written in the following form:

$$\frac{\partial^2 P}{\partial x^2} + \frac{\partial^2 P}{\partial y^2} + \frac{\partial^2 P}{\partial z^2} = 0, \tag{10.2}$$

where P is the potential of velocity.

The components of velocity are calculated as follows:

$$u = \frac{\partial P}{\partial x}, \quad v = \frac{\partial P}{\partial y}, \quad w = \frac{\partial P}{\partial z}.$$

The 2-D model of inviscid separated flows The governing equations of this model are the equation of vortex transport:

$$\frac{\partial \omega}{\partial t} + \frac{\partial u\,\omega}{\partial x} + \frac{\partial v\,\omega}{\partial y} = 0 \tag{10.3}$$

and Poisson's equation for flow function:

$$\frac{\partial^2 \psi}{\partial x^2} + \frac{\partial^2 \psi}{\partial y^2} = -\omega \tag{10.4}$$

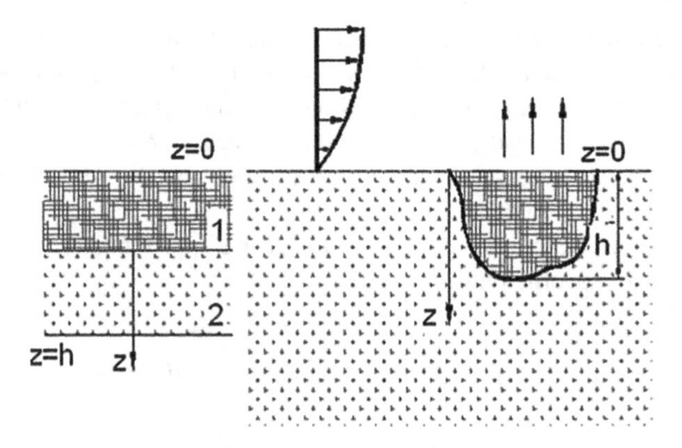

Fig. 10.1 Sketch of the computational domain in the problem of toxic substance evaporation from the soil. zone 1 – area where the liquid was evaporated; zone 2 – area with liquid

Components of wind velocity are calculated as follows:

$$u = \frac{\partial \psi}{\partial y}; \quad v = -\frac{\partial \psi}{\partial x}.$$

10.3.1 SubModel 1. Evaporation of Toxic Substance from the Soil After Spillage

The process of the toxic substance evaporation from the soil after spillage is simulated using the following equation:

$$\frac{\partial C}{\partial t} = \mu \frac{\partial^2 C}{\partial z^2}, \tag{10.5}$$

μ – diffusion coefficient in the zone 1 (Fig. 10.1); C – vapor concentration in zone 1; z – coordinate; t – time.

The Eq. 10.5 is solved together with transport model 10.1.

The speed of evaporation boundary in the soil is calculated using the following dependence:

$$\frac{dz}{dt} = \frac{Q}{S\rho},$$

where Q – mass speed of evaporation, S – square of evaporation, ρ – density.

10.3.2 SubModel 2. Air Pollution in the Room in the Case of Toxic Gas Infiltration Through the Ventilation System (Windows)

Change of the pollutant concentration in the room is calculated using the following balance model

$$VdC_{room} = L \cdot C \cdot dt - L \cdot C_{room} \cdot dt,$$

where V – volume of the room; C_{room} – outlet concentration; $C = f(t)$ – inlet concentration; t – time; L – air change rate.

Inlet concentration $C = f(t)$ is calculated using the transport model 10.1.

10.4 Numerical Model

The numerical integration of the governing Eqs. 10.1 and 10.2, 10.3 and 10.4 is carried out using rectangular grid. The main features of the difference schemes which are used are considered below.

To solve (Eq. 10.1) the implicit change-triangle difference scheme developed by Prof. Kchrutch V. and Prof. Biliaiev M. is used. In this case the time dependent derivative in (Eq. 10.1) is approximated as follows:

$$\frac{\partial C}{\partial t} \approx \frac{C_{ijk}^{n+1} - C_{ijk}^{n}}{\Delta t}.$$

At the first step the convective derivatives are represented in the following way:

$$\frac{\partial uC}{\partial x} = \frac{\partial u^{+}C}{\partial x} + \frac{\partial u^{-}C}{\partial x},$$

$$\frac{\partial vC}{\partial y} = \frac{\partial v^{+}C}{\partial y} + \frac{\partial v^{-}C}{\partial y},$$

$$\frac{\partial wC}{\partial z} = \frac{\partial w^{+}C}{\partial z} + \frac{\partial w^{-}C}{\partial z},$$

where $u^{+} = \frac{u+|u|}{2}$, $u^{-} = \frac{u-|u|}{2}$, $v^{+} = \frac{v+|v|}{2}$, $v^{-} = \frac{v-|v|}{2}$, $w^{+} = \frac{w+|w|}{2}$, $w^{-} = \frac{w-|w|}{2}$.

At the second step the convective derivatives are approximated as follows:

$$\frac{\partial u^+ C}{\partial x} \approx \frac{u^+_{i+1,j,k}\, C^{n+1}_{i\,jk} - u^+_{i\,jk}\, C^{n+1}_{i-1,j,k}}{\Delta x} = L^+_x\, C^{n+1},$$

$$\frac{\partial u^- C}{\partial x} \approx \frac{u^-_{i+1,j,k}\, C^{n+1}_{i+1,j,k} - u^-_{i\,jk}\, C^{n+1}_{i\,jk}}{\Delta x} = L^-_x\, C^{n+1},$$

$$\frac{\partial v^+ C}{\partial y} \approx \frac{v^+_{i,j+1,k} C_{ijk} - v^+_{ijk} C_{i,j-1,k}}{\Delta y} = L^+_y C^{n+1},$$

$$\frac{\partial v^- C}{\partial y} \approx \frac{v^-_{i,j+1,k} C_{i,j+1,k} - v^-_{ijk} C_{ijk}}{\Delta y} = L^-_y C^{n+1},$$

$$\frac{\partial w^+ C}{\partial z} \approx \frac{w^+_{i,j,k+1} C_{ijk} - w^+_{ijk} C_{i,j,k-1}}{\Delta z} = L^+_z C^{n+1},$$

$$\frac{\partial w^- C}{\partial z} \approx \frac{w^-_{i,j,k+1} C_{i,j,k+1} - w^-_{ijk} C_{i,j,k}}{\Delta z} = L^-_z C^{n+1}.$$

The second order derivatives are approximated as follows:

$$\frac{\partial}{\partial x}\left(\mu_x \frac{\partial C}{\partial x}\right) \approx \mu_x \frac{C^{n+1}_{i+1,j,k} - C^{n+1}_{ijk}}{\Delta x^2} - \mu_x \frac{C^{n+1}_{i,j,k} - C^{n+1}_{i-1,j,k}}{\Delta x^2} = \\ = M^-_{xx} C^{n+1} + M^+_{xx} C^{n+1},$$

$$\frac{\partial}{\partial y}\left(\mu_y \frac{\partial C}{\partial y}\right) \approx \mu_y \frac{C^{n+1}_{i,j+1,k} - C^{n+1}_{ijk}}{\Delta y^2} - \mu_y \frac{C^{n+1}_{i,j,k} - C^{n+1}_{i,j-1,k}}{\Delta y^2} = \\ = M^-_{yy} C^{n+1} + M^+_{yy} C^{n+1}$$

$$\frac{\partial}{\partial z}\left(\mu_z \frac{\partial C}{\partial z}\right) \approx \mu_z \frac{C^{n+1}_{i,j,k+1} - C^{n+1}_{ijk}}{\Delta z^2} - \mu_z \frac{C^{n+1}_{i,j,k} - C^{n+1}_{ij,k-1}}{\Delta z^2} = \\ = M^-_{zz} C^{n+1} + M^+_{zz} C^{n+1}.$$

In these expressions L^+_x, L^-_x, L^+_y, L^-_y, L^+_z, L^-_y, M^+_{xx}, M^-_{xx}, are the difference operators.

Solution of the transport equation in the finite-difference form is split in four steps on the time step of integration *dt*:

- at the first step ($k = \frac{1}{4}$) the difference equation is:

$$\frac{C^{n+k}_{i\,j} - C^{n}_{i\,j}}{\Delta t} + \frac{1}{2}\left(L^+_x\, C^k + L^+_y\, C^k + L^+_z\, C^k\right) + \frac{\sigma}{4} C^k_{i\,jk} = \\ = \frac{1}{4}\left(M^+_{xx}\, C^k + M^-_{xx} C^n + M^+_{yy}\, C^k + M^-_{yy}\, C^n + M^+_{zz}\, C^k + M^-_{zz}\, C^n\right) \tag{10.6}$$

– at the second step ($k = n + \frac{1}{2};\ c = n + \frac{1}{4}$) the difference equation is:

$$\frac{C^k_{i\,jk} - C^c_{i\,jk}}{\Delta t} + \frac{1}{2}\left(L^-_x\ C^k + L^-_y C^k + L^-_z C^k\right) + \frac{\sigma}{4} C^k_{i\,j} =$$
$$= \frac{1}{4}\left(M^-_{xx} C^k + M^+_{xx} C^c + M^-_{yy}\ C^k + M^+_{yy} C^c + M^-_{zz}\ C^k + M^+_{zz}\ C^c\right) \tag{10.7}$$

– at the third step ($k = n + \frac{3}{4}; c = n + \frac{1}{2}$) the expression (Eq. 10.7) is used;
– at the fourth step ($k = n + 1; c = n + \frac{3}{4}$) the expression (Eq. 10.6) is used.

At the fifth step (at this step the influence of the source of pollutant ejection is taken into account) the following approximation is used:

$$\frac{\overset{5}{C}{}^{n+1}_{i,j,k} - \overset{5}{C}{}^{n}_{i,j,k}}{\Delta t} = \sum_{l=1}^{N} \frac{Q_l\left(t^{n+1/2}\right)}{\Delta x\ \Delta y\ \Delta z}\ \delta_l.$$

Function δ_l is equal to zero in all cells accept the cells where the 'l' source of ejection is situated. This difference scheme is implicit and absolutely steady but the unknown concentration C is calculated using the explicit formulae at each step (so called "method of running calculation").

Instead of Laplace Eq. 10.2 the 'time-dependent' equation for the potential of velocity is used:

$$\frac{\partial P}{\partial \eta} = \frac{\partial^2 P}{\partial x^2} + \frac{\partial^2 P}{\partial y^2} + \frac{\partial^2 P}{\partial z^2}, \tag{10.8}$$

where η is the 'fictitious' time.

For $\eta \to \infty$ the solution of this equation tends to the solution of Laplace equation. To solve (Eq. 10.8), A.A. Samarskii's change-triangle difference scheme is used. According to this scheme the solution of equation is split in two steps:

– at the first step the difference equation is:

$$\frac{P^{n+1/2}_{i,j,k} - P^n_{i,j,k}}{0{,}5\Delta\eta} = \frac{P^n_{i+1,j,k} - P^n_{i,j,k}}{\Delta x^2} + \frac{-P^{n+1/2}_{i,j,k} + P^{n+1/2}_{i-1,j,k}}{\Delta x^2} + \frac{P^n_{i,j+1,k} - P^n_{i,j,k}}{\Delta y^2} +$$
$$+ \frac{-P^{n+1/2}_{i,j,k} + P^{n+1/2}_{i,j-1,k}}{\Delta y^2} + \frac{P^n_{i,j,k+1} - P^n_{i,j,k}}{\Delta z^2} + \frac{-P^{n+1/2}_{i,j,k} + P^{n+1/2}_{i,j,k-1}}{\Delta z^2},$$

– at the second step the difference equation is:

$$\frac{P^{n+1}_{i,j,k} - P^{n+1/2}_{i,j,k}}{0{,}5\Delta\eta} = \frac{P^{n+1}_{i+1,j,k} - P^{n+1}_{i,j,k}}{\Delta x^2} + \frac{-P^{n+1/2}_{i,j,k} + P^{n+1/2}_{i-1,j,k}}{\Delta x^2} + \frac{P^{n+1}_{i,j+1,k} - P^{n+1}_{i,j,k}}{\Delta y^2} +$$
$$+ \frac{-P^{n+1/2}_{i,j,k} + P^{n+1/2}_{i,j-1,k}}{\Delta y^2} + \frac{P^{n+1}_{i,j,k+1} - P^{n+1}_{i,j,k}}{\Delta z^2} + \frac{-P^{n+1/2}_{i,j,k} + P^{n+1/2}_{i,j,k-1}}{\Delta z^2}.$$

From these expressions the unknown value P is determined using the explicit formulae at each step (“method of running calculation”). To solve (Eq. 10.3) the implicit change-triangle difference scheme is used and A.A. Samarskii’s change-triangle difference scheme is used to solve (Eq. 10.4). On the base of the described difference schemes two codes were developed using FORTRAN language for programming.

10.5 Results of Numerical Simulation

Case 1 Assessment of the evacuation route safety.

The examples below concern the problem of the assessment of the evacuation route safety in the case when people leave the building after the accident (Figs. 10.2 and 10.3). To solve this problem, Eqs. 10.1 and 10.2 are integrated. To assess the risk of people hitting at the route of evacuation, the calculation of toxic doze for people is carried out:

$$TD = \int_0^t C\,(x, y, z, t)\,dt,$$

where t is the exposition time, C is the concentration of the toxic gas.

In the developed model the ‘skin’ toxic dose is also calculated using the following formula:

$$TD_S = \int_0^t \iint_S \mu \frac{\partial c}{\partial n}\,dsdt,$$

where $\mu \frac{\partial c}{\partial n}$ is the toxic substance transfer to the skin of the person; S is the square of the skin; t is the exposition time, C is the concentration of the toxic substance.

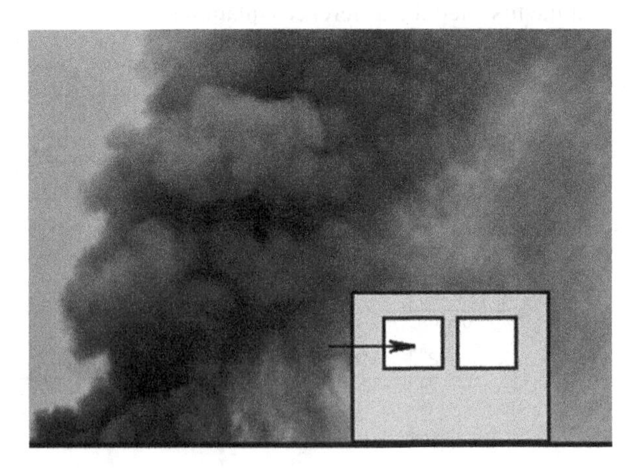

Fig. 10.2 Toxic gas cloud near building

Fig. 10.3 Evacuation from the room. Have you time to go out?

0	0	0	0	0	0	0	0	0	0	0	0	0	0	0	0	0	0	0	0	0
0	0	0	0	0	0	0	0	0	1	6	1	0	0	0	0	0	0	0	0	0
0	0	0	0	0	0	0	0	2	18	99	16	1	0	0	0	0	0	0	0	0
0	0	0	0	0	0	0	0	0	5	22	5	0		0	0	0	0	0	0	0
0	0	0	0	0	0	0	0	0	1	4	1	0	C	0	0	0	0	0	0	0
0	0	0	0	0	0	0	0	0	0	0	0	0		0	0	0	0	0	0	0
0	0	0	0	0	0	0	0	0	0	0	0	0	0	0	0	0	0	0	0	0
0	0	0	0	0	0	0	0	0		0	0	0	0	0	0	0	0	0	0	0
0	0	0	0	0	0	0	0	0	A	0	0	0	0	0	0	0	0	0	0	0
0	0	0	0	0	0	0	0	0	0	0	0	0	0	0	0	0	0	0	0	0
0	0	0	0	0	0	0	0	0	0	0	0	0	0	0	0	0	0	0	0	0
0	0	0	0		0	0	0	0	0	0	0	0	0	0	0	0	0	0	0	0
0	0	0	0	B	0	0	0	0	0	0	0	0	0	0	0	0	0	0	0	0
0	0	0	0	0	0	0	0	0	0	0	0	0	0	0	0	0	0	0	0	0
0	0	0	0	0	0	0	0	0	0	0	0	0	0	0	0	0	0	0	0	0
0	0	0	0	0	0	0	0	0	0	0	0	0	0	0	0	0	0	0	0	0
0	0	0	0	0	0	0	0	0	0	0	0	0	0	0	0	0	0	0	0	0
0	0	0	0	0	0	0	0	0	0	0	0	0	0	0	0	0	0	0	0	0
0	0	0	0	0	0	0	0	0	0	0	0	0	0	0	0	0	0	0	0	0
0	0	0	0	0	0	0	0	0	0	0	0	0	0	0	0	0	0	0	0	0
0	0	0	0	0	0	0	0	0	0	0	0	0	0	0	0	0	0	0	0	0
0	0	0	0	0	0	0	0	0	0	0	0	0	0	0	0	0	0	0	0	0
0	0	0	0	0	0	0	0	0	0	0	0	0	0	0	0	0	0	0	0	0
0	0	0	0	0	0	0	0	0	0	0	0	0	0	0	0	0	0	0	0	0
0	0	0	0	0	0	0	0	0	0	0	0	0	0	0	0	0	0	0	0	0

Fig. 10.4 Sketch of the computational domain: *A* – is the door; *B* – evacuation route; *C* – emission of NH_3

Figures 10.4, 10.5 and 10.6 present the results of numerical simulations of NH_3 cloud dispersion near the building from which the person got out and started to run along the evacuation route. It is clear that the toxic gas cloud very quickly covers the person at this route and creates the danger of hitting.

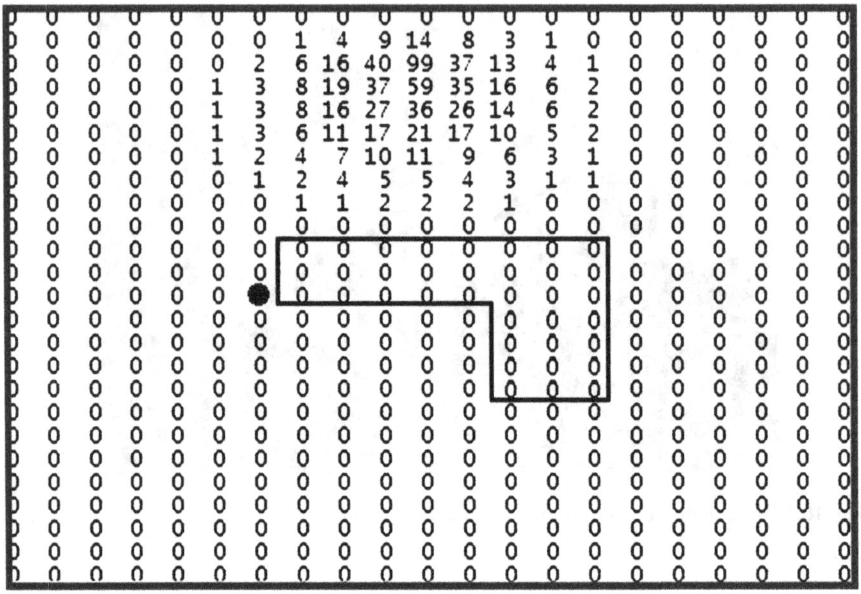

0	0	0	0	0	0	0	0	0	0	0	0	0	0	0	0	0	0	0	0
0	0	0	0	0	0	1	4	9	14	8	3	1	0	0	0	0	0	0	0
0	0	0	0	0	2	6	16	40	99	37	13	4	1	0	0	0	0	0	0
0	0	0	0	1	3	8	19	37	59	35	16	6	2	0	0	0	0	0	0
0	0	0	0	1	3	8	16	27	36	26	14	6	2	0	0	0	0	0	0
0	0	0	0	1	3	6	11	17	21	17	10	5	2	0	0	0	0	0	0
0	0	0	0	1	2	4	7	10	11	9	6	3	1	0	0	0	0	0	0
0	0	0	0	0	1	2	4	5	5	4	3	1	1	0	0	0	0	0	0
0	0	0	0	0	0	1	1	2	2	2	1	0	0	0	0	0	0	0	0
0	0	0	0	0	0	0	0	0	0	0	0	0	0	0	0	0	0	0	0
0	0	0	0	0	0	0	0	0	0	0	0	0	0	0	0	0	0	0	0
0	0	0	0	0	0	0	0	0	0	0	0	0	0	0	0	0	0	0	0
0	0	0	0	0	●	0	0	0	0	0	0	0	0	0	0	0	0	0	0
0	0	0	0	0	0	0	0	0	0	0	0	0	0	0	0	0	0	0	0
0	0	0	0	0	0	0	0	0	0	0	0	0	0	0	0	0	0	0	0
0	0	0	0	0	0	0	0	0	0	0	0	0	0	0	0	0	0	0	0
0	0	0	0	0	0	0	0	0	0	0	0	0	0	0	0	0	0	0	0
0	0	0	0	0	0	0	0	0	0	0	0	0	0	0	0	0	0	0	0
0	0	0	0	0	0	0	0	0	0	0	0	0	0	0	0	0	0	0	0
0	0	0	0	0	0	0	0	0	0	0	0	0	0	0	0	0	0	0	0
0	0	0	0	0	0	0	0	0	0	0	0	0	0	0	0	0	0	0	0
0	0	0	0	0	0	0	0	0	0	0	0	0	0	0	0	0	0	0	0
0	0	0	0	0	0	0	0	0	0	0	0	0	0	0	0	0	0	0	0
0	0	0	0	0	0	0	0	0	0	0	0	0	0	0	0	0	0	0	0
0	0	0	0	0	0	0	0	0	0	0	0	0	0	0	0	0	0	0	0

Fig. 10.5 Concentration of NH_3, $t = 8$ s. Black point in the figure depicts the man position at the route of evacuation

0	0	0	0	0	0	0	0	0	0	0	0	0	0	0	0	0	0	0	0
0	0	0	0	0	1	2	5	10	15	9	4	2	1	0	0	0	0	0	0
0	0	0	1	2	4	9	19	43	99	40	16	7	3	1	0	0	0	0	0
0	0	1	2	4	8	15	27	46	67	43	23	12	6	3	1	0	0	0	0
1	1	2	4	7	12	20	30	43	52	40	26	16	9	5	2	1	0	0	0
1	2	3	6	10	15	23	31	39	43	37	27	18	11	7	4	2	1	1	0
2	3	5	9	13	19	25	32	37	38	35	27	20	14	9	6	3	2	1	0
4	5	8	12	16	21	27	31	34	35	33	28	22	16	12	8	5	3	2	0
5	7	10	15	20	24	27	30	32	32	30	27	23	19	15	10	7	4	3	0
6	8	12	17	23	27	28	30	31	30	29	27	24	21	17	13	9	6	4	0
7	9	13	17	21	23	0	0	0	0	0	0	0	0	16	13	9	6	5	0
8	9	12	16	20	22	0	0	0	0	0	0	0	0	14	12	9	7	5	0
8	9	12	15	18	20	0	0	0	0	0	0	0	0	12	11	9	7	6	0
8	9	11	14	16	18	17	15	13	11	9	0	0	0	11	10	8	7	6	0
8	9	10	12	14	15	15	13	12	10	8	0	0	0	10	9	8	6	6	0
7	8	9	11	12	12	12	11	10	8	7	0	0	0	9	8	7	6	5	0
7	7	8	9	10	10	10	9	8	7	●	0	0	0	8	7	6	5	5	0
6	6	7	7	8	8	8	7	6	5	5	4	4	5	6	6	5	5	4	0
5	5	5	6	6	6	6	6	5	4	3	3	3	4	5	5	4	4	4	0
4	4	4	4	5	5	5	4	4	3	3	2	2	3	3	3	3	3	3	0
3	3	3	3	3	3	3	3	3	2	2	2	2	2	2	2	2	2	2	0
2	2	2	2	2	2	2	2	2	1	1	1	1	1	2	2	2	2	2	0
1	1	1	1	1	1	1	1	1	1	1	1	1	1	1	1	1	1	1	0
1	1	1	1	1	1	1	1	1	0	0	0	0	0	1	1	1	1	1	0
1	1	1	1	1	1	1	1	1	0	0	0	0	0	1	1	1	1	1	0

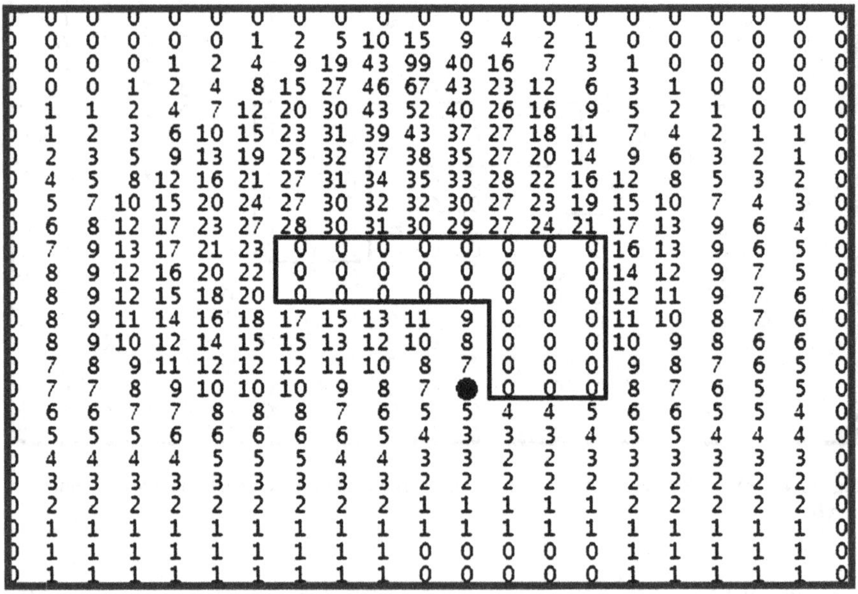

Fig. 10.6 Concentration of NH_3, $t = 17$ s. Black point in the figure depicts the man position at the route of evacuation

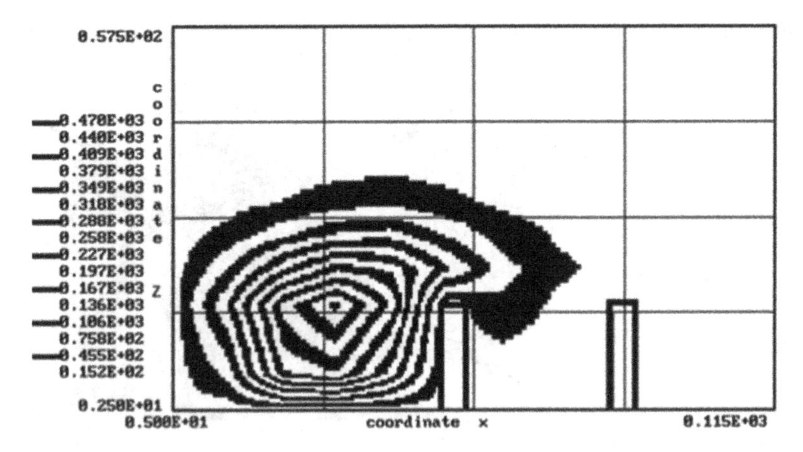

Fig. 10.7 Concentration of NH_3 $t = 5$ s

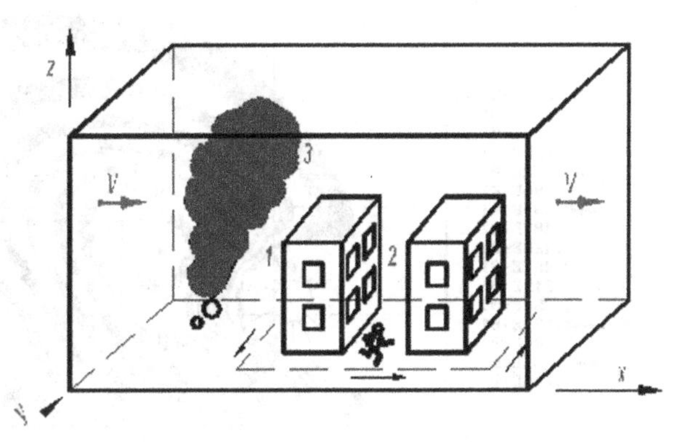

Fig. 10.8 Sketch of the computational domain: *1*, *2* –building; *3* – toxic gas cloud

In Fig. 10.7 the results of calculation of NH_3 cloud dispersion near two buildings are presented. The sketch of the computational domain and the route of evacuation are shown in Fig. 10.8. People leave the building in 8 s after the accident. The value of the toxic dose for the people running along the route of evacuation is as following:

t = 10 s TD = 1.11 mg · min/l
t = 12.5 s TD = 7.26 mg · min/l
t = 15 s TD = 12.55 mg · min/l
t = 17.5 s TD = 17.06 mg · min/l
t = 20 s TD = 21.10 mg · min/l
t = 22.5 s TD = 25.00 mg · min/l

Case 2 Modeling of NH_3 cloud neutralization.

The developed models allow assessing the efficiency of neutralization which is used to eliminate the toxic gas cloud. The reagent is supplied from the moving helicopter (Fig. 10.9). To solve the problem the additional equations which

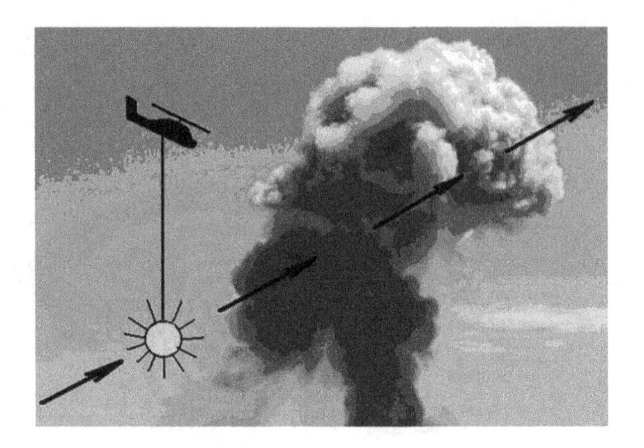

Fig. 10.9 Sketch of the helicopter route

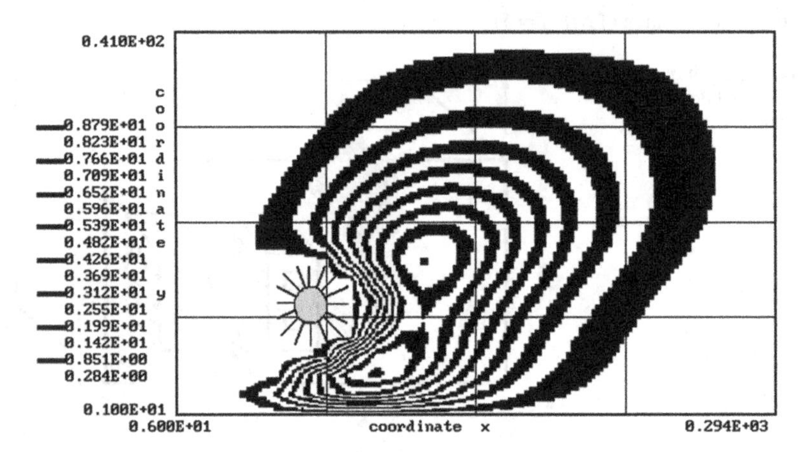

Fig. 10.10 Concentration of NH_3 (section $Y = 138$ m) $t = 9$ s (with neutralization)

describe the chemical reaction of the toxic gas and reagent are used together with (10.1–10.4).

The results of calculation of NH_3 concentration in case with neutralization are presented in Figs. 10.10 and 10.11.

The efficiency of the neutralization process can be estimated from Table 10.1.

In another study [2], the NH_3 toxic gas cloud was produced above the roof of the building. It was assumed that in the case of that accident, a hole in the roof was formed and toxic gas started to emit from the hole inside the building [2]. We tested in [2] the efficiency of supplying with a helicopter the reagent above the hole in the roof.

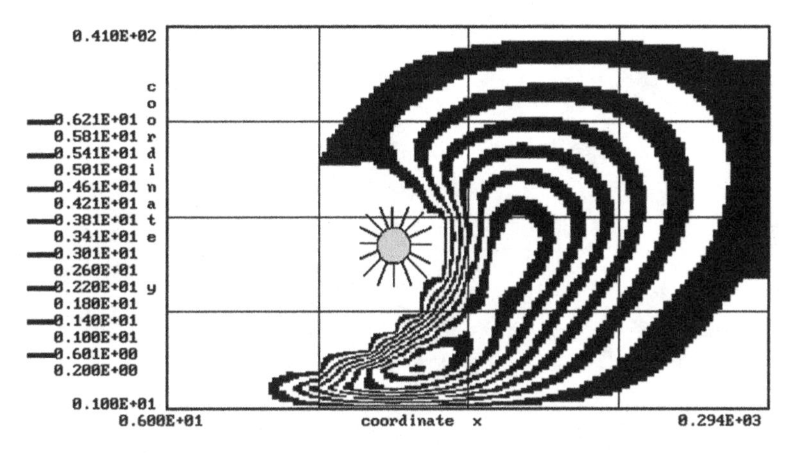

Fig. 10.11 Concentration of NH_3 (section $Y = 138$ m) $t = 12$ s (with neutralization)

Table 10.1 Mass of the eliminated NH_3

t, sec	12	16	20	26
Mass of NH_3, kg	129	213	271	317

10.6 Conclusions

We have presented the expert systems to predict the disaster impact on the person or environment in the case of accidents at chemical enterprises or storages. The expert systems are based on the numerical integration of the convective-diffusive equation and the equations of two fluid dynamic models. The developed numeric models allow predicting the atmosphere pollution among the buildings. Application of the expert systems for solving some problems in the field of the environmental safety was shown.

References

1. Belayev NN (1999) Computer simulation of the pollutant dispersion among buildings. In: Belayev NN, Kazakevitch MI, Khrutch VK (eds) Wind engineering into 21st century: proceedings of the tenth international conference on wind engineering, Copenhagen (Denmark). A. A. Balkema, Rotterdam/Brookfield, pp 1217–1220
2. Belyaev NN, Lisnyak VM (2012) Accident in chemically dangerous object: numeric modeling of the toxic gas neutralization process under building roof. News Dnipropetr Natl Univ Railw Transp 40:151–155 (in Russian)

Chapter 11
Fuzzy Logic and Data Mining in Disaster Mitigation

Abraham Kandel, Dan Tamir, and Naphtali D. Rishe

Abstract Disaster mitigation and management is one of the most challenging examples of decision making under uncertain, missing, and sketchy, information. Even in the extreme cases where the nature of the disaster is known, preparedness plans are in place, and analysis, evaluation, and simulations of the disaster management procedures have been performed, the amount and magnitude of "surprises" that accompany the real disaster pose an enormous demand. In the more severe cases, where the entire disaster is an unpredicted event, the disaster management and response system might fast run into a chaotic state. Hence, the key for improving disaster preparedness and mitigation capabilities is employing sound techniques for data collection, information processing, and decision making under uncertainty. Fuzzy logic based techniques are some of the most promising approaches for disaster mitigation. The advantage of the fuzzy-based approach is that it enables keeping account on events with perceived low possibility of occurrence via low fuzzy membership/truth-values and updating these values as information is accumulated or changed. Several fuzzy logic based algorithms can be deployed in the data collection, accumulation, and retention stage, in the information processing phase, and in the decision making process. In this chapter a comprehensive assessment of fuzzy techniques for disaster mitigation is presented. The use of fuzzy logic as a possible tool for disaster management is investigated and the strengths and weaknesses of several fuzzy techniques are evaluated. In addition to classical fuzzy techniques, the use of incremental fuzzy clustering in the context of complex and high order fuzzy logic system is evaluated.

A. Kandel (✉) • N.D. Rishe
School of Computer Science, Florida International University Miami,
11200 SW 8th St, Miami, FL, USA
e-mail: abekandel@yahoo.com; kandel@cse.usf.edu

D. Tamir
Department of Computer Science, Texas State University, San Marcos, TX, USA

H.-N. Teodorescu et al. (eds.), *Improving Disaster Resilience and Mitigation - IT Means and Tools*, NATO Science for Peace and Security Series C: Environmental Security,
DOI 10.1007/978-94-017-9136-6_11,

Keywords Fuzzy logic • Fuzzy functions • Fuzzy expectation • Black swan • Gray swan

11.1 Introduction

Disaster mitigation and management (DMM) is one of the most challenging examples of decision making under uncertain, missing, and sketchy, information. Even in the extreme cases where the nature of the disaster is known, preparedness plans are in place, and analysis, evaluation, and simulations of the disaster management procedures have been performed, the amount and magnitude of "surprises" that accompany the real disaster pose an enormous demand. In the more severe cases, where the entire disaster is an unpredicted event, the disaster management and response system might fast run into a chaotic state. Hence, the key for improving disaster preparedness and mitigation capabilities is employing sound techniques for data collection, information processing, and decision making under uncertainty.

Consequently, we must develop a formal set of tools to deal with the increasing number of potential disasters around the world on one hand and information dominance on the other hand. Moreover, we need a formal understanding of the relations between uncertainty, which is immersed in DMM programs, and fuzzy logic. Just as the formal notions of computer science have immensely benefited software and hardware design, a formal treatment of DMM programs should also lend to significant advances in the way that these programs and systems are developed and designed in the future.

Recently, the academic community and other agencies have presented spurring growth in the field of data mining in big-data systems. These advances are beginning to find their way into DMM programs and are redefining the way we address potential disaster and mitigate the effects of disasters. Nevertheless, academia, industry, and governments need to engage as a unified entity to advance new technologies as well as applied established technologies in preparation and response to the specific emerging problems of disasters. Since research and development is seldom conducted in national isolation it must include international collaboration and global activities. This will bring many countries and scientists together in order to enjoy broad access to all leading edge technologies and tools to be used for prevention and remedies of disasters.

Analysis of disasters shows three types of challenges, the first is the ability to predict the occurrence of disasters, the second is the need to produce a preparedness plan, and the third is the actual real time response activities related to providing remedies for a currently occurring disaster. Inspired by the metaphor of black swan coined by Taleb [1, 2] we show that every disaster falls into the category of a gray level swan (with black swan being a special case of gray level swan). Furthermore, we show that each of the three challenges involves dealing with uncertainty and can be addressed via fuzzy logic based tools and techniques.

Experience shows that swans come in different shades of gray, spawning the range from black-to-white. There is no need to expand on white swans. Those are adorable animals and under the Taleb metaphor, they represent pleasant events, highly predictable pleasant events, or pleasant surprises. Our collective knowledge, however, tells us that white swans are rare. Moreover, sometimes white swans change abruptly without advanced notice into black or gray swans. The black swans are the other extreme. They represent completely unpredictable highly adverse events (disasters). Again, common experience shows that black swans are rare. This is due to two reasons: (1) there is almost always some information about the upcoming black swan. This is evident from the term "connecting the dots," which refers to the fact that information about disasters might be available; but, our ability to process this information is limited. The second is related to the authors' personal observation that often there is some amount of "good news" that accompanies any type of "bad news." Following these observation we classify all the swans as gray level swans; where the level of gray is related to the amount of surprise and the severity of the disastrous event represented by the swan. We further elaborate on gray swans in Sect. 11.2. In Sect. 11.3, we review methods for improving our ability to identify and "bleach" gray swans using fuzzy logic-based tools.

Predicting the encounter with gray swans is an important component of DMM. It can enable early setting of a mitigation plan. Nevertheless, the fact that a disaster (gray swan) is somewhat predictable does not completely reduce the amount of surprise that accompany the actual occurrence of the disaster. Hence, any mitigation plan; that is, a set of procedures compiled in order to address the adverse effects of disasters, should be flexible enough to handle additional surprises. We refer to these surprises as second generation swans. Finally, the real time ramification program is the actual set of procedures enacted and executed as the disaster occurs. There are two preconditions for successful remedy of the disaster effect. First, the leadership of the authorities who have to attempt following the original mitigation plan as close as possible while instilling a sense of trust and calmness in the people that experience the disaster and the mitigation provider teams. Second, and as a part of their leadership traits, the authorities must possess the ability to adapt their remedy procedures to the dynamics of the disaster; potentially providing effective improvisations. Again, this relates to the notion of second generation gray swans which are secondary disastrous events that evolve from the main disaster. A simple example for such swans is events of looting and violence that might accompany a major disaster. For this end, fast automatic assessment of the dynamics is paramount. Again, numerous tools including fuzzy logic based tools can assist the leaders in getting a good grasp on the disaster dynamics. A related notion is the notion of unknown unknowns (dark black swans) and known unknowns (gray swans).

Fuzzy logic based techniques are some of the most promising approaches for disaster mitigation. The advantage of the fuzzy-based approach is that it enables keeping account on events with perceived low possibility of occurrence via low fuzzy membership/truth-values and updating these values as information is accumulated or changed. Several fuzzy logic based algorithms can be deployed in the data collection, accumulation, and retention stage, in the information processing phase, and in the decision making process. In this chapter a comprehensive assessment of fuzzy techniques for disaster mitigation is presented. The use of fuzzy logic as a possible tool for disaster management is investigated and the strengths and weaknesses of several fuzzy techniques are evaluated. In addition to classical fuzzy techniques, the use of incremental fuzzy clustering in the context of complex and high order fuzzy logic system is evaluated.

The rest of the chapter elaborates on the unknown unknowns (that is, black swans) and known unknowns (i.e., gray swans) their relation to DMM and uncertainty which can be addressed via fuzzy logic based tools. This is elaborated in Sect. 11.2. Section 11.3, provides an overview of several fuzzy logic based tools for dealing with the uncertainty of the two sorts of unknowns and Sect. 11.4 provides brief conclusions as well as proposals for future enhancements of this research.

11.2 Uncertainty in Disaster Mitigation and Management

Disasters occur with different degrees of unpredictability and severity which is manifested in two main facets. First, the actual occurrence of the disaster might be difficult (potentially impossible) to predict. Second, regardless of the level predictability of the disaster, it is very likely that the disaster will be accompanied by secondary effects. Hence, disasters are a major source of "surprise" and uncertainty and their mitigation and management requires sound automatic and intelligent handling of uncertainty. Often, the stakeholders of DMM programs are classifying the unpredictability of disasters as two types of unknowns: unknown unknowns and known unknowns.

11.2.1 Unknown Unknowns and Known Unknowns

In his excellent books [1, 2], Nassim Nicholas Taleb describes the features of black swan events. We provide a slightly different set of features.

1. The black swan is an outlier; lying outside of the space of our regular expectations.
2. It has very low predictability and it carries an extremely adverse impact
3. The unknown component of the event is far more relevant than the known component.
4. Finally, we can explain, de facto, the event and through those explanations make it predictable in retrospect.

In this sense, the black swan represents a class of problems that can be referred to as the "unknown unknowns." However, a thorough investigation of many of the events that are widely considered as black swans, e.g., the September 11, 2001 attack in NYC, shows that there was available information concerning the attack; yet, this information did not affect the decision making and response prior to the attack.

This suggests that the term black swan is a bit too extreme and one should consider using the term gray swan. In this chapter, a gray swan represents an unlikely event that can be anticipated and carries an extremely adverse impact. In this respect the gray swans represent a two dimensional spectrum of information. The first dimension represents the predictability of the event where black swans are highly unpredictable and white swans are the norm. The second dimension represents the amount of adverse outcome embedded in the event; with black swans representing the most adverse outcomes. Consequently, the black swan is a special case of a gray swan. The following is a partial list of well-known gray swans each of which has a different degree of surprise as well as different degree of severity.

1. Bangladeshi factory collapse; Iceland volcano eruption
2. Fukushima – tsunami followed by nuclear radiation and risk of meltdown of nuclear facilities in the area.
3. 9/11 NYC attack followed by the collapse of the Twins
4. Y2K (was this a wolf-wolf-wolf ignited by grid?)
5. Financial Markets (1987, 2008), Madoff
6. 1998 Long-Term Capital Management
7. Yom Kippur War
8. December 7, 1941 - Pearl Harbor
9. Lincoln, Kennedy, Sadat, and Rabin assassinations

One type of gray swans relates to a set of events that can be considered as known unknowns. For example, a hurricane occurring in Florida during the hurricane season should not be considered as a part of the set of unknown unknowns. It should not surprise the responsible authorities. Moreover, often, there is a span of a few days between the identification of the hurricane and the actual landfall. Regardless, even a predictable hurricane land-fall carries numerous secondary disastrous events that are hard to predict.

We refer to these secondary events as second generation gray swans. Second generation swans are generated and/or detected while the disaster is in effect. A gray swan might (and according to the Murphy lows is likely to) spawn additional gray swans. Generally, second generation swans evolve late and fast. Hence, they introduce a more challenging detection problem and require specialized identification tools such as dynamic clustering.

11.2.2 Preparedness and Real Time Ramification

Following the discussion above, disaster mitigation and management has two key components. The first is referred to as preparedness. In preparedness, authorities consider as many known unknowns as possible, and compile procedures for early prediction, detection, and alert of disasters as well as sound remedy procedures for these disasters. The second component of DMM is referred to as real time remedies.

Considering real time remedies, there is no recipe (algorithm) for success. One can note that *leadership* is paramount and trying to adhere to a set of "best practices" which have been set ahead of time and used in training and ramification exercises can help reducing cost and risk. Nevertheless, a great amount of ability to improvise, rapidly change plans as dictated by changing circumstances, and significant amount of flexibility is required.

Even in the extreme cases where the nature of the disaster is known, preparedness plans are in place, and analysis, evaluation, and simulations of the disaster management procedures have been performed, the amount and magnitude of "surprises" that accompany the real disaster pose an enormous demand. Detecting relatively slow evolving [first generation] gray swans before the disaster occurs and relatively fast evolving second generation gray swans requires an adequate set of uncertainty management tools. In addition, existing swans might (are likely to) spawn unanticipated second generation swans. The collapse of the Twins in 9/11 is an example of a second generation gray swan.

Fuzzy logic is one of the suggested tools that can help creating a better understanding of DMM tools including, but no limited to intelligent robotics, learning and reasoning, language analysis and understanding, and data mining. While NATO continues to lead a strong technical agenda in DMM technologies, academia and industry must assume a preeminent position in driving a variety of leading-edge technologies and tools in order to address and mitigate disasters. We believe that the role of our research in fuzzy logic and uncertainty management in achieving these goals is critical for producing a successful DMM programs.

This emerging field must therefore drive a novel set of research directions for the USA and NATO assisting the scientific community and the private sector to develop a science and tools for anti-terror management.

11.3 Tools for Predictions and Evaluations of Fuzzy Events

Fuzzy logic based techniques are some of the most promising approaches for disaster mitigation. The advantage of the fuzzy-based approach is that it enables keeping account on events with perceived low possibility of occurrence via low fuzzy membership/truth-values and updating these values as information is accumulated or changed. Several fuzzy logic based algorithms can be deployed in the data collection, accumulation, and retention stage, in the information processing phase, and in the decision making process. Therefore, in this section we describe several possible fuzzy tools to try and predict disasters and cope with evolving disasters via sound DMM programs. We consider the following fuzzy logic based tools:

1. Fuzzy Switching Mechanisms
2. Fuzzy Expected Value
3. Fuzzy Relational Data Bases, Fuzzy Data-Mining and Fuzzy social Network Architectures (FSNA)

4. Complex and Multidimensional Fuzzy Sets, Logic, and Systems
5. Neuro-Fuzzy-Based Logic, and Systems
6. Dynamic and Incremental Fuzzy Clustering

11.3.1 Making Decisions with No Data

As an example for this idea we will use the fuzzy treatment of the transient behavior of a switching system and its static hazards [3]. Perhaps the major reason for the ineffectiveness of classical techniques in dealing with static hazard and obtaining a logical explanation of the existence of static hazard lies in their failure to come to grips with the issue of fuzziness. This is due to the fact that the hazardous variable implies imprecision in the binary system, which does not stem from randomness; but, from a lack of sharp transition between members in the class of input states. Intuitively, fuzziness is a type of imprecision which stems from a grouping of elements into classes that do not have sharply defined boundaries - that is, in which there is no sharp transition from membership to non-membership. Thus, the transition of a state has a fuzzy behavior during the transition time, since this is a member in an ordered set of operations, some of which are fuzzy in nature.

Any fuzzy-valued switching function can be expressed in disjunctive and conjunctive normal forms, in a similar way to two-valued switching functions. As before, fuzzy-valued switching functions over n variables can be represented by the mapping $f: [0, 1]^n \rightarrow [0, 1]$. We define a V-fuzzy function as a fuzzy function $f(x)$ such that $f(\xi)$ is a binary function for every binary n-dimensional vector ξ. It is clear that a V-fuzzy function f induces a binary function F such that $F: [0, 1]^n \rightarrow [0, 1]$ determined by $F(\xi) = f(\xi)$ for every binary n-dimensional vector ξ.

If the B-fuzzy function f describes the complete behavior of a binary combinational system, its steady-state behavior is represented by F, the binary function induced by f. Let $f(x)$ be an n-dimensional V-fuzzy function, and let ξ and ρ be adjacent binary n-dimensional vectors. The vector $T_{\xi_j}^{\rho}$ is a static hazard of f iff $f(\xi) = f(\rho) \neq f\left(T_{\xi_j}^{\rho}\right)$.

If $f(\xi) = f(\rho) = 1$, $T_{\xi_j}^{\rho}$ is a 1-hazard. If $f(\xi) = f(\rho) = 0$, $T_{\xi_j}^{\rho}$ is a 0-hazard. If f is B-fuzzy and $T_{\xi_j}^{\rho}$ is a static hazard, then $f\left(T_{\xi_j}^{\rho}\right)$ has a perfect fuzzy value, that is, $f\left(T_{\xi_j}^{\rho}\right) \in (0, 1)$. Consider the static hazard as a malfunction represented by an actual or potential deviation from the intended behavior of the system. We can detect all static hazards of the V-fuzzy function $f(x)$ by considering the following extension of Shannon normal form. Let $f(\overline{x})$, $\overline{x} = (x_1, x_2, \ldots, x_n)$, be a fuzzy function and denote the vector

$$\left(x_1, x_2, x_{j-1}, x_{j+1}, \ldots, x_n,\right) \text{ by } x^j,$$

By successive applications of the rules of fuzzy algebra, the function $f(x)$ may be expanded about, say, x_j as follows:

$$f(x) = x_j f_1\left(x^j\right) + \overline{x}_j f_2\left(x^j\right) + x_j\overline{x}_j f_3\left(x^j\right) + f_4\left(x^j\right),$$

where f_1, f_2, f_3, and f_4 are fuzzy functions. It is clear that the same expansion holds when the fuzzy functions are replaced by B-fuzzy functions of the same dimension. Let ξ and ρ be two adjacent n-dimensional binary vectors that differ only in their j th component. Treating ξ_j as a perfect fuzzy variable during transition time implies that $T_{\xi_j}^{\rho}$ is a 1-hazard of f iff $f(\xi) = f(\rho) = 1$ and $f\left(T_{\xi_j}^{\rho}\right) \in [0, 1)$. We show that the above conditions for the vector $T_{\xi_j}^{\rho}$ to be 1-hazard yield the following result.

Theorem 1 ([3]): The vector $T_{\xi_j}^{\rho}$ is a 1-hazard of the B-fuzzy function $f(x)$ given above iff the binary vector ξ_j is a solution of the following set of Boolean equations:

$$f_1\left(x^j\right) = 1, \quad f_2\left(x^j\right) = 1, \quad f_4\left(x^j\right) = 0.$$

Proof

State 1: $\xi_j = 1$ and $\overline{\xi}_j = 0$ imply $f_1(\xi^j) + f_4(\xi^j) = 1$.
State 2: $\xi_j = 0$ and $\overline{\xi}_j = 1$ imply $f_2(\xi^j) + f_4(\xi^j) = 1$.

Transition state: $\xi_j \in (0, 1)$ [which implies $\overline{\xi}_j \in (0, 1)$], and thus:

$$0 \leq max\left\{min\left[\xi_j, f_1\left(\xi^j\right)\right], min\left[\overline{\xi}_j, f_2\left(\xi^j\right)\right], min\left[\xi_j, \overline{\xi}_j, f_3\left(\xi^j\right)\right], f_4\left(\xi^j\right)\right\} < 1.$$

It is clear from the transition state that $f_4(\xi_j)$ cannot be equal to one, and thus:

$$f_4\left(\xi^j\right) = 0, \quad f_1\left(\xi^j\right) = f_2\left(\xi^j\right) = 1.$$

Several items must be pointed out. The system is not a fuzzy system. It is a Boolean system. The modeling of the system as a fuzzy system, due to the lack of knowledge regarding the behavior of x_j during the transition provided us with a tool to make decisions (regarding the Boolean values of f_1, f_2 and f_4) with no data whatsoever regarding x^j. Thus, we were able to make non-fuzzy decisions in a deterministic environment with no data. The interesting question is whether or not we can apply this idea to DMM programs.

11.3.2 Fuzzy Expectations

Ordinarily, imprecision and indeterminacy are considered to be statistical, random characteristics and are taken into account by the methods of probability theory.

In real situations, a frequent source of imprecision is not only the presence of random variables, but the impossibility, in principle, of operating with exact data as a result of the complexity of the system, or the imprecision of the constraints and objectives. At the same time, classes of objects that do not have clear boundaries appear in the problems; the imprecision of such classes is expressed in the possibility that an element not only belongs or does not belong to a certain class, but that intermediate grades of membership are also possible. The membership grade is subjective; although it is natural to assign a lower membership grade to an event that have a lower probability of occurrence. The fact that the assignment of a membership function of a fuzzy set is "non-statistical" does not mean that we cannot use probability distribution functions in assigning membership functions. As a matter of fact, a careful examination of the variables of fuzzy sets reveals that they may be classified into two types: statistical and non-statistical.

Definition 1 ([3]) Let B be a Borel field (σ -algebra) of subsets of the real line Ω. A set function $\mu(\cdot)$ defined on B is called a fuzzy measure if it has the following properties:

1. $\mu(\Phi) = 0$ (Φ is the empty set);
2. $\mu(\Omega) = 1$;
3. If $\alpha, \beta \in B$; with $\alpha \subset \beta$, then $\mu(\alpha) \leq \mu(\beta)$;
4. If $\{\alpha_j | 1 \leq j < \infty\}$ is a monotone sequence, then

$$\lim_{j\to\infty} [\mu(\alpha_j)] = \mu\left[\lim_{j\to\infty} (\alpha_j)\right].$$

Clearly, $\Phi, \Omega \in B$; also, if $\alpha_j \in B$ and $\{\alpha_j | 1 \leq j < \infty\}$ is a monotonic sequence, then $\lim_{j\to\infty} (\alpha_j) \in B$. In the above definition, (1) and (2) mean that the fuzzy measure is bounded and nonnegative, (3) means monotonicity (in a similar way to finite additive measures used in probability), and (4) means continuity. It should be noted that if Ω is a finite set, then the continuity requirement can be deleted. (Ω, B, μ) is called a fuzzy measure space; $\mu(\cdot)$ is the fuzzy measure of (Ω, B). The fuzzy measure μ is defined on subsets of the real line. Clearly, $\mu[\chi_A \geq T]$ is a non-increasing, real-valued function of T when χ_A is the membership function of set A. Throughout our discussion, we use ξ^T to represent $x | \chi_A(x) \geq T$ and $\mu(\xi^T)$ to represent $\mu[\chi_A \geq T]$, assuming that the set A is well specified. Let $\chi A : \Omega \to [0, 1]$ and $\xi^T = x | \chi_A(x) \geq T$. The function χ_A is called a B-measurable function if $\xi^T \in B, \forall\, T \in [0, 1]$. Definition 2 defines the fuzzy expected value (*FEV*) of χ_A when $\chi_A \in [0, 1]$. Extension of this definition when $\chi_A \in [a, b], a < b < \infty$, is presented later.

Definition 2 ([3]): Let χ_A be a B -measurable function such that $\chi_A \in [0, 1]$. The fuzzy expected value (*FEV*) of χ_A over a set A, with respect to the measure

$\mu(\cdot)$, is defined as $\sup_{T\in[0,1]}\{min\left[T,\mu\left(\xi^{T}\right)\right]\}$, where $\xi^{T}=\{x|\chi_{A}(x)\geq T\}$. Now, $\mu\{x|\chi_{A}(x)\geq T\}=f_{A}(T)$ is a function of the threshold T. The actual calculation of $FEV(\chi_{A})$ then consists of finding the intersection of the curves $T=f_{A}(T)$. The intersection of the two curves will be at a value $T=H$, so that $FEV(\chi_{A}=H\in[0,1]$. It should be noted that when dealing with the $FEV(\eta)$ where $\eta\in[0,1]$, we should not use a fuzzy measure in the evaluation but rather a function of the fuzzy measure, η', which transforms η under the same transformation that χ and T undergo to η and T', respectively. In general the *FEV* has the promise and the potential to be used as a very powerful tool in developing DMM technologies.

11.3.3 Fuzzy Relational Databases and Fuzzy Social Network Architecture

The Fuzzy Relational Database (FRDB) model which is based on research in the fields of relational databases and theories of fuzzy sets and possibility is designed to allow representation and manipulation of imprecise information. Furthermore, the system provides means for "individualization" of data to reflect the user's perception of the data [4]. As such, the FRDB model is suitable for use in fuzzy expert system and other fields of imprecise information-processing that model human approximate reasoning such as FSNA [5, 6].

The objective of the FRDB model is to provide the capability to handle imprecise information. The FRDB should be able to retrieve information corresponding to natural language statements as well as relations in FSNA. Although most of these situations cannot be solved within the framework of classical database management systems, they are illustrative of the types of problem that human beings are capable of solving through the use of approximate reasoning. The FRDB model and the FSNA model retrieve the desired information by applying the rules of fuzzy linguistics to the fuzzy terms in the query.

The FRDB as well as the FSNA development [4–6] were influenced by the need for easy-to-use systems with sound theoretical foundations as provided by the relational database model and theories of fuzzy sets and possibility. They address the following issues:

1. representation of imprecise information,
2. derivation of possibility/certainty measures of acceptance,
3. linguistic approximations of fuzzy terms in query languages,
4. development of fuzzy relational operators (IS, AS . . . AS, GREATER, . . .),
5. processing of queries with fuzzy connectors and truth quantifiers,
6. null-value handling using the concept of the possibilities expected value,
7. modification of the fuzzy term definitions to suit the individual user.

The fuzzy relational data base and the FSNA are collections of fuzzy time-varying relations which may be characterized by tables, graphs, or functions, and manipulated by recognition (retrieval) algorithms or translation rules.

As an example let us take a look at one of these relations, the similarity relation. Let D_i be a scalar domain, $\in D_i$. Then $s(x, y) \in [0, 1]$ is a similarity relation with the following properties: Reflexivity: $s(x, x) = 1$; Symmetry: $s(x, y) = s(y, x)$; Θ -transitivity: where Θ is most commonly specified as max-min transitivity. If, $y, z \in U$, then $s(x, z) \geq max(y \in D_i)\ min(s(x, y), s(y, z))$. Another example is the proximity relation defined below. Let D_i be a numerical domain and, $y, z \in D_i$. Here $p(x, y) \in [0, 1]$ is a proximity relation that is reflexive, and symmetric with transitivity of the form

$$p\,(x, z) \geq max\,(y \in D_i)\ p\,(x, y) * p\,(y, z)\,.$$

The generally used form of the proximity relations is $p(x, y).\, e^{-\beta|x-y|}$, where $\beta > 0$. This form assigns equal degrees of proximity to equally distant points. For this reason, it is referred to as absolute proximity in the FRDB and FSNA models. Similarity and proximity are used in evaluation of queries of the general form: "Find X such that $X.A \ominus d$" Where $X.A$ is an attribute of X, $d \in D$ is a value of attribute A defined on the domain D, and $\ominus$ is a fuzzy relational operator. Clearly both FRDS and FSNA may have numerous applications in black swan as well as gray swan prediction.

In many DMM programs and disaster models the amount of information is determined by the amount of the uncertainty – or, more exactly, it is determined by the amount by which the uncertainty has been reduced; that is, we can measure information as the decrease of uncertainty. The concept of information itself has been implicit in many DMM models. This is both as a substantive concept important in its own right and as a consonant concept that is ancillary to the entire structure of DMM

11.3.4 Complex Fuzzy Membership Grade

Several aspects of the DMM program can utilize the concept of complex fuzzy logic [3, 7–14]. Complex fuzzy logic can be used to represent the two dimensional information embedded in the description of a disaster; namely, the severity and uncertainty. In addition, inference based on complex fuzzy logic can be used to exploit the fact that variables related to the uncertainty that it a part of disasters is multi-dimensional and cannot be readily defined via single dimensional clauses connected by single dimensional connectives. Finally, the multi-dimensional fuzzy space defined as a generalization of complex fuzzy logic can serve as a media for clustering of disaster in a linguistic variable-based feature space.

Tamir et al. introduced a new interpretation of complex fuzzy membership grade and derived the concept of pure complex fuzzy classes [13]. This section introduces the concept of a pure complex fuzzy grade of membership, the interpretation of this concept as the denotation of a fuzzy class, and the basic operations on fuzzy classes.

To distinguish between classes, sets, and elements of a set we use the following notation: a class is denoted by an upper case Greek letter, a set is denoted by an upper case Latin letter, and a member of a set is denoted by a lower case Latin letter.

The Cartesian representation of the pure complex grade of membership is given in the following way:

$$\mu(V, z) = \mu_r(V) + j\mu_i(z)$$

Where $\mu_r(V)$ and $\mu_i(z)$, the real and imaginary components of the pure complex fuzzy grade of membership, are real value fuzzy grades of membership. That is, $\mu_r(V)$ and $\mu_i(z)$ can get any value in the interval $[0, 1]$. The polar representation of the pure complex grade of membership is given by:

$$\mu(V, x) = r(V)e^{j\sigma\phi(z)}$$

Where $r(V)$ and $\phi(z)$, the amplitude and phase components of the pure complex fuzzy grade of membership, are real value fuzzy grades of membership. That is, they can get any value in the interval $[0, 1]$. The scaling factor, σ is in the interval $(0, 2\pi]$. It is used to control the behavior of the phase within the unit circle according to the specific application. Typical values of σ are $\{1, \frac{\pi}{2}, \pi, 2\pi\}$. Without loss of generality, for the rest of the discussion in this section we assume that $\sigma = 2\pi$.

The difference between pure complex fuzzy grades of membership and the complex fuzzy grade of membership proposed by Ramot et al. [11, 12], is that both components of the membership grade are fuzzy functions that convey information about a fuzzy set. This entails different interpretation of the concept as well as a different set of operations and a different set of results obtained when these operations are applied to pure complex grades of membership. This is detailed in the following sections.

11.3.4.1 Complex Fuzzy Class

A fuzzy class is a finite or infinite collection of objects and fuzzy sets that can be defined in an unambiguous way and complies with the axioms of fuzzy sets given by Tamir et al. and the axioms of fuzzy classes given by *Běhounek* [9, 15–20]. While a general fuzzy class can contain individual objects as well as fuzzy sets, a *pure fuzzy class of order one* can contain only fuzzy sets. In other words, individual objects cannot be members of a pure fuzzy class of order one. A pure fuzzy class of order M is a collection of pure fuzzy classes of order $M - 1$. We define a *Complex Fuzzy Class* Γ to be a pure fuzzy class of order one i.e., a fuzzy set of fuzzy sets. That is,

$\Gamma = \{V_i\}_{i=1}^{\infty}$; or $\Gamma = \{V_i\}_{i=1}^{N}$ where V_i is a fuzzy set and N is a finite integer. Note that despite the fact that we use the notation $\Gamma = \{V_i\}_{i=1}^{\infty}$ we do not imply that the set of sets $\{V_i\}$ is enumerable. The set of sets $\{V_i\}$ can be finite, countably infinite, or uncountably infinite. The use of the notation $\{V_i\}_{i=1}^{\infty}$ is just for convenience.

The class Γ is defined over a universe of discourse T. It is characterized by a pure complex membership function $\mu_\Gamma(V, z)$ that assigns a complex-valued grade of membership in Γ to any element $z \in U$ (where U is the universe of discourse). The values that $\mu_\Gamma(V, z)$ can receive lie within the unit square or the unit circle in the complex plane, and are in one of the following forms:

$$\mu_\Gamma(V, z) = \mu_r(V) + j\mu_i(z)$$

$$\mu_\Gamma(z, V) = \mu_r(z) + j\mu_i(V)$$

Where $\mu_r(\alpha)$ and $\mu_i(\alpha)$, are real functions with a range of [0,1]. Alternatively:

$$\mu_\Gamma(V, z) = r(V)e^{j\theta\phi(z)}$$

$$\mu_\Gamma(z, V) = r(z)e^{j\theta\phi(V)}$$

Where $r(\alpha)$ and $\phi(\alpha)$, are real functions with a range of [0, 1] and $\theta \in (0, 2\pi]$.

In order to provide a concrete example we define the following pure fuzzy class. Let the universe of discourse be the set of all the hurricanes that hit the East Coast of the USA (in any time in the past) along with a set of attributes related to hurricanes such as wind speed, rain, movement of the hurricane eye, and related surges. Let M_i denote the set of hurricanes that hit the East Coast of the USA in the last i years. Furthermore consider a function (f_1) that associates a number between 0 and 1 with each set of hurricanes. For example, this function might reflect the severity in terms of average wind gust of all the hurricanes in the set. In addition, consider a second function (f_2) that associates a number between 0 and 1 with each specific hurricane. For example, this function might be a normalized value of level of destructiveness of the hurricane. The functions (f_1,f_2) can be used to define a pure fuzzy class of order one. A compound of the two functions in the form of a complex number can represent the degree of membership in the pure fuzzy class of "destructive" (e.g., catastrophic) hurricanes in the set of hurricanes that occurred in the last 10 years.

Formally, let U be a universe of discourse and let 2^U be the power set of U. Let f_1 be a function from 2^U to [0, 1] and let f_2 be a function that maps elements of U to the interval [0, 1]. For $V \in 2^U$ and $z \in U$ define $\mu_\Gamma(V, z)$ to be:

$$\mu_\Gamma(V, z) = \mu_r(V) + j\mu_i(z) = f_1(V) + jf_2(z)$$

Then, $\mu_\Gamma(V, z)$ defines a pure fuzzy class of order one, where for every $V \in 2^U$, and for every $z \in U$,

$\mu_\Gamma(V,z)$; is the degree of membership of z in V and the degree of membership of V in Γ. Hence, a complex fuzzy class Γ can be represented as the set of ordered triples:

$$\Gamma = \left\{V, z, \mu_\Gamma (V,z) \middle| V \in 2^U, z \in U\right\}$$

Depending on the form of $\mu_\Gamma(\alpha)$ (Cartesian or polar), $\mu_r(\alpha)$, $\grave{\imath}_i(\alpha)$, $r(\alpha)$, and $\phi(\alpha)$ denote the degree of membership of z in V and/ or the degree of membership of V in Γ. Without loss of generality, however, we assume that $\mu_r(\alpha)$ and $r(\alpha)$ denote the degree of membership of V in Γ for the Cartesian and the polar representations respectively. In addition, we assume that $\mu_i(\alpha)$ and $\phi(\alpha)$ denote the degree of membership of z in V for the Cartesian and the polar representations respectively. Throughout this chapter, the term complex fuzzy class refers to a pure fuzzy class with pure complex-valued membership function, while the term fuzzy class refers to a traditional fuzzy class such as the one defined by *Běhounek* [15].

Degree of Membership of Order *N*

The traditional fuzzy grade of membership is a scalar that defines a fuzzy set. It can be considered as degree of membership of order 1. The pure complex degree of membership defined in this chapter is a complex number that defines a pure fuzzy class. That is, a fuzzy set of fuzzy sets. This degree of membership can be considered as degree of membership of order 2 and the class defined can be considered as a pure fuzzy class of order 1. Additionally, one can consider the definition of a fuzzy set (a class of order 0) as a mapping into a one dimensional space and the definition of a pure fuzzy class (a class of order 1) as a mapping into a two dimensional space. Hence, it is possible to consider a degree of membership of order N as well as a mapping into an N -dimensional space. The following is a recursive definition of a fuzzy class of order N. Note that part 2 of the definition is not really necessary it is given in order to connect the terms pure complex fuzzy grade of membership and the term grade of membership of order 2.

Definition 3 ([13]):

1. A fuzzy class of order 0 is a fuzzy set; it is characterized by a degree of membership of order 1 and a mapping into a one dimensional space.
2. A fuzzy class of order 1 is a fuzzy class; that is, set of fuzzy sets. It is characterized by a pure complex degree of membership. Alternatively, it can be characterized by a degree of membership of order two and a mapping into a two dimensional space.
3. A fuzzy class of order N is a fuzzy set of fuzzy classes of order $N-1$; it is characterized by a degree of membership of order $N+1$ and a mapping into an $(N+1)$ -dimensional space.

Table 11.1 Basic propositional fuzzy logic connectives

Operation	Interpretation
Negation	$f('P) = 1 + j1 - f(P)$
Implication	$f(P \rightarrow Q) = \min(1, 1 - p_R + q_R) + j \times \min(1, 1 - p_I + q_I)$
Conjunction	$f(P \otimes Q) = \min(p_R, q_R) + j \times \min(p_I, q_I)$
Disjunction	$f(P \oplus Q) = \min(p_R, q_R) + j \times \min(p_I, q_I)$

Generalized Complex Fuzzy Logic

A general form of a complex fuzzy proposition is: "$x \ldots A \ldots B \ldots$" where A and B are values assigned to linguistic variables and '...' denotes natural language constants. A complex fuzzy proposition P can get any pair of truth values from the Cartesian interval $[0, 1] \times [0, 1]$ or the unit circle. Formally a fuzzy interpretation of a complex fuzzy proposition P is an assignment of fuzzy truth value of the form $p_r + jp_i$, or of the form $r(p)e^{j\theta(p)}$, to P. In this case, assuming a proposition of the form "$x \ldots A \ldots B \ldots$," then p_r $(r(p))$ is assigned to the term A and p_i $(\theta(p))$ is assigned to the term B.

For example, under one interpretation, the complex fuzzy truth value associated with the complex proposition: "x is a *destructive hurricane with high surge*." can be $0.1 + j0.5$. Alternatively, in another context, the same proposition can be interpreted as having the complex truth value $0.3e^{j0.2}$. As in the case of traditional propositional fuzzy logic we use the tight relation between complex fuzzy classes/complex fuzzy membership to determine the interpretation of connectives. For example, let C denote the complex fuzzy set of "destructive hurricanes with high surge," and let $f_C = c_r + jc_i$, be a specific fuzzy membership function of C, then f_C can be used as the basis for interpretations of P. Next we define several connectives along with their interpretation.

Table 11.1 includes a specific definition of connectives along with their interpretation. In this table P, Q, and S denote complex fuzzy propositions and $f(S)$ denotes the complex fuzzy interpretation of S. We use the fuzzy Łukasiewicz logical system as the basis for the definitions [16, 19]. Hence, the max t-norm is used for conjunction and the min t-conorm is used for disjunction. Nevertheless, other logical systems such as Gödel fuzzy systems can be used [19, 21].

The same axioms used for fuzzy logic are used for complex fuzzy logic, and Modus ponens is the rule of inference.

Complex Fuzzy Propositions and Connectives Examples

Consider the following propositions (P and Q respectively):

1. "x is a *destructive hurricane with high surge*."
2. "x is a *destructive huricane with fast moving center*."

Let A be the term "*destructive hurricane*," Hence, P is of the form: "x is A in B," and Q is of the form "x is A in C." In this case, the terms "*destructve hurricane*," "*high surge*," and "*fast moving center*", are values assigned to the linguistic variables $\{A, B, C\}$. Furthermore, the term "*destructve hurricane*," can get fuzzy truth values (between 0 and 1) or fuzzy linguistic values such as "*catastriphic*," "*devastating*," "*and disastrous*." Assume that the complex fuzzy interpretation (i.e., degree of confidence or complex fuzzy truth value) of P is $p_r + jp_i$, while the complex fuzzy interpretation of Q is $q_r + jq_i$ $(q_r = p_r)$. Thus, the truth value of "*x is a devastating hurricane*," is p_R, the truth value of "*x is in a high surge*," is p_i, the truth value of "x is a *catastriphic huricane*," is q_r, and the truth value of "x is a *fast moving center*," is q_i, Suppose that the term "*moderate*" stands for "*non* − destructive" which stands for "*NOT destructive*," the term "*low*" stands for "*NOT high*," and the term "*slow*" stands for "*NOT fast*." In this context, *NOT* is interpreted as the fuzzy negation operation. Note that this is not the only way to define these linguistic terms and it is used to exemplify the expressive power and the inference power of the logic. Then, the complex fuzzy interpretation of the following composite propositions is:

1. $f('P) = (1 - p_r) + j(1 - p_I)$

That is, $'P$ denotes the proposition "*x is a non − destructive hurricane with a low surge*." The confidence level in $'P$ is $(1 - p_r) + j(1 - p_i)$; where the fuzzy truth value of the term "*x is a non − destructive hurricane*," is $(1 - p_r)$ and the fuzzy truth value of the term "low surge," is $(1 - p_i)$

2. $'P \rightarrow 'Q = \min(1, q_r - p_r) + j \times \min(1, q_i - p_I)$

Thus, $('P \rightarrow 'Q)$ denotes the proposition *If* "x is a *non − destructive* hurricane with a low surge,"

THEN x is a *non − destructive huricane with low moving center*." The truth values of individual terms, as well as the truth value of $'P \rightarrow 'Q$ are calculated according to Table 11.1.

3. $f(P \oplus 'Q) = \max(p_r, 1 - q_r) + j \times \max(p_i, 1 - q_i)$

That is, $(P \oplus 'Q)$ denotes a proposition such as "x is a *destructive hurricane with high surge*." OR

"x is a *non − destructive huricane with low moving center*." The truth values of individual terms, as well as the truth value of $P \oplus 'Q$ are calculated according to Table 11.1.

4. $f('P \otimes Q) = \min(1 - p_r, q_r) + j \times \min(1 - p_i, q_i)$.

That is, $(P \otimes Q)$ denotes the proposition "*x is a devastating hurricane with high surge*." AND "*x is a devastating huricane with fast moving center*." The truth values of individual terms, as well as the truth value of $'P \otimes Q$ are calculated according to Table 11.1.

Complex Fuzzy Inference Example

Assume that the degree of confidence in the proposition $R='P$ defined above is r_r+jr_i, let $S='Q$ and assume that the degree of confidence in the fuzzy implication $T=R\rightarrow S$ is t_r+jt_i. Then, using Modus ponens

$$\begin{array}{l} R \\ \underline{R\rightarrow S} \\ S \end{array}$$

One can infer S with a degree of confidence $\min(r_r,t_r)+j\times\min(r_i,t_i)$.
In other words if one is using:

"*x is a non − destructive hurricane with a low surge,*"
IF "x is a non − destructive hurricane with a low surge," THEN "x is non − destructive huricane with slow moving center."

"*x is non − destructive hurricane with slow moving center.*"
Hence, using Modus ponens one can infer:
"x is non − destructive hurricane with slow moving center." with a degree of confidence of $\min(r_r,t_r)+j\times\min(r_i,t_i)$.

11.3.5 Neuro-Fuzzy Systems

The term neuro-fuzzy systems refers to combinations of artificial neural networks and Fuzzy logic. Neuro-fuzzy systems enable modeling human reasoning via fuzzy inference systems along with the modeling of human learning via the learning and connectionist structure of neural networks. Neuro-fuzzy systems can serve as highly efficient mechanisms for inference and learning under uncertainty. Furthermore incremental learning techniques can enable observing outliers and the Fuzzy inference can allow these outliers to coexist (with low degrees of membership) with "main-stream" data. As more information about the outliers becomes available, the information, and the derivatives of the rate of information flow can be used to identify potential black swans that are hidden in the outliers The classical model of Neuro-Fuzzy systems can be extended to include multidimensional Fuzzy logic and inference systems in numerical domains and in domains characterized by linguistic variables. We plan to address this in future research.

11.3.6 Incremental Fuzzy Clustering

Clustering is a widely used mechanism for pattern recognition and classification. Fuzzy clustering (e.g., the Fuzzy C-means) enables patterns to be members of more than one cluster. Additionally, it enables maintaining clusters that represent outliers through low degree of membership. These clusters would be discarded in clustering of hard (vs. fuzzy) data. The incremental and dynamic clustering (e.g., the incremental Fuzzy ISODATA) enable the clusters' structures to change as information is accumulated. Again, this is a strong mechanism for enabling identification of unlikely events (i.e., black swans) without premature discarding of these events. The clustering can be performed in a traditional feature space composed of numerical measurements of feature values. Alternatively, the clustering can be performed in a multidimensional fuzzy logic space where the features represent values of linguistic variables The combination of powerful classification capability, adaptive and dynamic mechanisms, as well as the capability to consider uncertain data, maintain data with low likelihood of occurrence, and use a combination of numerical and linguistic values makes this tools one of the most promising tools for detecting black swans. We are currently engaged in research on dynamic and incremental fuzzy clustering and it is evident that the methodology can serve as a highly efficient tool for identifying outliers. We plan to report on this research in the near future.

11.4 Conclusions

In this chapter, we have outlined features of disasters using the metaphor of gray swans. We have shown that an important part of the challenges related to disaster are identifying slow evolving uncertain data that points to the potential of occurrence of disaster before it occurs and fast evolving data concerning the secondary effect of disasters after the occurrence of a major disaster. We have outlined as set of fuzzy logic based tools that can be used to address these and other challenges related to DMM.

While the USA and NATO continue to lead the technical agenda in DMM technologies, recent disasters are showing that there is still a lack of technology-based tools in specific decision support tools for addressing disaster, mitigating their adverse impact and managing disaster response programs. Thus, the USA and NATO must develop additional DMM capabilities. Additional activities that will assist in DMM programs include [22]:

1. Accelerated delivery of technical capabilities for DMM
2. Development of world class science, technology, engineering and mathematics (STEM) capabilities for the DOD and the Nation.

On the top of those important tasks, one should never forget that in the development of DMM programs we do not have the luxury of neglecting human intelligence [23]. In any fuzzy event related to a gray swan an investigation after the fact reveals enough clear data points which have been read correctly but were not treated properly.

In the future, we intend to investigate the DMM utility of several other fuzzy logic based tools including:

1. Value-at-Risk (VaR) under Fuzzy uncertainty
2. Non-cooperative Fuzzy games
3. Fuzzy logic driven web crawlers and web-bots
4. Fuzzy Expert Systems and Fuzzy Dynamic Forecasting (FDEs)

Finally, we plan to expand our research on complex fuzzy logic based neuro-fuzzy systems as well as the research on incremental and dynamic fuzzy clustering. Both of these research threads are expected to provide significant advancement to our capability to identify and neutralized (as much as possible) first generation and second generation gray swans.

Acknowledgments This material is based in part upon work supported by the National Science Foundation under Grant Nos. CNS-0821345, CNS-1126619, HRD-0833093, IIP-0829576, CNS-1057661, IIS-1052625, CNS-0959985, OISE-1157372, IIP-1237818, IIP-1330943, IIP-1230661, IIP-1026265, IIP-1058606, IIS-1213026.

References

1. Taleb NN (2004) Fooled by randomness. Random House, New York
2. Taleb NN (2007) The black swan. Random House, New York
3. Kandel A (1986) Fuzzy mathematical techniques with applications. Addison-Wesley, Reading
4. Zemankova-Leech M, Kandel A (1984) Fuzzy relational data bases – a key to expert systems. Verlag TUV Rheinland, Koln
5. Last M, Kandel A, Bunke H (eds) (2004) Data mining in time series databases, series in machine perception and artificial intelligence, vol 57. World Scientific, Singapore
6. Mikhail RF, Berndt D, Kandel A (2010) Automated database application testing, series in machine perception and artificial intelligence, vol 76. World Scientific, Singapore
7. Zadeh LA (1975) The concept of a linguistic variable and its application to approximate reasoning – Part I. Inform Sci 7:199–249
8. Klir GJ, Tina A (1988) Fuzzy sets, uncertainty, and information. Prentice Hall, Upper Saddle River
9. Tamir DE, Kandel A (1990) An axiomatic approach to fuzzy set theory. Inform Sci 52:75–83
10. Tamir DE, Kandel A (1995) Fuzzy semantic analysis and formal specification of conceptual knowledge. Inform Sci Intell Syst 82(3–4):181–196
11. Ramot D, Milo R, Friedman M, Kandel A (2002) Complex fuzzy sets. IEEE Trans Fuzzy Syst 10(2):171–186
12. Ramot D, Friedman M, Langholz G, Kandel A (2003) Complex fuzzy logic. IEEE Trans Fuzzy Syst 11(4):450–461
13. Tamir DE, Lin J, Kandel A (2011) A new interpretation of complex membership grade. Int J Intell Syst 26(4):285–312
14. Dick S (2005) Towards complex fuzzy logic. IEEE Trans Fuzzy Syst 13:405–414
15. Běhounek L, Cintula P (2005) Fuzzy class theory. Fuzzy Set Syst 154(1):34–55
16. Fraenkel AA, Bar-Hillel Y, Levy A (1973) Foundations of set theory, 2nd edn. Elsevier, Amsterdam
17. Mundici D, Cignoli R, D'Ottaviano IML (1999) Algebraic foundations of many-valued reasoning. Kluwer Academic, Boston

18. Hájek P (1995) Fuzzy logic and arithmetical hierarchy. Fuzzy Set Syst 3(8):359–363
19. Casasnovas J, Rosselló F (2009) Scalar and fuzzy cardinalities of crisp and fuzzy multisets. Int J Intell Syst 24(6):587–623
20. Cintula P (2003) Advances in LΠ and LΠ1/2 logics. Arch Math Log 42:449–468
21. Montagna F (2005) On the predicate logics of continuous t-norm BL-algebras. Arc Math Log 44:97–114
22. Lemnios ZJ, Shaffer A (2009) The critical role of science and technology for national defense, Computing Research News, a publication of the CRA, vol 21, no 5
23. Last M, Kandel A (eds) (2005) Fighting terror in cyberspace, series in machine perception and artificial intelligence, vol 65. World Scientific, Singapore

Chapter 12
Decision Support to Build in Landslide-Prone Areas

Irina Lungu, Anghel Stanciu, Gabriela Covatariu, and Iancu-Bogdan Teodoru

Abstract Soil mass movements are potential effects of insufficient soil shear strength (when construction development is to be decided). Landslides regarded as first-time or reactivated soil movements become disasters in relation to significant human and material losses. Landslide prone areas related to urban slopes are identified as such and characterized according to various criteria. The public perception related to construction decisions is first-hand provided by mapping the landslide potential, hazard and risk of each country/county/city zone. The paper presents a comparative analysis between methods to provide decision-making support for the local authorities to build on urban slopes with previous soil mass movements.

12.1 Resilience to Disaster in the Context of Landslide Events

The urban land slopes have been proved to be prone to landslide events of various frequencies, and in consequence, to be the cause of community losses in human lives and/or property damage, augmented by medium and long-term social secondary effects. The landslide is thus considered both from the geomorphologic perspective, and from the potential/susceptibility and the hazard of the event triggering. Furthermore, there are certain risk levels associated with the event and the vulnerability of the affected community. All these concern a future event, while the event's categorization as a disaster is possible only post-factum.

The consequences of a landslide, regarding both the natural and built environment and the local community, are generating either an emergency situation response or a calamity response, the difference residing in the amount of involved resources to

I. Lungu (✉) • A. Stanciu • G. Covatariu • I.-B. Teodoru
Faculty of Civil Engineering and Building Services, "Gheorghe Asachi" Technical University of Iasi, D. Mangeron Bd., 700050 Iasi, Romania
e-mail: ilungu@ce.tuiasi.ro; irina_lungu2003@yahoo.com

H.-N. Teodorescu et al. (eds.), *Improving Disaster Resilience and Mitigation - IT Means and Tools*, NATO Science for Peace and Security Series C: Environmental Security,
DOI 10.1007/978-94-017-9136-6_12,

restore the affected area to safety. The landslide – as a disaster triggering event – is often the immediate consequence of other disaster generating natural events, such as floods and/or earthquakes.

The term of landslide vulnerability is often associated with a specific community (as an integrated system with social and physical elements), but it may also be considered independently as a characteristic of a geomorphologic complex or it may simply be associated with the rock mass as a natural system. The vulnerability indicates the intensity of the event, from the prejudice perspective in the first case, and the physical modifications resulted from the rock mass mechanical instability, in the second case.

The affected community may react during the event by actively adapting to minimize the impact, thus indicating a coping capacity. This requires previously established adaptation strategies that become effective during the event. Although the adaptation strategies are apparently concerning the social element, the adaptive development of the built environment in the landslide susceptible areas [3], bearing in mind the geotechnical risks, may also be considered an adaptation strategy. From these perspectives, the communities are required to be disaster resilient, the term also covering the landslide event. The resilience consists in the community's ability to adapt in real time to the intensity of the event, sustaining minimized losses and dysfunctionalities, and also the ability to fully recover and evolve, by means of post-event activities, towards proper preparation concerning a future event. This will considerably reduce the effects otherwise to be considered a disaster.

The resilience to disaster is considered the opposite of vulnerability [14], also bearing a constructive meaning and not being limited to ascertaining the effects. The location on a time axis is different for the two terms, as vulnerability is evaluated before the event while the resilience is evaluated pot-event. The resilience is thus permanently brought up to date by the possibility and the ability to learn the post-event lessons. In the case of landslide, the vastly varied social and physical context, as well as the association with other hazardous natural agents (floods and earthquakes), are creating community responses that are requiring "education" on multiple levels. It is because of these that the inter-disciplinary approach of community disaster resilience is necessary. The frequency of disastrous landslide events in a certain area or community is paradoxically a potent motivator to generate resilience developing actions, both on local level and on national and international levels.

If the majority of the countries outside the European continent are ad-dressing the infrastructure developing strategies in an a real specific manner, most of the European countries are sharing a joint built environment approaching strategy from the perspective of the building standards and from the point of view of the regulations concerning the natural disaster risks. The structural design of buildings may thus be uniformly supported on a European scale. However, the execution, the monitoring, and the maintenance of the constructions vary, even within the boundaries of the same country, with the offered financial support and more, with the attitude and the mentality in the history of a specific community.

The economic development of a landslide susceptible area increases the community vulnerability and therefore the need to initiate and develop social projects that will improve the landslide post-event resilience. The financial investments in improving the material quality of life are relatively ineffective if not doubled by appropriate training strategies in natural disaster situations.

12.2 Rock Mass Instability: Landslides

The geomorphological instability, by the rock volume that falls and/or topples (hard rocks) or moves (translations and/or rotations, in case of soft rocks) either downward or spreading laterally, can predictably affect the built environment and the community in a specific zone.

The rock massifs bordered by man-man or natural slopes become unstable when the shear stress (τ) –developed by self-weight, other external mechanical actions, as well as from the dynamics of the underground water and weathering – equals the rock/soil shear strength (τ_f), thus generating a sliding surface for the material situated above it.

Man-made slopes are built to border the rock mass within construction works (embankments for transportation infrastructure, dams and dykes), or to protect civil engineering works made on natural slopes, in this way altering the ground surface with new slopes. The design of man-made slopes results in the maximum slope (1:m), that provides a minim safety as a stability factor ($F_{S.\,\mathrm{effective}} \geq F_{S.\,\mathrm{min}}$), relative to the required level difference (ΔH) and rock type.

Urban slopes are naturally inclined surfaces that border rock/soil massifs as support for existing constructions, new constructions being allowed under the same constraint for the safety factor against landsliding, marginally considered as 1.

The instability/failure/sliding is considered as triggered when the corresponding safety factor $F_{S.\,\mathrm{min}} = 1.00$, that makes valid the statement the landsliding mass is at limit equilibrium. That signifies that the variation of the safety factor against landsliding towards 1.00 quantitatively assesses the stability or instability in the zone of interest – Fig. 12.1.

By definition, although equally as meaning, two formulations are the most representatives in terms of geotechnical significance, characteristic for discrete and not continuum materials [18]:

- the safety factor against landsliding represents the ratio between the available shear strength (mobilized at maximum value) and the mean shear strength necessary to be mobilized when the limit equilibrium is developed in the soil mass along the given failure surface ($F_S = F_{S.\,med} = 1.00$);
- the safety factor against landsliding is defined as the factor that applied to the soil shear strength, along the failure surface, would develop the limit equilibrium state for the potential sliding mass.

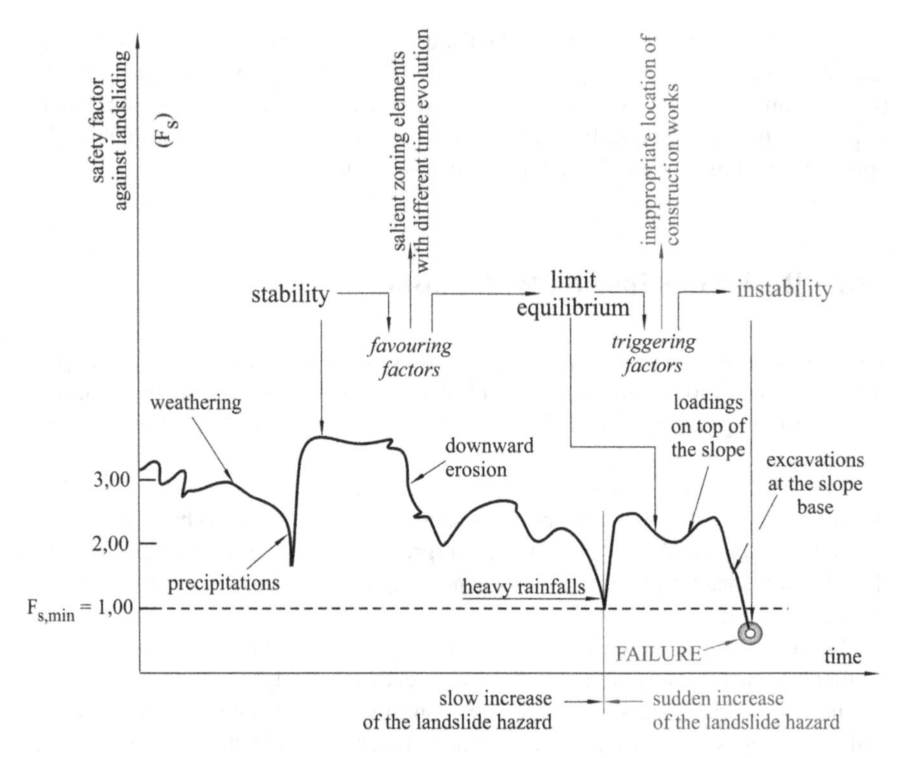

Fig. 12.1 The dynamic of the safety factor against landsliding by intersecting the climatic and the anthropic indicators

The land mass instability in the vicinity of a construction may be locally generated, or it may be at a larger scale. In this latter case, the community from a specific area of a slope is affected and the instability is considered a landslide. The scenarios generated when designing a new building include the land mass stability analyses before and after the construction, taking into account the long or short term effects of drainage conditions. As a result, in certain cases, additional landslide prevention and/or protection measures concerning the building are needed.

The recorded development in the stability analysis methodology in the last decades may be summarized in the following results:

- The development of deterministic methods in considering the simultaneous action of static and dynamic forces generated by various building types, seismic loads, hydrostatic and hydrodynamic loads, their intensity dynamic and the modification of the calculus involved geotechnical parameters; however, complex rock behavior laws are included, which require geotechnical parameters difficult to estimate for undisturbed rock samples.
- The development of bidimensional software for slope stability analysis (e.g. GeoSLOPE, Geo5, Oasis, Galena), with dedicated subroutines for numerous

cases of building design. Optimized design versions are generated according to relevant criteria.
- The determination of the spatial landslide effect, by weighted summing the safety factors from the bidimensional analyses of the geometrical surfaces of the site planning, or by computing the land mass interaction resulted from the 3D discretization of the landslide susceptible mass.
- The development and improvement of laboratory and in situ equipment able to record with satisfactory accuracy the rock/soil parameters, that are required in the stability analyses.
- The development and permanent improvement of rock mass movement monitoring equipment, able to record surface or deep movements, as well as the intensity of climatic and water infiltration factors, thus generating evolution models, useful in stability prediction and in the validation of technical stability solutions.
- Starting with the theoretical limitations of the slope stability analyses by statistical methods (stability factors variation lows that are not consistent with the property dynamic of the same factors), the concept of Artificial Neuronal Network (ANN) has been developed – effective in the determination of the safety factor against landsliding. The ANN "black box" model may operate only with a broad geotechnical date base, monitored on the same site, or the same category of civil engineering work.

From the perspective of trusting the stability calculations, the stability analyses producing a stability indicator as safety factor against landsliding were deemed sufficient up until the eighties when generating a stability model of a site. Since the development of the GIS technology in mapping a region for landslide potential, susceptibility, vulnerability, and landslide risk (by overlapping the landslide hazard and vulnerability of an area), the slope stability analysis based on the stability indicator value is applied for each practical situation, when urban planning decisions or building expertise are required.

12.3 Landslide Management: Awareness, Prevention, Real Time Action, Post-event Recovery, the Improvement of Existing Management

The prejudices recorded in every new disaster have triggered over time the reaction of the affected community, leading to post-event processing and to the development of appropriate planning meant to decrease the negative effects and the economic recovery period. Thus, it has been developed the emergency and disaster situations management, with those caused by natural agents – floods, landslides and earthquakes – being the foremost concern.

When important land masses from the built area are moving with increased speed, the landslides are becoming emergency situations (involving responses on local, regional, or national level) or disaster situations (also involving responses on international level). Bearing in mind the vulnerability increment by building/developing infrastructure doubled by the climatic change effects in the latest period, the frequency and magnitude of landslides have increased, while the life disturbances in the affected communities have not been easily removed. Certain authorities have underlined a need for a more organized approach of similar situations, on the one hand, and the need for improved economic development strategies that are addressing the landslide as a possible situation and its effects reduced from the design phase, on the other hand. A dedicated terminology has been introduced – potential, hazard, vulnerability, and landslide risk, and furthermore, new graphic mapping technology – GIS, and a new science branch that integrates these concepts and improves the knowledge in the field – landslide management.

In order to put the theory into practice, this new type of management was supported by the governments by creating a specific legal framework, and by the authorities by specific implementation guides.

The process of initiating, development and implementation of landslide management has evolved differently at international level, being permanently supported by the academic environment with research producing spectacular results at an early phase. This attracted financial interest and support and the discipline keep on developing at present. Dedicated master and doctoral papers have created appropriate approaches adapted for each region, thus creating a specific and efficient management.

In general, the researchers have become specialists in a segment of the management area, as integrating the concerned parties remains the most difficult objective:

- Geologists, who are approaching the landslide phenomenon from the geomorphological and hydrological perspective and who elaborate potential and hazard maps;
- Geotechnicians, who quantitatively determine the stability reserve from the perspective of a new building in the area; they also design specific protection installations and they are able to provide predictions regarding the volume of potentially sliding landmass;
- Sociologists, who are appreciating the vulnerability by quantifying the damages/losses on all the social levels and who are able to provide predictions concerning individual and collective behaviors;
- Managers, who objectively perceive the phase flowchart and who are able to provide new approaches in difficult process situations.

The main effect of triggering a mass interest in the landslide management was – depending on the frequency and magnitude of landslide events in the last 20 years – the decrease of losses. This was also the international objective of the UNISDR – The United Nations Office for Disaster Risk Reduction programs [9].

However, in certain countries, the assigned funding was insufficient to properly address the losses, negative medium and long-term socio-economic effects being observed. It was acknowledged the importance of individual response in case of landslides and of the potential disaster they may trigger. More important than resistance against landslides (commonly perceived by durable stabilization works), the community needs to adapt in real time to the event – as coping ability and furthermore, to recover and improve the reaction to a following hazard – that is, resilience.

International fora, symposiums, workshops, and conferences dedicated to increase the disaster resilience, the natural ones among them – floods, landslides and earthquakes – have brought together specialists in various fields, to acknowledge the need for interdisciplinarity, to share specific knowledge, and to establish joint work teams in order to address current needs in the general context.

12.4 Decision Support to Build in Landslide Prone Areas in Romania: Case Study

Romania makes not exception from the occurrence of natural disasters and calamities. The national strategy for sustainable development of Romania, elaborated in 1998, acknowledges the calamities of earthquakes, landslides, and floods that are to be addressed by adopting appropriate prevention measures. In the volume published by the World Bank, Cristoph Pisch, Preventable Losses: Saving Lives and Propriety Through Hazard Risk Management, The World Bank, Washington DC, October 2004, it is recorded that up to 57 % of annual economic losses in Romania are caused by floods and landslides [5].

Also, the volume Natural Disaster Hotspots, Disaster Risk Management Series, Washington DC 2005, page 90, shows that in Romania 37,4 % of the country surface is in state of risk, 45,8 % of the population lives in the state of risk, while 50,3 % of the country annual budget is affected by natural catastrophe [4].

A national program meant to identify and delimitate natural hazards in each county has been initiated, including the landslides in this category. Its main objective is the elaboration of risk maps, distinctly formulated for each type of natural hazard, in order to issue recommendations and enable decision making based on them regarding real estate destinations and location. These considerations are to be included in urban and territory planning.

In Romania, landslide risk maps regarding a particular zone reflect the stress state within soil massifs/natural slopes and are elaborated based on the values of the safety factors against landsliding, in representative soil profiles [8]. Based on these values, the timing of reaching the critical state/limit equilibrium as the starting point of the landslide could be estimated. Prior to these evaluations and corresponding zoning of the landslide risk, mapping of the same regions are developed regarding the potential and probability of the landsliding and its hazard, respectively. The coefficients ($K_{a \ldots h}$) for the eight hazard zoning criteria [15] are assessed quantitatively and individually over the zone of interest. They are superposed to result in a mean value as hazard mean coefficient mapped over the same zone, without specifying the type and spreading of the soil/rock movement.

Table 12.1 Qualitative assessment of the landslide hazard

Landslide hazard	Ranging domain of the mean hazard coefficient K_m
Low	<0.100
Medium	0.100–0.300
Medium – high	0.310–0.500
High	0.510–0.800
Very high	>0.800

The mean hazard coefficient K_m, corresponding to each surface resulted from the superposition of the eight individual mappings, is computed as:

$$K_m = \frac{K_a \times K_b}{6} \left(K_c + K_d + K_e + K_f + K_g + K_h\right) \tag{12.1}$$

where K_a is the coefficient value of the lithological factor; K_b – the coefficient value of the geomorphological factor; K_c – the coefficient value of the structural factor; K_d – the coefficient value of the hydro-climatic factor; K_e – the coefficient value of the hydrogeological factor; K_f – the coefficient value of the seismic factor; K_g – the coefficient value of the forestation factor; K_h – the coefficient value of the anthropic factor.

The following legal categories [15] are used to assess the landslide hazard for a specific zone in Romania, according to Table 12.1.

Figures in the papers [2], [4], and [1], all related to previous work in our group, present examples of hazard maps for landsliding at the national-, Iasi county-, and metropolitan-level. The implementation of this hazard assessment is required in particular cases for real estate developments in a metropolitan area prone to landsliding, as presented in [2].

In case of building in that area, the landslide risk is assessed by the safety factor against landslide for the selected representative soil profiles. The succession of the operation procedure to elaborate the landslide risk map in that area is the following:

- Materialization on the topographic map the zone limits for which the detailed calculation of the landslide risk will be initiated;
- Selecting the relevant soil profile alignments on the topographic map, before construction, and for the proposed real estate development, along which the earth pressures are calculated at the soil block boundaries and the safety factor is computed using the Janbu's method [18].
- Designation along each soil profile of the calculated earth pressure values (active forces) and of the resistances (passive forces);
- Correlation of the points of isovalue separately for the active and passive forces and drawing the corresponding maps, which in assembly represents the landsliding risk map of this zone prior to construction;

Application of the previous steps for the relevant design scenarios, resulting in the influence of the new constructions, of the water level increase and seismic action in this area.

Once the landslide risk has been evaluated and analyzed, the treatment deals with reducing the risk of the landslide event. In this respect, typical options would include:

- Accept the risk, when risk is confined to the acceptable domain.
- Avoid the risk, when a new construction is in question in the potential area of a landslide risk, abandoning the project and looking for alternative construction sites where risk is acceptable.
- Reduce the likelihood that would require consolidation/stabilization measures – Fig. 12.2 – to control triggering aspects [18], [11]; after implementation of the stabilization work, the risk would fall within acceptable domain.
- Reduce consequences by setting defensive measures of stabilization.
- Introduce monitoring and warning systems in the area of risk on a permanent or temporary basis; this can be regarded as measures to reduce consequences as well (Romanian Law 260, [16]).
- Transfer, compensate, or share the risk, involving insurance companies.

All the above mentioned options are regarded via their relative costs and benefits. Combinations of options are also appropriate to consider, especially since large reductions of risk may be achieved.

Slope management as emerged from the societal need to control geotechnical risks, landslide risk included, is mainly visible when deciding over granting construction permits in a landslide prone area. Slope management can be seen as an integrated approach of the landslide management related to the civil engineering project, as presented in [10].

At national level, and subsequently at regional and municipal level, a Land Management Plan is issued, with the main purpose of correlating the development of that territory. It has a prospective character, underlining possible advantages as well as limitations or issues. Although it is not an operational plan to build a construction, the Land Management Plan establishes guidelines and mainly determines spatial development policies (regulations) and not limitations or practical objectives achievable by buildings design.

As an instrument of synthesis and perspective concerning the development of a settlement, the General Urban Plan (GUP) is instituted – the documentation that regulates the objectives, actions and development measures for a future or existing settlement, for a predetermined period, on the basis of multi-criteria analysis of the existing situation; it directs the implementation of certain policies concerning a settlement land management. To this documentation, the Urban Regulations of the specific settlement is associated.

The general urban plans are elaborated in correlation with the land management plan of the territory in the settlement's administration. The General Urban Plan and the associated Urban Regulations, as approved, are the documentations supporting the issuing of urban certificates and construction permits in the territory of the

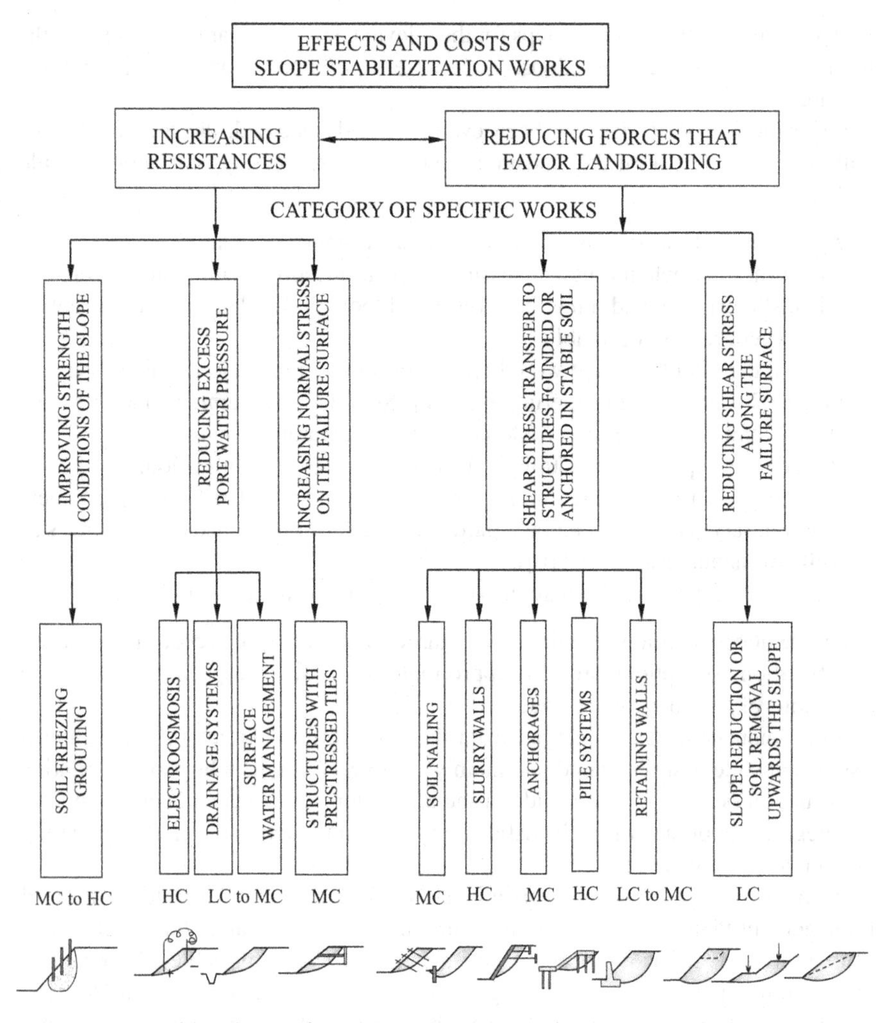

Fig. 12.2 Interventions to consider in slope stabilization

respective settlement. The term of validity of the general urban plan is, as a rule, of 5–10 years, with the exception of cases when exceptional situations occur that are calling for a documentation update.

As a functional model to implement the risk maps on the Mayor's Office level in the specialty departments, the fields where these maps are a decisional support are as follows:

- Urbanism and land management, to elaborate the urban documentations (GUP – General Urban Plan, ZUP – Zonal Urban Plan, LUP – Local Urban Plan), the

planning and establishment of residential areas, urban infrastructure, building services networks, urban rehabilitation etc.;
- Urban slopes, to issue building permits on slopes, determining the landslide risk areas, slope monitoring and stabilization etc.;
- Civil protection, for assisting in elaborating disaster situations protocols etc.

12.5 The Role of the Neuronal Networks in Slope Stability Pre-dictions

The artificial neuron as defined by analogy to the nerve-cell represents the basic working unit for the neuronal calculation. This unit processes a set of values as informational input and produces a result as informational output, considered as the activity level of the neuron. In general terms, Artificial Neural Networks (ANN) represents an assembly of artificial neurons linked by specific connections that are provided with a weighing functionality and a path for the traffic information.

At present, from a theoretic perspective, the ANNs are superior to other methods in slope stability analysis, man-made or natural slopes, due to their capacity to learn by training, based on existing data. Learning for an ANNs refers to any modification of the multitude of parameters (weights associated to connections and thresholds associated to units) that provides the adaptation of the ANN behavior to the problem in question [6]. A formalization of the problem is implicitly embedded in the network parameters by the learning process (knowledge of the phenomena by the regular/classic theories and methods), to be used during service. The necessity to obtain stable predictions when changing the initial conditions is provided by a important feature of the ANN – the generalization capacity, which means to provide good results for data that have not been yet trained for.

A trained ANN can generate results to the following problems for complex distribution of the zones prone to landslides over a specific territory:

- Approximation and estimation – by creating a functional dependency between two relevant parameters for that specific zone, based on an existent set of values (e.g. precipitation volume – underground water level);
- Prognosis – approximation of the next value within an initiated series (e.g. previously established safety factors against landsliding);
- Optimization – in determining one value that meets certain constraints and optimize and objective function (e.g. the slope for embankments);
- Modeling and adaptive control – based on a reference model, the system can be controlled by determining a parameter u(t) that ensures the resulting parameter as required v(t) (e.g. construction loadings that maintains the stability, as safety factor against landsliding over unity).

The neural networks, being connecting models, function mainly based on their connections. Their weighing limit the result of the training process and represents the long term memory of the model. In addition to that, three important features of the neuronal networks are vital in slope stability analysis [6]:

- the parallel workability – during data processing, many neurons "act" simultaneously;
- the robustness – if some neurons are "malfunctioning", the system can still function unaffected [13];
- accommodation, necessary as in many cases, the existing data are different from the updated ones as new situations develop.

During the last decades, the neuronal networks have solved two main issues:

- mapping zones prone to landslides by susceptibility/potential and hazard [7], with the main focus on geological data, GIS technologies and remote sensor recordings;
- stability assessment for a relatively small area, by safety factors against landsliding [17], based on geotechnical knowledge, with regular soil parameters from the soil investigation report; the ground surface geometry is simpler in man-made slopes than for natural ones, but the construction/urban planning decision requires the stability of the affected natural slopes.

We refer the reader to [12] for pictures of unfortunate situations of landslides triggered upward from a stabilization work as rehabilitation solution of the road zone with "chronic" landslides, as reactivated frequently.

A network trained with a consistent database (soil parameters, existing stabilization works and topographic monitoring) for that region could offer immediate technical solutions for multiple adaptation, with the advantage of shorter working time and implicit robustness compared to the classical methods.

12.6 Conclusions Regarding Urban Planning Decisions in Landslide Prone Areas

The urban planning decisions to build in landslide susceptible areas are addressing the new building constructions as real estate, or as economic development, depending on the economic and social policy and on the moral standards of the involved community. Thus, simply ensuring the stability of the land plots in the light of increasing building number is but an incipient phase that is maintaining the area's vulnerability rather high. The public transportation system, with additional access routes in case of necessity, the water supply and sewage system monitored in terms of time losses, public utility buildings to serve the distant areas from the communication center are but a few examples that are scarcely meet in reality, although in the urban planning, documentations are always mentioned. The implementation deficiencies become visible in the light of the next event that triggers an emergency or a disaster situation in the community.

In parallel with responsible decision making concerning building construction on local/ regional/ national level, the individual responsibility to increase disaster resilience, landslides included, is a human trait that has begun to be taught in schools, at all levels. Behaviors and responses are trained in simulations, while information is becoming accessible on various communication channels. The landslides are studied both as a phenomenon, and from the perspective of proper management, enabling cutting negative effects on the community. The increase in volume of official documents, standards, and civil engineering regulations on national and international levels, as well as their permanent modification in the latest years, may become reasons for ineffective implementation of theory into practice. To compensate, professional associations in the field, on national and international levels, are coagulating the experience and good practices, generously disseminating knowledge in the community. The knowledge transfer, both vertically and horizontally, is a key element in creating the desired disaster resilience of a community.

References

1. Adomniței CM (2010) Fundamentation of decisions for urban planning based on hazard landsliding maps. Doctoral thesis, Gheorghe Asachi, Technical University of Iasi
2. Bejan F (2010) Mapping the slope stability for an urban development in the N-W zone of the Iasi city. Dissertation, Gheorghe Asachi, Technical University of Iasi, Romania
3. Chang Y, Wilkinson S, Seville E, Potangaroa R (2010) Resourcing for a resilient post-disaster reconstruction environment. Int J Disaster Resilience Built Environ 1(1):65–83
4. Copilau J (2008) Contributions regarding the inventory, monitoring and rehabilitation works of natural slopes. Doctoral thesis, Gheorghe Asachi, Technical University of Iasi, Romania
5. Copilau J, Stanciu A, Lungu I (2008) The use of landslide risk in the rehabilitation of terrestrial transportation infrastructure. In: Proceedings of the national conference of geotechnical and foundation engineering, Timișoara, pp 545–552
6. Covatariu G (2010) Applications of neural networks in civil engineering. Doctoral thesis, Gheorghe Asachi, Technical University of Iasi, Romania
7. Ermini L, Catani F, Casagli N (2005) Artificial neural networks applied to landslide susceptibility assessment. Geomorphol J 66:327–343
8. GT 019–98 Elaboration Guide for landslide risk mapping of slopes to provide construction stability. Constructions Bulletin, vol 6, Bucharest, 2000
9. (HFA) Hyogo Framework for Action 2005–2015: Building the Resilience of Nations and Communities to Disasters http://www.unisdr.org/we/coordinate/hfa
10. Lungu I, Stanciu A, Boti N (2012) Integrated geotechnical risk approach in urban decisions to increase the quality of life. Ann Acad Rom Sci 4(1):63–71
11. Manea S (1998) Evaluation of the slopes landslide risk. Conspress, Bucharest
12. Manea S, Stanciu C, Ungureanu DA (2008) Landslide of the left slope of the Buzau Valley, at thekm. 81 + 000 on DN 10 Buzău – Brașov. In: Proceedings of the national conference of geotechnical and foundation engineering, Timișoara, pp 774–784
13. Melchiorre C, Castellanos, Abella EA, van Westen CJ, Matteucci M (2011) Evaluation of predictibility capability, robustness, and sensitivity in non-linear landslide susceptibility models, Guantanamo, Cuba. Comput Geosci J 37:410–425
14. Ollet EJ (2008) Flash flood and landslide disasters in the Philippines: reducing vulnerability and improving community resilience. Dissertation, The University of Newcastle, Australia

15. Romanian Decision 447 (2003) Methodological norms regarding the content and procedure to elaborate the natural risk maps for landslides. The Romanian Official Monitor, part 1, no. 305/2003
16. Romanian Law no. 260/2008 regarding the mandatory insurance of dwellings against earthquakes, landslides or floods. The Romanian Official Monitor, part 1, no. 757/2008
17. Sakellariou MG, Ferentinou MD (2005) A study of slope stability prediction using neural networks. J Geotech Geol Eng 23:419–445
18. Stanciu A, Lungu I (2006) Foundations I – physics and mechanics of soils. Technical Publishing House, Bucharest

Part II
Tools and Applications

Chapter 13
Mitigation using Emergent Technologies

Leon J.M. Rothkrantz

Abstract Mitigation is the effort to reduce loss of life and property by lessening the impact of disasters. In this paper we discuss a communication network for crisis situations based on smart phones. A special icon language enables users to report about their position and context situation and to receive information how to reach safe areas. Messages are ordered by the intensity of the emotional content. Users can be tracked by their GPS position and this information is used as input of a special dynamic routing algorithm based on ideas from Artificial Life, called Ant Based Control.

13.1 Introduction

After the onset of a crisis situation caused by a tsunami, hurricane, flooding, or explosion, first responders try to localize victims and evacuate them to safe areas. To get full context awareness of what is going on, communication plays a vital role. Internet, mobile Internet and wireless communication technologies enable communication of short and large distances. Traditionally detection, response and recovery are coordinated by the crisis management. The starting point is a crisis plan prepared long time before the onset of a crisis with a description of the crisis management and a schedule of planned actions. Usually the crisis centre is localized on a specific place. But it proves that such an organization lacks agility, flexibility and adaptively.

L.J.M. Rothkrantz (✉)
Intelligent Interaction, Delft University of Technology, Mekelweg 4, 2628 CD Delft, The Netherlands

SEWACO, The Netherlands Defence Academy, Het Nieuwe Diep 8, 1781 AC Den Helder, The Netherlands
e-mail: l.j.m.rothkrantz@tudelft.nl

H.-N. Teodorescu et al. (eds.), *Improving Disaster Resilience and Mitigation - IT Means and Tools*, NATO Science for Peace and Security Series C: Environmental Security,
DOI 10.1007/978-94-017-9136-6_13,

Recent studies show that a decentralized, bottom up approach is much more flexible. Social media play a central role. In the process of information sharing that can provide useful information for managers but also in the process of dissemination of information to the public. It proves that people in a crisis situation build up a virtual community by their smart phones employing "social software." Actions emerge spontaneously and are not initiated or controlled by the central authorities. The challenge of the authorities is how to support these emergent actions. From a mitigation viewpoint it is important to improve the collaborative resilience of organizations, cities and communities by means of social media. This requires an improvement of the IT infrastructures for crisis management and disaster response.

In this paper we focus on the use of social media in crisis management and response. The focus of our research is how social media enable citizens and communities to participate and collaborate in all stages of a disaster. The following research questions will be discussed:

RQ1: How to design a network of smart phones supporting communication between users and virtual agents in the network?

RQ2: How to design an icon based language supporting communication about crisis events for users unable to use current languages?

RQ3: How to detect emotions in messages and select those messages based on strong negative emotions?

RQ4: How to design a routing algorithm supporting citizens to escape to a safe area, based on emergent crowd movements?

To solve those questions we will describe a communication system based on social media (section Architecture). During a crisis it proves that affected people are able to provide in continuous way information about their position, about their observations (section Icon language). We also discuss the emotional content of messages. User information is used as input of a dynamic routing system based on ideas from artificial life (section Ant Based routing). The system supports users to reach safe areas.

13.2 Related Work

In this paper we discuss the application of location awareness and navigation during crisis situations. GPS is one of the most efficient positioning technologies. Thanks to the reduction in the size of the GPS receivers and the integration of GPS with mobile phones, users can be localized and tracked. In [1] the authors present a distributed trajectory similarity search framework. It focuses on GPS trace search in smart phone networks by decentralized and in-situ data.

Another mobile application based on GPS positioning is to locate family members and alert when friends are nearby [2]. We developed a system based on the GPS technology to navigate car drivers along the shortest route to their destination [3]. An adapted version of that system was developed to navigate pedestrians to a destination for example a safe area [4]. The localization process is running in the background, no action of the user is required. The smart phone sends at regular times an update of its position. The system is able to track users and this provides important information about the movement of individual persons and crowds of people. The system is able to check in which direction people move, which roads are still open or apparently blocked and is able to support movements of people in the right direction for example to safe areas. Local information provided by human observers in the crisis area is assimilated to global information by the system and in this way the system has real time context awareness.

An important issue is if people are able and willing to use their smart phones to share their observations with others. Social media had become an integral part of disaster response during the Hurricane Sandy not only because of the loss of cell phone service during the peak of the storm. Twitter and Facebook were used by citizens to keep informed, locate friends and family, notify authorities and express support. Researchers from the University of Colorado and the University of California-Irvine participate in the Project HEROIC [5]. The goal of the project is a better understanding of the dynamics of informal online communication in response to extreme events. Through a combination of data collection and modeling of conversation dynamics, the project team aims to understand the relationship between hazard events, informal communication and emergency response.

Studies of how humans behave during times of collective stress shows that the affected population in a disaster area are not helpless victims, they are actually capable humans who tend to act rationally and exhibit a great deal of pro-social behavior [6, 7]. The research challenge is how to utilize the potential capacity of affected people. Emergent technology will be used to support people reaching a safe area based on the tracking information provided by them.

13.3 Architecture of the ABC Routing System

In [3, 8–11] we presented our research on dynamic routing systems based on the ABC algorithm. The goal of the system was to route cars in the shortest time from start to their destination. A well-known routing algorithm is Dijkstra algorithm which routes cars via the shortest path from start to destination. In current traffic situations the shortest path is usually not equal to the shortest path in time due to traffic jams. In rush hours and during traffic incidents car drivers can be delayed for hours. Dynamic routing algorithms take care of the current traffic flow and to be expected traffic flows. In case of traffic jams cars can be rerouted along less congested routes. We developed a system providing travel advices to car drivers via their smart phones. The input of the system is the current traffic. Car drivers will be tracked via their smart phones. At regular times a car sends his GPS position to the Routing system. From the car tracks the speed of cars can be computed. To be expected delays along the routes can be computed via prediction models using a historic database. We tracked cars along the routes in the area around Delft for a

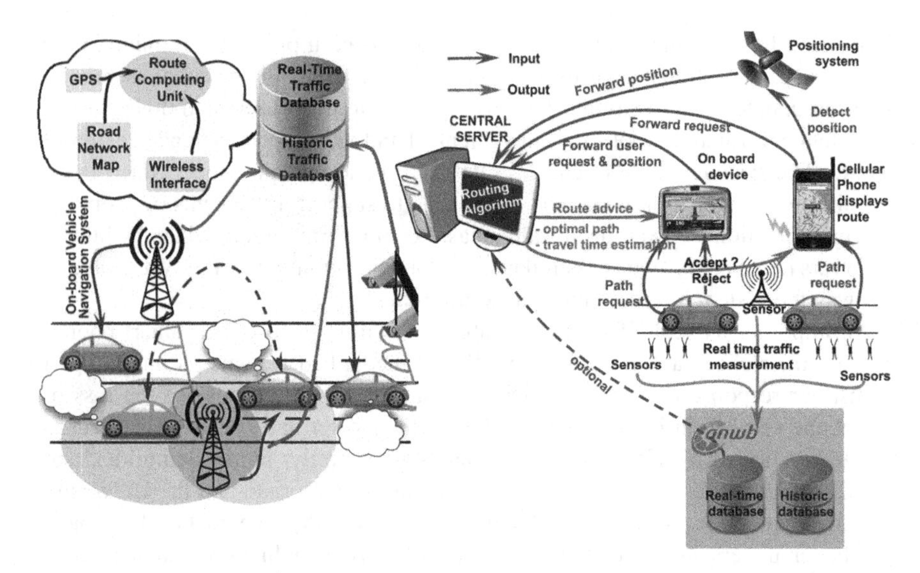

Fig. 13.1 Infrastructure of a decentralized and a centralized routing system

year and computed prediction models of traffic flows depending of the time of the day, time of the week, weather prediction and special incidents.

A road network can be presented as a graph. Along the edges of the graph we attach the travel time from along the edge which is updated all the time. The travel time from A to B can be computed and we used the ABC algorithm to compute the shortest route in time.

In [8, 9] we adapted the developed dynamic routing system to an evacuation system in crisis situations. In the next section we will discuss the main aspects and components of the system displayed in Fig. 13.1.

13.3.1 Centralized or Decentralized Systems

In crisis situations the ICT infrastructure can be damaged. Also the street network can be changed. Roads can be blocked, bridges destroyed. A decentralized approach guarantees that at least in some parts of the crisis area the ICT network and street network is still intact. We tested our system in a simulated crisis situation in the city of Rotterdam. We used a decentralized approach and planned crisis centre all over the city on places in to be expected safe areas.

In a centralized architecture the results can be analyzed much easier. A central system is vulnerable in case the system's server breaks down. On the other hand, when evaluating a routing algorithm in a decentralized architecture, namely a mobile ad-hoc network (comprised of cell phones and handheld devices), the results are scarcely repeatable and reliable. Nevertheless, it involves real-time access to restricted traffic information databases for each user. Another possible disadvantage of the decentralized architecture is that traditional algorithms have to be modified in order to be used. What is more, a decentralized architecture involves some extra communication. The amount of information that has to be transferred to a part of the decentralized system is variable. Because of this factor a decentralized architecture can be faster or slower in processing user requests. One advantage brought by the decentralized system is the possible increase in speed plus memory space. A decentralized system results in the diminution of the susceptibility to failure.

13.3.2 Street Network

Disasters like tsunamis, hurricanes, flooding or earthquakes are able to destroy a complete infrastructure. Our system will be limited to those areas where the communication ICT infrastructure is still working and the street network may be damaged but still exists. The basic assumption of our system is that there is a map of the crisis environment. In case the street network is completely damaged it is difficult or even impossible to drive by cars. In that situation it is still possible to route pedestrians to safe areas by providing only the direction to safe areas (compass) and adapted maps. But our system has only been tested in areas with a street network. We selected OpenStreetMap as our basic maps. Communication architecture OpenStreetMap (OSM) is a collaborative project to create a free editable map of the world. OSM follows a similar concept as Wikipedia does, but for maps and other geographical facts. An important fact is that the OSM data does not resume to streets and roads. Anybody can gather location data across the globe from a variety of sources such as recordings from GPS devices, from free satellite imagery or simply from knowing an area very well, for example because they live there. This information then gets uploaded to OSM's central database from where it can be further modified, corrected and enriched by anyone who notices missing facts or errors about the area. OSM creates and provides free geographic data such as street maps to anyone who wants to use them.

In order to implement the graphical user interface of the system and to construct an initial database out of intersections and highways from The Netherlands, we embedded an OpenStreetMap map viewer in the application. This was a Java panel which allowed several listeners and functions to be redefined. Due to the modular design of Swing component library, the integration was an easy task.

13.4 Icon Language

Many cities in the Netherlands have a sizeable population of emigrants who do not master the Dutch language. To enable communication during a crisis between all users and the crisis centre, a special icon based language has been developed [12–14]. Before the written language invention, images have been used for communication. Visual language refers to communication that occurs through visual symbols as opposed to words. People use such symbols to represent objects, relations and events. The arrangement of the symbols creates information that represents contexts. Although the meaning of these symbols may be ambiguous due to different cultural backgrounds, people are able to learn and understand them because they represent concepts those are commonly known and actually interacted with. Interaction using visual language can bring visual awareness, evoke readiness to respond and fast exchange of information. Its usage particularly makes any application accessible for large individual difference present in the population, e.g. in language independent context, for mobile use and for speech and hearing impaired users.

According to semiotics, visual symbols can be interpreted based on their perceivable form (syntax), their relation between their form and their meaning (semantics), and by their usage (pragmatics). A "sentence" in visual language has a completely different syntax from spoken language. The latter is based on a sequential ordering of words, while visual language has a simultaneous structure with a parallel temporal and spatial configuration of symbols. Its meaning can be derived as a result of combination of these symbols by a global semantic analysis of the sentence using shared common contexts and world model.

Inspired by the way children make drawings to represent their visual thinking, our research aimed at a free way of creating visual language-based messages using a spatial arrangement of graphical symbols, such as icons, lines, arrows, and ellipses. A communication interface in the field of crisis management has been developed as a proof of concept that allows people involved in a crisis event to report observations using visual language. Its knowledge representation was designed to represent a limited number of concepts in this domain. Similar methodology can be applied for the development in different domains.

13.4.1 The Concept-Based Visual Language

The (grammar-) free visual language provides a way of representing the world phenomena using a spatial arrangement of icons, lines, arrows, and ellipses. The icons represent concepts (see Figs. 13.2a and 13.6). They can also represent structural units of the visual perception, such as direction, color, motion, texture, pattern, orientation, scale, angle, space, and proportion. The arrow can specify the direction of the relationship, such as causal, temporal, and possession. The lines can also be used to specify non-directed relationships between icons, e.g. correlation and conjugation. Grouping icons that represent a close relationship, can be expressed using ellipses around the related icons.

Another way is by placing the icons close to each other and away from other "unrelated" icons. These free arrangements can create meaningful "sentences" that represent different contexts (see Figs. 13.2a, b).

The law of proximity defined by Gestalt psychologist has been explored to understand the way people create certain concepts to describe events. The law states

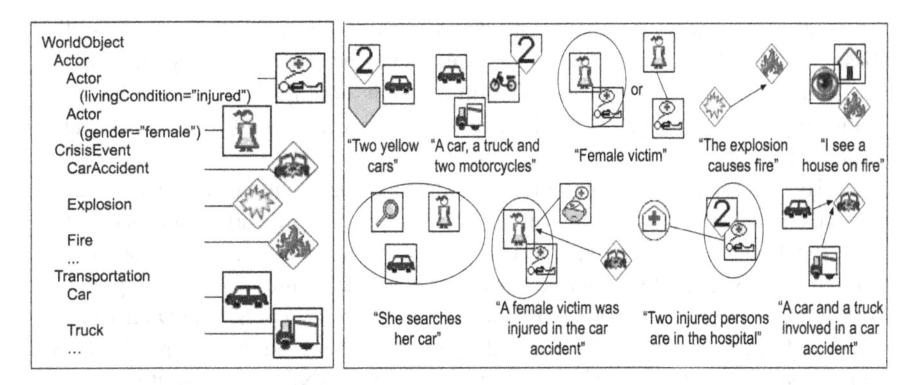

Fig. 13.2 (**a**) A part of the crisis management ontology; an icon is an instance of a class/concept in the ontology, and (**b**) examples of the free visual language-based sentences

that the brain more closely associates objects close to each other than it does when two objects are far apart.

Spatial or temporal proximity of elements may induce the mind to perceive a collective or totality. One of the learning points from this theory is that not only because human's inclination of wanting to be neat and organized, by grouping "related" objects together, we are also making room for other interesting information to be added to a layout that would not fit otherwise. Visual messages being conveyed can be understood because of the relationships underlying the grouping of elements within the messages.

13.5 Emotion in Text and Speech

Micro blogging sites such as Twitter can play a vital role in spreading information during "natural" or man-made disasters. But the volume and velocity of tweets posted during crises today tend to be extremely high, making it hard for disaster-affected communities and professional emergency responders to process the information in a timely manner. Furthermore, posts tend to vary highly in terms of their subjects and usefulness; from messages that are entirely off-topic or personal in nature, to messages containing critical information that augments situational awareness. Finding actionable information can accelerate disaster response and alleviate both property and human losses. In [15] Imran et al., describe automatic methods for extracting information from microblog posts. Specifically, they focus on extracting valuable "information nuggets," brief, self-contained information items relevant to disaster response.

To process the data the authors use NLP technology, such as n-grams, Part of Speech (POS), and Verbnet for example. "Naive Bayesian classifiers" were used to automatically classify a tweet in one of the following classes "caution and advice," "information source," "donation," "causalities and damage." The authors used a huge dataset of tweets posted during the Joplin 2011 tornado that struck Joplin, Missouri. The 206,764 unique tweets were selected by monitoring the Twitter Streaming API using the hashtag #joplin a few hours after the tornado hit. This monitoring process continued until the number of tweets about the tornado became particularly sparse.

In [16] we focused on the semantic and emotional content of tweets. Speech Valence and Speech Arousal, as well as the Word Topics are part of our method for operationalizing Speech Semantics. There are no ground truths available for valence and arousal of speech. Our approach is to use a simple technique based on previous work for creating lists of words with valence and arousal scores. The Dictionary of Affect in Language (DAL) [17] and Affective Norms for English Words (ANEW) [18] are options that we considered. We chose to use ANEW because it contains a smaller word list but with a focus on emotion related words.

To maximize the matching between our words and the words from the ANEW list, we have applied stemming for all words. The score of each utterance is initialized to neutral valence and arousal values which in our case are represented by the value 5. All the words from the utterance are looked up in the ANEW list. If matches are found, a new score is computed by averaging over the valence and arousal scores of the matching words. We find that only 16 % of the utterances contained words from the ANEW list, therefore the majority of the scores still indicated neutral valence and arousal.

In [19] we developed a model to extract special characteristics from speech, such as loudness, pitch, jitter, speech rate etc. Using these features we were able to detect stress in speech using a special database of speech recordings. We are especially interested to detect emotions/stress in speech recordings in emergency calls. We want to test our models on this type of recordings. So we would like to have some hours of recordings of 112- emergency calls.

The final goal of our research is a supporting tool for operators to assess the urgency of calls. In case of crises operators can become overloaded and automatically prioritizing the calls based on their urgency is needed to handle in real time the huge amount of calls. We want to stress that such a system would not replace an operator but it will be a supporting tool. For the success of our system it is very important to have real data as opposed to using actors because their emotions are not genuine and do not preserve the real characteristics.

13.6 Ant Based Routing

In [3, 8–11] we presented our Ant Based Control Algorithm. This algorithm mimics the food searching behavior of Argentinean ants. Ants traveling between a food source and their colony deposit a pheromone (an odor-molecule) trail that marks the trail they have taken and which evaporates over time. At each obstacle encountered, an ant decides whether to go left or right. Each path is taken with a probability proportional with the strength of the pheromone trail (Table 13.1). Thus, initially approximately fifty percent of the ants will choose to divert by going left around an

Table 13.1 A probability table for node 2

(2) Destination	Next	1	3	5
1		0.90	0.02	0.08
3		0.03	0.90	0.07
4		0.44	0.19	0.37
5		0.08	0.05	0.87
6		0.06	0.30	0.64
7		0.05	0.25	0.70

obstacle while the others will go right. If, however, the right route is shorter than the left route, the pheromone trail of the ants following that route creates a denser pheromone trail than the ants traveling via the left route. The higher density will entice other ants in following the right route, which thereby reinforces itself until the trail on the left route has evaporated. In the end, only a single trail around all obstacles remains. It has been shown that, under certain conditions, the path marked by this trail is the shortest path from food to colony.

In traffic networks, obstacles are formed by congestions, road works or accidents. Avoidance of these obstacles is paramount when attempting to reach the destination in the shortest time possible. The ant based routing algorithm enables human drivers to cooperate in a manner equal to ants in order to form time optimal shortest paths throughout the city, circumventing the obstacles. Vehicles provide the system with a constant stream of up-to-date data concerning obstacles. These data enable intelligent agents called ants to approximate the shortest routes through a city. This routing information is then relayed real time to vehicles. The following subsections describe the infrastructure and present the algorithm in more detail.

13.6.1 Updating Routing Tables

It is extremely important how the information coming from the ants is combined in the routing table of each node i (Fig. 13.3 and Table 13.1). The Pheromone deposit will exert great influence over the path advice and over the distribution of the traffic. So, for a backward ant B_{ds}, arriving at a new node i, all entries of the routing table R_i that correspond to the destination d will be updated. First the backward ant B_{ds} that arrived in i and which has popped from the stack the pair (i, t_{si}) selects the time interval $t_{si} \in I_k$ for which the routing table R_i is going to be updated, j is between i and d. This is going to be realized in 4 steps:

1. Get the estimated travelling time on the segment starting in node i and ending in node j (i, j successive nodes) when starting in the interval I_k at node i:

$$t_{ij} = t_{sj} - t_{si} .$$

   The two durations t_{sj}, t_{si} are coming from the stack memory of the ant agent.

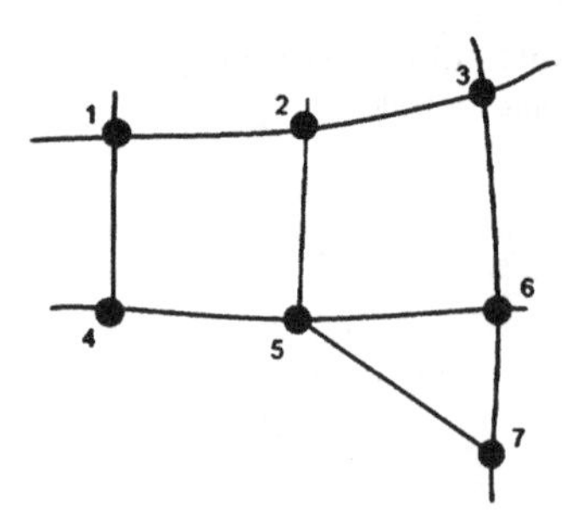

Fig. 13.3 A simple network with seven nodes

2. Update the average travelling time $\mu_j(I_k)$ of node j:

$$\mu_j\left(I_k\right) = \mu_j\left(I_k\right) + \eta\left(t_{ij} - \mu_j\left(I_k\right)\right), \eta = 0.1.$$

   The value of η is 0.1. This variable represents what impact, the recent trip times samples, have over the average value $\mu_j(I_k)$. In this way only 50 observations will impact on the average travel time measurement.

3. Compute the reinforcement r to be used to update the routing table. This is a function of the time t_{ij} and its mean value $\mu_j(I_k)$.

$$r = \begin{cases} \frac{t_{ij}}{a\mu_j(I_k)}; \frac{t_{ij}}{a\mu_j(I_k)} < 1, a = 1.1 > 1 \\ r = 1 \text{ otherwise.} \end{cases}$$

4. Update the routing table for destination j.

Changes will be performed also on the entries corresponding to $f \in S_{id}, f \neq d$, for all $f \in V$ on the subpaths followed by ant F_{sd} after visiting the node i. Because all the forward ant decisions depended on the destination node, the subpaths are side effects and are sub-optimal. After all, even though suboptimal paths, these were discovered by the ants at no additional cost.

Considering for example one forward ant F_{AD} and the corresponding backward ant B_{AD} we are going to exemplify how the pheromone matrix is going to be updated. If the destination is $d' = i + 1, \ldots, C$ and the current node is $i = C, \ldots, A$, the pheromone matrix R_i is updated by incrementing the pheromone $P_{d'j}, f = i + 1$ of the last node f the ant B_{AD} came from (the pheromone suggesting to choose neighbor f when destination is d') and decrementing by normalization the pheromone values of the other neighbors $P_{d'j}$, $j \in N_i$, $j \neq f$.

The pheromone mainly depends on the measure of goodness associated with the trip time $t_{id'}$ (I_k) experienced from i to d' by the forward ant. This time is the only available feedback to score paths and the best chosen cost since it integrates the number of hops in the network and the delay suffered because of congestion. Nevertheless, the optimal trip time can not be determined because is dependant of the entire network status. A path with a good trip time during a period of congestion might not be so good in case of a fluid traffic. But this is only a disadvantage that ant agents need to overcome. It is a typical problem frequent in the reinforcement learning field.

Update of the routing table for the destination j is done accordingly to the formulas:

$$P'_{jn}(I_k) = P_{jn}(I_k) + (1\text{–}r)\ \left(1\text{–}P_{jn}(I_k)\right) \text{ for the link } i \rightarrow n.$$

otherwise

$$P'_{jl}(I_k) = P_{jl}(I_k) - (1\text{–}r)\ P_{jl}(I_k)\Big) \text{ for } l \neq n.$$

When the backward ant B_{ds} arrives in s, it is deleted.

13.7 Experiments

In this section we describe three experiments testing our dynamic navigation system. At regular times practice exercises are organized to train first responders as firemen, police, and medical people. In 2010 a huge disaster simulation was simulated in the city of Rotterdam. A controlled explosion was created in the harbor; a ship was set on fire, a terroristic attack in the Underground and an explosion of a car transporting toxic liquid in a tunnel. Volunteers, most students were requested to play the role of victims, observers or just citizens. Professionals were involved to rescue people and flush the fire.

13.7.1 Experiment 1

Non professionals were requested to use the Icon Lang tool (see Figs. 13.4 and 13.5) and to report about the events they observed. In the crisis room incoming messages were placed on a digital map and operators were assumed to give a semantic interpretation of the messages. It proves that nobody has problems using the tool. But we have to notice that students were used as probants. The operators in the room had problems with the right semantic interpretation of two events. Messages about the terroristic attack in the underground were interpreted as "there is a fire in the underground". A terroristic attack fortunately never happened before and was misinterpreted by the operators. Secondly the ship on fire causes problems because the reports localized the ships on different positions. It took some time before the solution was found. The ship was drifting away on the river and was observed on different locations.

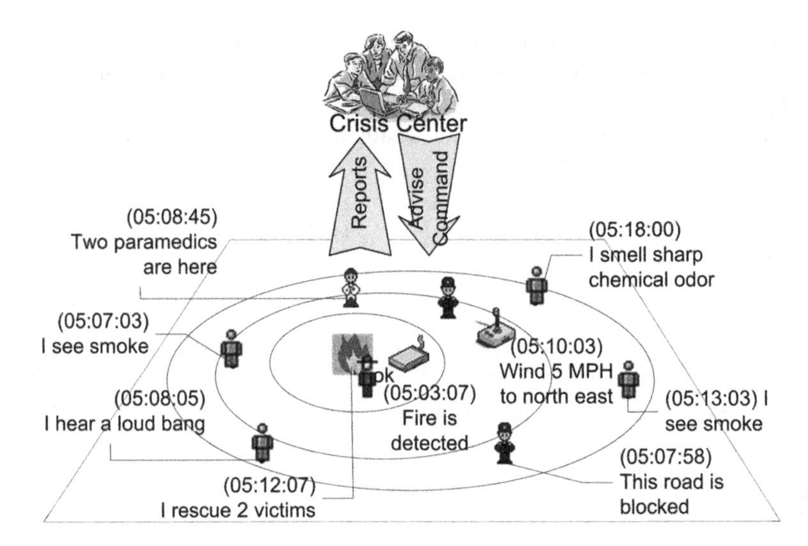

Fig. 13.4 Annotated map of the area4

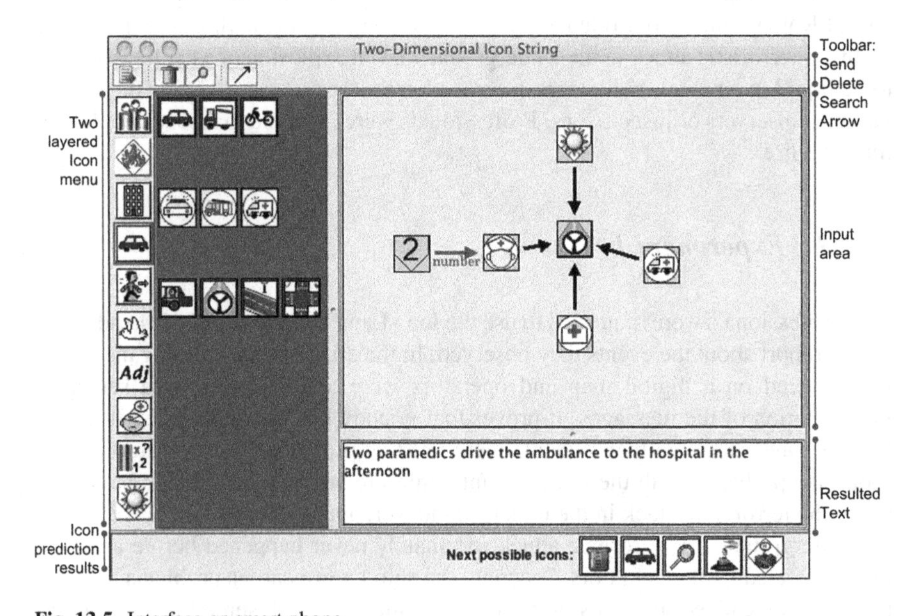

Fig. 13.5 Interface on smart phone

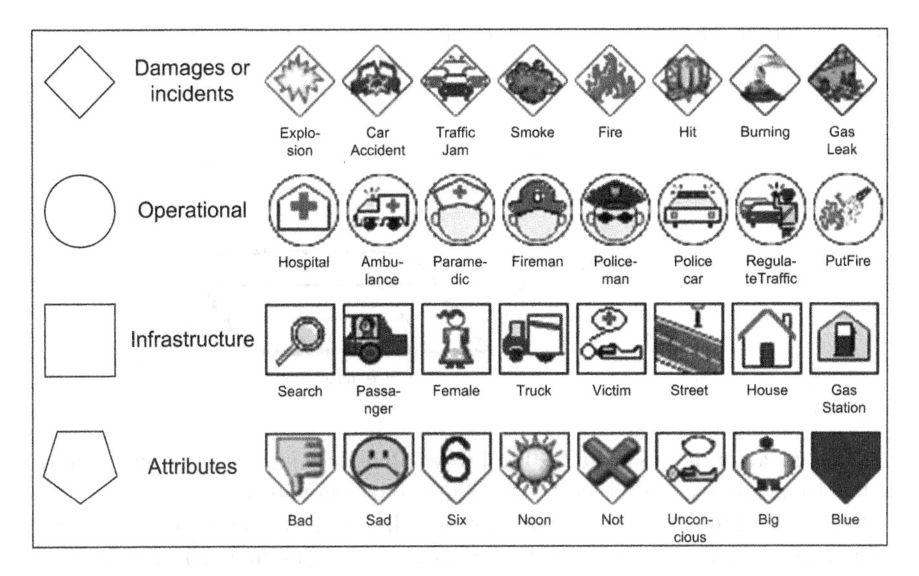

Fig. 13.6 Selection of icons

13.7.2 Experiment 2

In [8, 9] we described our first experiments using a hierarchical version of our dynamic routing system. It splits traffic networks into several smaller and less complex networks by introducing a hierarchy between the roads. At each intersection our dynamic ABC-system maintains a routing table used for guiding the cars. This model is supplemented with actual data from the traffic by the vehicles themselves. In order to test our system we built a simulation environment with a routing system that guides the vehicles between cities or sectors in cities using the fastest way in time and taking into account the load on the roads. The simulator makes possible to play different accident scenarios, like a crisis game where, because of a disaster, multiple roads become unavailable or heavily congested, and most of the drivers are disoriented.

The main idea of hierarchical routing is to divide a large network in zones corresponding to local and global sectors on a map. In this way a hierarchical traffic network based on the human's behavior to plan a route is created. A separate network exists for every sector or city. The nodes situated at the border of a sector and which have connection with other sectors are called routing nodes. As we will see later these nodes will play a special role, and their activity is different from the one of an inner sector node. At such nodes the transition between lower and higher layers takes place. An example of such a network is shown in Fig. 13.7. The network of global points is a routing table on the highest level.

A common feature of all ant based routing algorithms is the presence in every network node of a data structure, called routing table, holding all the information

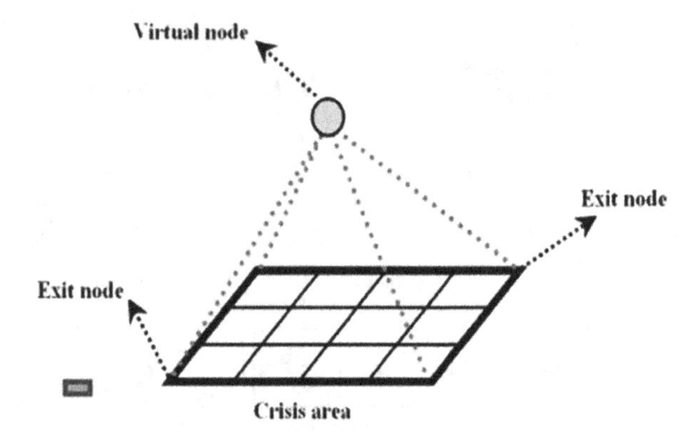

Fig. 13.7 Virtual node

used by the algorithm to make the local forwarding decisions. The routing table is both a local database and a local model of the global network status.

Each node *i* in the network has a probability table for every possible final destination *d*. The tables have entries for each next neighboring node *n*, P*dn*. This expresses the goodness of choosing node n as its next node from the current node *i* if the packet has to go to the destination node *d*. For every sector that contains a city network a virtual node is introduced. This can be understood as an abstraction for all the nodes of the sector (see Fig. 13.7). Each virtual node will have an entry in the data structures of every node. They will be used to route the data between different sectors. The concept virtual node is also used in crisis situations.

A service provided by the system is escaping from a dangerous area. The goal is to route the traffic as soon as possible out of a certain zone via some safe exit points. The problem is similar with dynamic routing using multiple destinations which can be any of the exit points. A virtual node (VNode) is introduced in the ant network. It represents the collection of all exit points. Each routing table will contain, for this virtual node, an entry as possible destination. All the traffic inside the crisis area will be evacuated. Each vehicle will have the virtual node as destination.

In similar way can be solved the evacuation of pedestrians. Instead of vehicles we will have people carrying smart phones connected to the escaping system.

Considering the map with three sectors from Fig. 13.8. Table 13.2 is an example of a routing table of the node 34, which is part of sector 1 and has three neighbors. Every node of sector 1 is a possible destination and has a corresponding row with probabilities in the table. There are also two virtual nodes, one for sector 2 and one for sector 3. For our sample experiment, we choose a skeleton road map with 4 cities connected by motorways (Fig. 13.9). We split the vehicles in four categories:

- Vehicles that have no communication with the routing system. They are guided according with Dijkstra's algorithm,
- Vehicles that give information about the state of the traffic to the system, but doesn't use it for navigation (ex. buses and other public transport vehicles),
- Vehicles that give no information to the system but they use it for routing,
- Vehicles that use the routing system but also provides it with information.

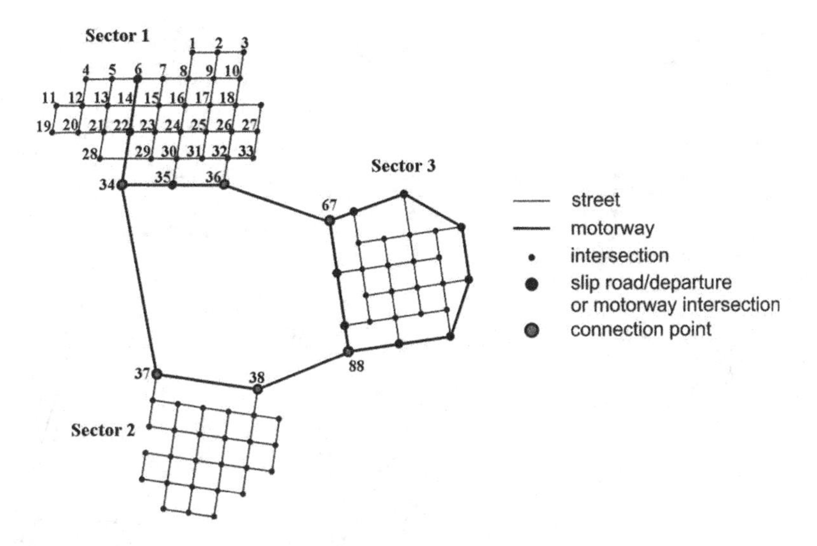

Fig. 13.8 Network model with three sectors

Table 13.2 Routing table in node 34

Neighbour			
Destination	N22	N35	N37
D1	0.58	0.42	0.00
...	...	...	0.00
D33	0.35	0.65	0.00
D35	0.96	0.04	0.00
D36	0.93	0.07	0.00
V2	0.02	0.89	0.09
V3	0.03	0.20	0.77

The first two categories we called them 'standard vehicles' and the ones which use the routing system 'smart vehicles'. We generate vehicles in the sectors 1, 2 and 3 and we choose sector 4 as destination for them. The speed on the ring roads was set 80 km/h and 50 km/h inside the cities. The total simulation period was 25,000 'time steps' = 5,000 s. After 2,500 time steps, we generate an accident on the road between sector 1 and 3 and the speed on this road gets less than 20 km/h. This was the default road to sector 4 for most of the cars starting in sector 1.

To measure the performance of the system, we counted the travel time necessary for the cars which started in sector 1 to reach the destination sector 4. In Fig. 13.10 we display the average time that was necessary for the standard vehicle and in Fig. 13.11 the average time a 'smart vehicle' needed to reach the destination. Because of the delay on the link with the accident, the cars using the routing system are routed via sector 2. We can observe that the cars which encounter the delay arrive after 5,000 time steps. In the graph corresponding to the smart cars (Fig. 13.11) the values start to increase earlier. This is because the cars which were rerouted via sector 2 will arrive faster than the 'standard' ones which took the road with the congestion.

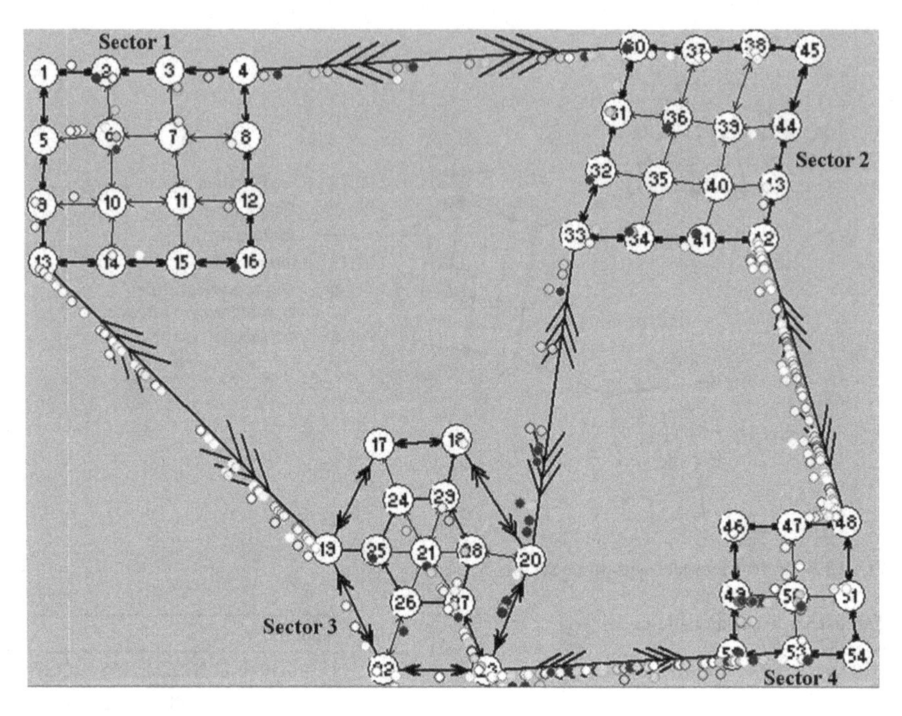

Fig. 13.9 Traffic simulator

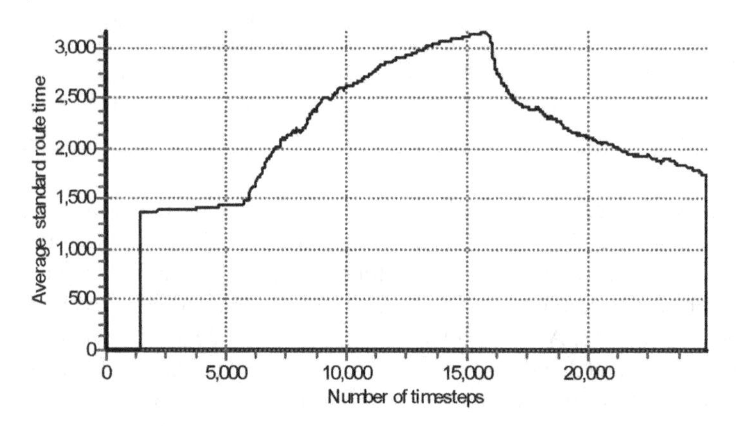

Fig. 13.10 Standard vehicle trip time

The average 'smart car' time grows up to 1,800 time steps and then it stabilize around 1,700 time steps. So the rerouting will cost the vehicle using the routing system about 300 time steps. In the meanwhile the 'standard vehicle' delay continues to grow. It was also probably going to a stabile value somewhere above 2,500 time steps. But at the time 10,000 we decided to put delay on another road – the one connecting sector 2 with 4. Because of this the 'standard vehicle' graph

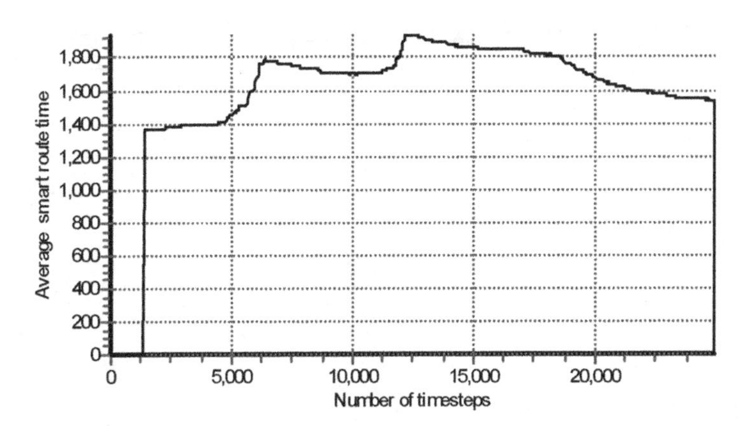

Fig. 13.11 Smart vehicle trip time

follows the ascendant slope. The 'smart cars' will get only about 150 units more delay. This is because the routing system reacts to the new event and finds an alternative route via sector 3.

At time 15,000, we release the link we blocked first, between sector 1 and 3. In this way, the road used by the standard vehicles gets clear and the average trip time will decrease fast. The new available road is noticed also by the routing system, which stops to divert the users via sector 2. Both graphs go smooth to the initial values (Figs. 13.10 and 13.11).

13.7.3 Experiment 3

It is difficult to get permission of a city counsel to test systems in real life situations. Fortunately we developed a city parking service for the city of Rotterdam (see Fig. 13.12). The system monitors free parking places in the Park houses of the city and reserves parking places for participants of the system. The City Based Parking Routing System (CBPRS) guides vehicles using our ant based distributed hierarchical routing algorithm to their reserved parking place. We considered the Park houses as safe areas and the rest of the city was threatened by a toxic cloud. So car drivers have to look for a rescue to the safe areas in the Park houses.

Participating vehicles can be characterized along two dimensions. The first is the routing algorithm employed, which can either be Dijkstra's algorithm or the ant based algorithm. (We do not simulate vehicles that do not employ routing.) Since Dijkstra's algorithm is non-adaptive we expect it to perform inferior when many traffic jams occur. The second dimension concerns parking behavior: Vehicles can either park or not, and if they do, they can use the CBPRS parking service or they can search for a place randomly. This leads to six possible vehicle types. We experimented with various relative frequencies for these vehicle types, as shown in Table 13.3. Our wish was to test a higher percentage of Ant Parking vehicles up

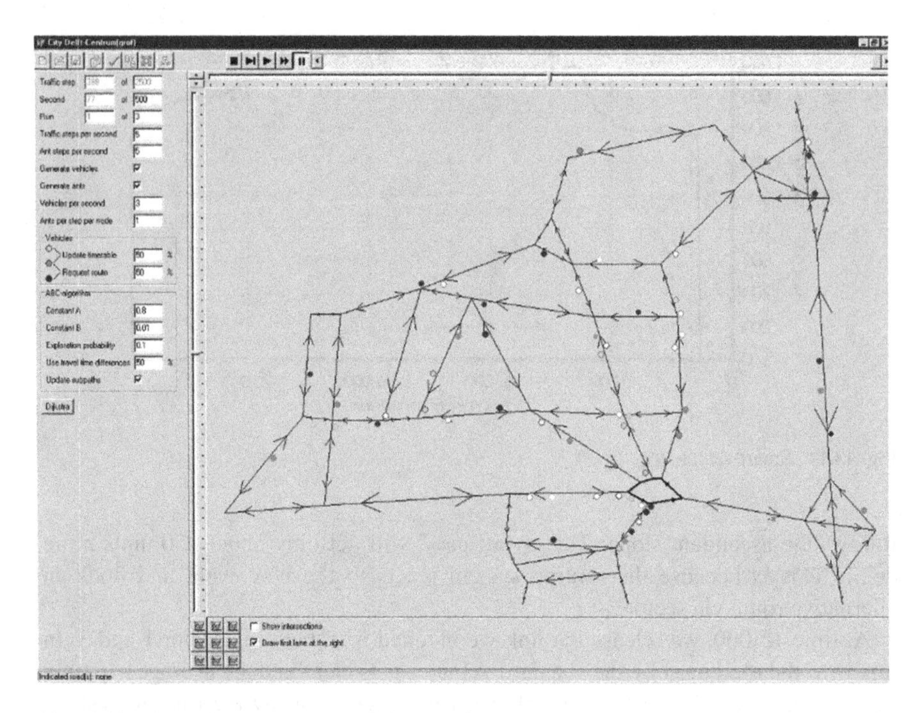

Fig. 13.12 Simulation environment of the city of Rotterdam

Table 13.3 Relative frequencies of participating vehicles during experiments

	Vehicles	Dijkstra vehicle (%)	Dijkstra parking vehicle (%)	Ant vehicle (%)	Ant parking vehicle (%)
Experiment one	7,500	90	10	0	0
Experiment two	7,500	90	0	0	10
Experiment three	7,500	0	0	90	10
Experiment four	7,500	40	0	50	10
Experiment five	7,500	90	10	0	0

to 100 %, comparable to a crisis situation. But this was not an option in the given realistic not crisis related test environment. It is also not clear how many people find their own way during a crisis and how many people follow the guidelines of the authorities.

Experiment one served as our benchmark situation where all vehicles were routed via Dijkstra's algorithm. Experiment two measured the effectiveness of the CBPRS when 10 % of all vehicles, the minimal participation limit set in this study, used the CBPRS services. Experiment three shows the maximum possible level of effectiveness of the CBPRS when all vehicles participate in the system. Other experiments illustrate upper and lower limits of effectiveness found in this simulation environment.

Table 13.4 Results of experiments with Parking (P) and non-parking vehicles (N)

	Exp one		Exp two		Exp three		Exp four		Exp five	
	P	N	P	N	P	N	P	N	P	N
Samples	666	14,162	692	14,353	630	14,261	666	13,785	585	14,379
Roads traveled	18	10	10	10	9	11	10	11	16	10
Average	544	295	278	281	198	227	234	241	384	277
St dev	680	349	288	319	103	114	105	135	490	320

The city environment can be distinguished from the smaller environments through the fact that traffic volumes are reasonable high and remain high for almost the entire duration of the day sometimes stretching well into the evening. The ring road allows vehicles to move at an increased maximum speed of 80 km/h giving vehicles the opportunity to travel quickly from one side of the city to another. This prevents overcrowding of the roads within the inner city that have a lower maximum speed. While such a ring road has certain benefits, it can also be the source of congestions within a city.

During experiment one all vehicles within the simulation environment used Dijkstra's algorithm in order to follow the shortest path to their destination. The shortest path towards most destinations however includes traveling along the ring road for a certain distance. Taking into account that all other vehicles also use Dijkstra's algorithm it is clear that the traffic volume on the ring road is liable to increase beyond its capacity and therefore causing traffic jams as is described in the following paragraphs. This leads to a situation where the parking vehicles of experiment one are forced to wait in the traffic jams caused mainly by the non-parking vehicles thereby increasing the travel time significantly in comparison to the other three experiments.

Experiments two and three are conducted using ant based parking vehicles. The difference between these experiments is that experiment two made use of non-parking Dijkstra oriented vehicles while experiment three relied solely on ant based vehicles. The CBPRS in experiment two is faced with a situation in which dynamic routing information is scarce. The ant based algorithm is constantly struggling to find optimal routes. As the simulation progresses we see that the minimal information received is insufficient and travel times start to increase steadily. In experiment three the CBPRS is supplied with an abundance of dynamic routing information and is able to quickly find the most optimal paths through the city.

The optimal distribution of vehicles in experiment four was found by methodically trying different combinations and comparing results. The results of this process are shown through experiment four that represents the most optimal distribution of non-parking Dijkstra vehicles and ant based vehicles. Experiment five compares the performance of Dijkstra's algorithm and the ant based algorithm by allowing Dijkstra orientated parking vehicles to use the CBPRS parking service and so duplicating the settings of experiment two. The results of the experiments clearly show a reduction in travel times for Dijkstra orientated parking vehicles when comparing to the situation in experiment one (Table 13.4).

13.8 Summary and Conclusion

In this paper we discussed different emergent technologies to improve mitigation in crisis situation. To answer the first research question we introduced an architecture of smart phones. This architecture was tested during several training exercises for first responders. For the second research question we introduced a special nonverbal language based on icons and special syntax and grammar. The language has been tested during several crisis exercises and proved to be very useful. Especially in cities where more than 30 % of the people doesn't master the Dutch language the icon language was very successful. The icon language is very close to the incremental use of multimedia with adaptive languages full of emoticons and acronyms.

The third research question was about the search for emotional content in speech and text messages, such as Tweets. We were able to detect the emotional content but the relation between messages with a strong emotional content and senders who need urgent help was not as strong as expected. It proves that even in dangerous situation most people were able to analyze the situation in a rational way. The relation between emotions as anger and need for help was evident.

T answer the last research question we proposed the use of a special dynamic routing device to evacuate people from a dangerous situation. We were not able to test the technology in a real crisis situation but the test results during exercises were promising.

All the research questions we discussed are examples from emergent technologies. The focus is on a bottom up approach and not top down. To employ information from people in a crisis situation proved to be very useful and is supported by the use of social media on a large scale.

References

1. Gunawan LT, Alers H, Brinkman W-P, Neerincx M (2011) Distributed collaborative situation-map making for disaster response. Interact Comput 23(4):308–316
2. Al-Suwaidi GB, Zemerly MJ (2009) Locating friends and family using mobile phones with global positioning system (GPS). Proceedings of 7th ACS/IEEE International conference on computer systems and applications program, Rabat, 10–13 May 2009
3. Radu AA, Rothkrantz LJM, Novak M (2012) Digital traveler assistant. Inform Control Autom Robot 174:101–114
4. Gunawan LT, Fitrianie S, Brinkman W-P, Neerincx M Utilizing the potential of the affected population and prevalent movile technology during disaster response. ISCRAM2012
5. Spiro E, Sutton J, Johnson B, Fitzhugh S, Butts C (2012) Superstorm sandy: looking at the twitter response. Online Research Highlight. http://heroicproject.org
6. Quarantelli EL, Dynes RR (1972) When disaster strikes (it isn't much like what you've heard and read about). Psychol Today 5:66–70
7. Lomnitz, C, Fisher HW (1999) III Response to disaster: fact versus fiction and its perpetuation. The sociology of disaster. Nat Hazards 19:79–80

8. Tatomir B, Rothkrantz LJM (2004) Dynamic traffic routing using ant based control. International conference on systems, man and cybernetics IEEE SMC, 2004, The Hague, Netherlands, pp 3970–3975
9. Tatomir B, Rothkrantz LJM (2005) H-ABC: a scalable dynamic routing algorithm. Adv Nat Comput 3:279–293
10. Rothkrantz LJM, Boehle JL, van Wezel M (2013) A rental system of electrical cars in Amsterdam. Transport Lett Int J Transport Res 5(1):38–48
11. Tatomir B, Rothkrantz LJM, Suson AC (2009) Travel time prediction for dynamic routing using ant based control. In: Proceedings of the 2009 Winter simulation conference Austin, pp 1069–1078
12. Fitrianie S, Rothkrantz LJM (2005) Language-independent communication using icons on a PDA. In: Proceedings of TSD'05, LNCS 3658, Springer, pp 404–411
13. Fitrianie S, Rothkrantz (2006) Two-dimensional visual language grammar. In: Proceedings of TSD'06, LNCS 4188, Springer, pp 573–580
14. Fitrianie S, Rothkrantz LJM (2009) Computed ontology-based situation awareness of multi-user observations. In: Proceedings of ISCRAM'09, Gothenburg, Sweden
15. Imran M, Elbassuoni S, Castillo C, Diaz F (2013) Meier extracting information nuggets from disaster-related messages in social media. In: Proceedings of the 10th international ISCRAM conference – Baden-Baden, Germany, May 2013
16. van Willigen I, Rothkrantz LJM, Wiggers P (2009) Lexical affinity measure between words. Text Speech Dialogue 5729:234–241
17. Whissell CM (1989) The dictionary of affect in language, J Emotion: Theory, Res Experience 4:113–131
18. Bradley MM, Lang PJ (1999) Affective norms for english words (Anew)' NIHM center for the study of emotion and attention. University of Florida
19. Lefter I, Rothkrantz LJM, van Leeuwen DA, Wiggers P (2011) Automatic stress detection in emergency (telephone) calls. Int J Int Def Support Syst 4(2):148–168

Chapter 14
Geographic Information for Disaster Management – An Overview

Wolfgang P. Reinhardt

Abstract The importance of geographic information (GI) for disaster management (DM) has been recognized pretty well although there is still a need to establish knowledge about GI more widely. Also, there are quite a number of research issues related to DM, where GI plays a significant role. In this paper, the role of GI for DM is demonstrated mainly through an overview on research-oriented projects and other activities related to different tasks within DM, which have been carried out during the last years.

14.1 Introduction and Overview

The increase of the number of natural disasters in the last decades and the associated tremendous damages to properties and people has led to a high number of activities on different organizational levels. Examples of activities are DM policies, action plans, directives etc., as well as research and development projects. Only a few can be mentioned here, like the United Nations International Strategy for Disaster Reduction [1], which is of worldwide importance. In Europe, disaster management has been addressed in the entire framework programs (FP), e.g., within the safety and security agenda of FP7 program. IT-related topics of interest were e.g.: in-situ monitoring and smart sensor networks, risk information infrastructures and generic services, public safety communication, alert systems and rapidly

W.P. Reinhardt (✉)
AGIS/Institut für Angewandte Informatik, University of the Bundeswehr Munich, Werner-Heisenberg-Weg 39, 85579 Neubiberg, Germany
e-mail: Wolfgang.Reinhardt@unibw.de

H.-N. Teodorescu et al. (eds.), *Improving Disaster Resilience and Mitigation - IT Means and Tools*, NATO Science for Peace and Security Series C: Environmental Security, DOI 10.1007/978-94-017-9136-6_14,

deployable emergency telecommunications systems, emergency management and rescue operations, as well as distributed early warning and alert systems. Of very high importance are the activities in the space-based applications at the service of the European society which are based on development projects and realizations of directives of FP7 such as

COPERNICUS, formerly known as **GMES,** Global Monitoring for Environment and Security, [2] and

INSPIRE – Infrastructure for Spatial Information in Europe [3].

The tsunami in 2004 in the Indian Ocean initiated a research program in Germany called "Early Warning Systems" under the umbrella of the "Geotechnologien" (Geotechnologies) program [4], funded by the German Ministry of Research and Education and the German Research foundation. It was mainly a research-oriented program to improve the methodology within early warning systems. Geo and IT scientists cooperated in all of the project consortia. The author of this paper worked in the consortium of one of the funded project within this program (see Sect. 14.3.3).

The importance of geographic information and related methods and tools were emphasized very often, e.g. in a report of the US National Research Council [5], cited after Goodchild and Glennon [6] which states:

> The (report)'s central conclusion is that geospatial data and tools should be an essential part of all aspects of emergency management – from planning for future events, through response and recovery, to the mitigation of future events ......

In consequence, GI methods and tools can be considered as one example of the very important information and communication technology (ICT) tools for DM.

The remainder of this paper is organized in the following way: The second section gives a short introduction into the field of geographic information. In the third part an overview of the DM phases and examples of the usage of geographic information for disaster management (GI4DM) are given, along with some comments on spatial simulations and crowd sourcing for DM. Section 14.4 includes an overview on the project "TranSAFE-Alp" which has been finished recently and demonstrates how geographic information is used in a specific application within DM. Finally, the paper is concluded by some remarks on GI4DM in general.

The focus of this paper is GI4DM with special emphasis on issues where still research related to ICT is ongoing or needed. It is obvious that the trial to give such an overview on this very wide field can almost be "disastrous" due to high ambition, but nevertheless the trial is made, of course biased through the author's personal experience.

14.2 Basics of Geographic Information

14.2.1 Geographic Information and Geographic Information Systems

14.2.1.1 Geographic Data

Geographic data refers to the earth's surface and sometimes also to the geosphere and the underground, and are available in different forms (see examples in Fig. 14.1). Most important are:

- Geographic features with geometric properties (in vector form, in two dimensions x, y (2D) or in x, y, z (3D) and thematic properties (attributes). Aspects of time are not described here, but also can be treated.
- Coverages, which are discrete continuous fields in the form of
  - Raster data, which is pixel based, like satellite or airborne imageries
  - Point based data like grids or triangular irregular networks (TIN), mainly used to represent the height structure of the terrain

Geographic data is widely available through governmental and other mapping agencies, private data providers like the navigation data vendors or satellite data providers and also in an open form, as Open Street Map [7].

For readers who notice the hopping between the terms *data* and *information* in this section, a short note that the understanding of these terms in this field is in line with the general understanding of the terms in information management.

14.2.1.2 Geographic Information System

The origin of the Geographic Information System (GIS) goes back to the 50s/60s of the last century and became a well-known and widely used technology since that time.

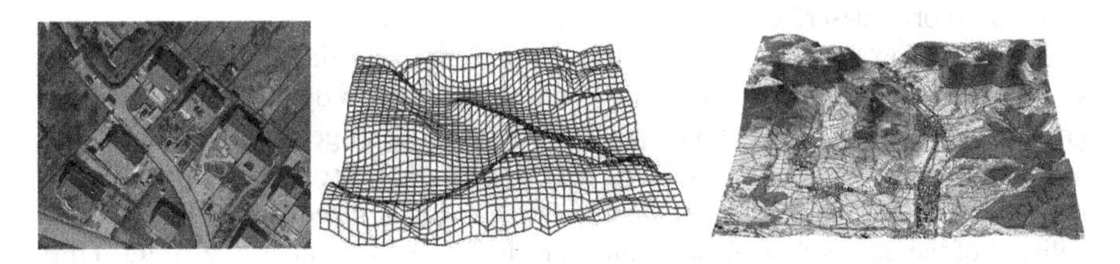

Fig. 14.1 Examples of geographic information: aerial imagery and geographic features in vector form (*left*), gridded data (*middle*), overlay of gridded data and vector data (*right*)

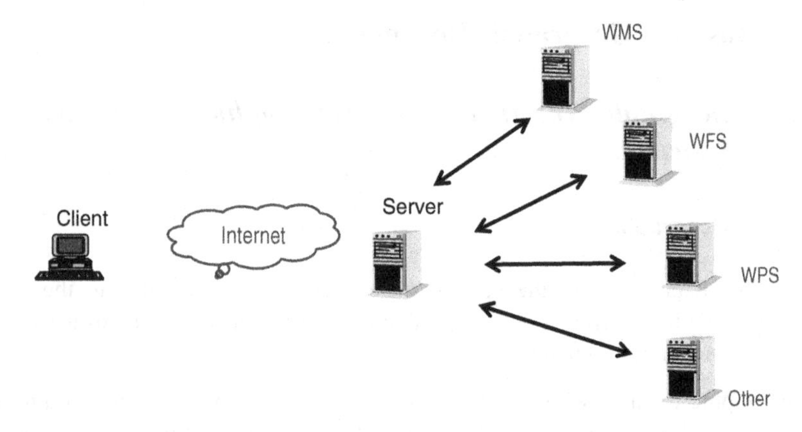

Fig. 14.2 Geo Web Services

According to NCGIA [8] a GIS is a system of hardware, software and procedures to facilitate the management, manipulation, analysis, modeling, representation and display of geo-referenced data to solve complex problems regarding planning and management of resources.

As the location of things is very important in many fields of applications and also in disaster management, Geographic Information became very important also for DM. The main benefit of GIS lies in the possibility to exploit various different layers of data with the help of proper analysis methods to generate knew knowledge. For a further study of the field lots of text books are available like Longley et al. [9].

14.2.2 Geo Web Services

Geo Web Services (GWS) are web services dedicated to the delivery/access to or processing of geographic information. Special classes of Geo Web Services are the GWS with interfaces standardized by the Open Geospatial Consortium [10] which partly are also ISO Standards [11], see examples in Fig. 14.2. The most common GWS are shortly described below:

The Web Map Service (WMS) is used to present maps via the Internet through standardized interfaces. It renders maps with spatial content dynamically from geographic information. Map is understood as the portrayal of geographic information as a digital image file, which is suitable to be displayed on a computer screen. Please notice that only the parameters of the interface are standardized for all GWS and the delivery of the data is done in well-known raster or vector formats like JPEG, TIFF or SVG. The processing on the server side, in case of WMS to generate the map, is done in a proprietary environment.

The Web Feature Service (WFS) is used to deliver geographic feature data encoded in the ISO Standard GML (Geographic Markup Language). An extension, the WFS-T (Web Feature Service Transactional), also allows the modification (insert, update, delete) of geographic features.

The Web Processing Service (WPS) is used to provide GI methods as Services. In this case, the level of standardization is somehow lower as knowledge of the used method has to be known to be used in as a WPS. More details on this issue can be found e.g. in Stollberg and Zipf [12].

Web Services in general follow the Publish-Find-Bind principle [13], which means:

- Descriptions of the resource (the service) are provided through metadata. (*publish*)
- Potential users of the service are able to *find* the necessary information to be able to use the service within their environment (*bind*).

That means that remote services can be included in service-based applications. As many mapping agencies and satellite imagery providers offer such services this concept gives the opportunity to include relevant information also in disaster management applications with low effort. The issue of license costs for the usage of services will not be discussed here.

Geo Web Services are widely used within Spatial Data Infrastructures (SDI) like INSPIRE which already has been mentioned Sect. 14.1.

14.3 Use of Geographic Information in Disaster Management

The use of geographic information for disaster management is reasonably related to the different phases of DM. Unfortunately, phases and terms used in DM are not used in a common way across countries and regions and the different disciplines involved. There is no space to discuss this issue in this paper. Only a short outline about phases is given in Sect. 14.3.1. A very good introduction in the field of DM, the terms used and also the phases is given by UNISDR [14] as well as by the German Center for Disaster Management and Risk Reduction Technology [15].

14.3.1 Phases in Disaster Management

As already indicated, the use of phases is very common in DM, but unfortunately there is no generally accepted model of phases are available. A very detailed description of the phases used in different countries, disciplines and organizations

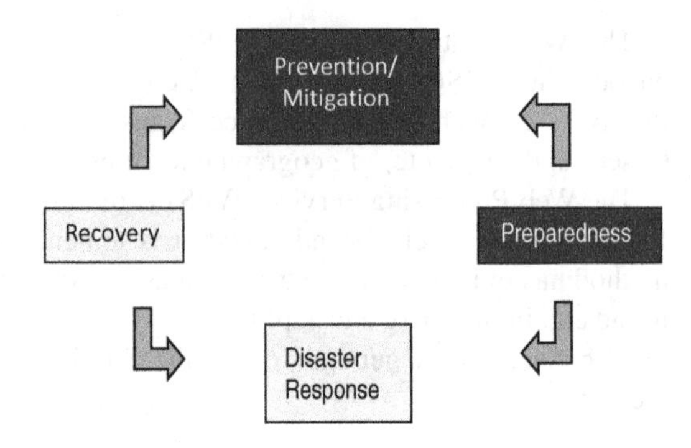

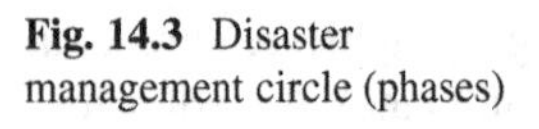
Fig. 14.3 Disaster management circle (phases)

is given (in German language) by Stangl and Stollenwerk [16]. For this paper, the model depicted in Fig. 14.3 is used which is used in a similar way by other organizations. The phase prevention/mitigation and preparedness are sometimes subsumed under the term disaster protection.

According to the latter literature the assignment of actions also are pretty different in different regions and disciplines. Therefore, the usage of GI in the following is related to civil protection as an aggregation of prevention/mitigation and preparedness. The recovery phase is not treated in this paper.

14.3.2 Examples of Geographic Information Used Within DM Phases

In the phase of disaster protection GI plays important roles for the following tasks:

- Hazard prediction and modeling
- Susceptibility mapping
- Risk assessment and mapping
- Public Awareness and education
- Scenarios development
- Emergency planning and training
- Real time monitoring and forecasting
- Early warning and alerting
- ...

With respect to the restricted length of this paper, only a few of these tasks are discussed shortly in this section.

The mapping of susceptible, endangered, hazardous, risky or vulnerable (Terms!) areas to natural and/or technical disasters is a task usually done by responsible public administrations and their subcontractors. But there are still research scopes in this field to optimize the used methods, for example in landslide applications, see Gallus et al. [17] or Pradhan et al. [18].

From an ICT perspective the tasks of real time monitoring, early warning and alerting have been treated intensively in the last decade especially in conjunction with so-called Geo-Sensor-Networks (GSN) which are according to Nittel [19] able to deliver geographic or geo-referenced data with importance for the tasks mentioned. Since the Open Geospatial Consortium (OGC) has introduced the Sensor Web Enablement [20], a concept to model, connect and access various kinds of sensors (which became a standard also), there application within DM has been investigated in various applied research projects. Kandawasvika [21] as well as Walter [22] discussed different aspects of the suitability of the concept for landslide applications in their dissertations.

Examples of tasks where GI is used in the response phase of DM:

- Dispatching of resources
- Situational awareness
- Command & control coordination
- Information dissemination

To improve the post disaster situational awareness up-to-date satellite imagery is very helpful. Therefore the national aeronautics and space research center of the Federal Republic of Germany (DLR) has established a Center for Satellite based Crisis Information [23]. Reznik [24] investigated within the framework of the ESS project [25], the use of Unmanned Arial Vehicles (UAV) to provide up-to-date information as well as the usage of SWE (see above). Reznik's work was in conjunction with concepts for an open, standards based command and control systems. Wang and Yuvan [26] also discuss the usage of UAVs for the same purpose.

More information and examples about GI4DM can be found in the literature, e.g. in a special issue of a journal edited by Konecny and Reinhardt [27, 28].

14.3.3 Geographic Information and Simulation

In almost all scientific disciplines, simulation models are important ways to study systems which are inaccessible to scientific experimental and observational methods.

Within disaster management, simulation plays a major role to study the effects of disasters. Consequently there are a tremendous number of publications available discussing methods for the simulation of different disaster types. A comprehensive overview on simulation processes using geographic information – also called spatial simulations – is given by O'Sullivan and Perry [29].

A connection of the simulation software to a GIS is beneficial for simulation requiring geographic data. Figure 14.4 gives an overview on this connection. As all spatial simulations need geographic data, the GIS can be used for providing and preparing the geographic data and to combine it with functional data, which is often needed for the specific simulation, e.g. material parameters in case of a landslide. This data has to be transferred in a suitable form to the simulation system,

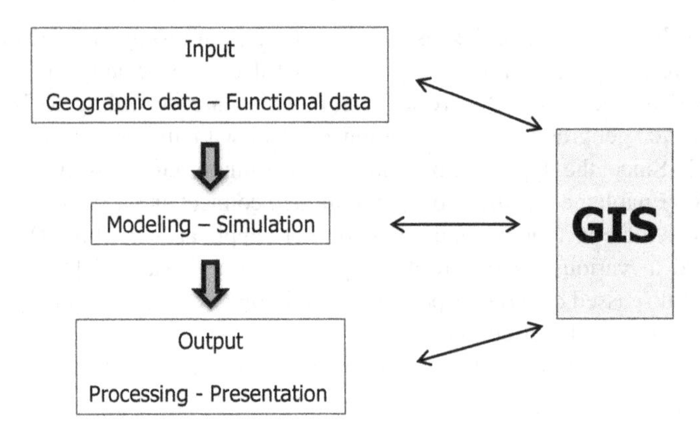

Fig. 14.4 GIS and modeling/simulation

which might be a GIS component or a separate software system. The way how both components communicate also depends from the latter, but this issue shall not be discussed further in this paper. As the output also refers to locations, the GIS is very suitable for presenting them and for optimizing the presentation as indicated e.g. in Dransch et al. [30].

The approach presented here was applied in the project "EGIFF", which was funded in the framework of the Geotechnologien program (Sect. 14.1). In this case a Finite Element (FE) approach was used to simulate landslide processes, see Nuhn and Reinhardt [31]. In the first step the project test area was modelled in 3D and the model was incorporated in the FEM Software. After that within the FEM software loads, simulating for example heavy rain, were introduced and the effects of different magnitudes of loads to the landslide were studied intensively. Figure 14.5 depicts a cross section of the investigated area which shows the results of the simulation (the red areas indicate a high risk of sliding). In case of FE approach the advantages of a GIS – and additional methods like spatial clustering – were demonstrated very well as the results of a FE simulation include ten, or even hundreds, of thousands of vectors showing the direction and the amount of a potential displacement of each finite element. Only after a suitable processing this results can be visually interpreted and presented to decision makers as shown in Nuhn et al. [32]. For the project study area, lots of historical data were available which allowed for a simulation of historical events. The results showed clearly the suitability of the developed approach but according to Trauner et al. [33] it was also clearly shown that the transferability and applicability to other areas, with different geology and soils, is very limited as very good knowledge about that are necessary for such an approach. More details about the EGIFF project can be found in Breunig et al. [34].

Many other examples of the simulation of natural disasters can be found in the literature, especially for flood simulation, e.g. Merz et al. [35] but also for technical disasters like explosions, e.g. Fischer et al. [36].

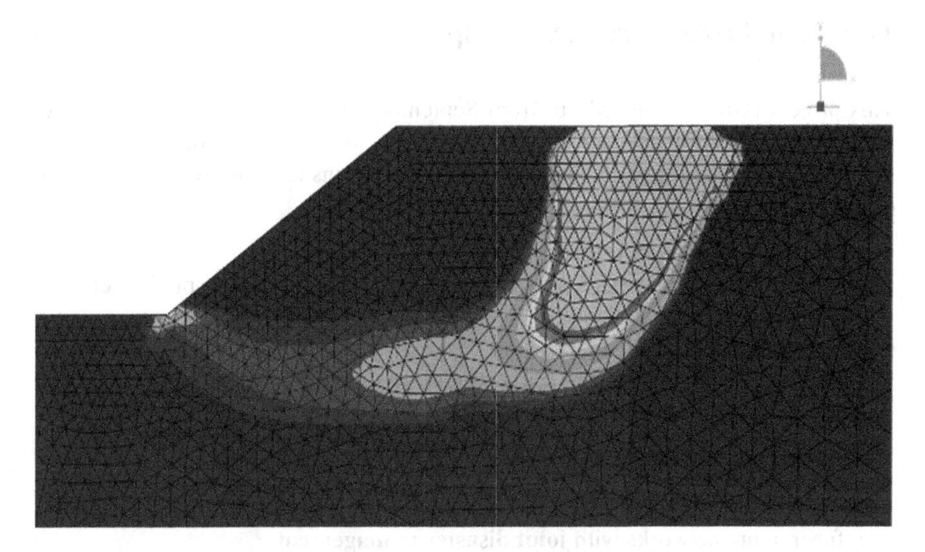

Fig. 14.5 Modeling and simulation of a landslide (cross section)

14.3.4 Geographic Information and Crowd Sourcing

The concept of crowd sourcing (CS) is probably much older but the term was introduced by Howe [37]. It is related to the practice of obtaining needed ideas or content by soliciting contributions from a large group of people. CS in conjunction with GI and DM is discussed in Goodchild and Glennon [6].

The usage of CS within disaster management became popular through the Ushahidi platform [38], which was applied for all major disasters in the last years. An analysis of this platform applied for the Haiti disaster is available on the same web site [39]. The main purpose of the platform is the documentation of any relevant incidents/damages and their location through users to make this information available to other people in the area. Of course, there is a number of issues that have to be considered in this context, like:

- The reliability of the captured information available on the platform ("Trust in information")
- The information is usually captured by untrained persons and therefore poorly structured and not standardized.

Ongoing activities try to overcome these obstacles through different approaches like

- Fusion of the crowd sourcing data with other data (e.g. from sensors)
- Exploitation of social networks
- Definition and inclusion of quality indicators

14.4 The Project TranSAFE-Alp

This project [40] was carried out from September 2011 till August 2013, funded through the EU Alpine Space Program [41]. The partners are from Austria (2), Germany (2), Italy (7) and Slovenia (1). The involved institutions are from different organizations like:

- Operators of Transport Infrastructures (3)
- Organizations responsible for civil protection, traffic/transport, public security and/or disaster management (5)
- Research institutes/Universities (4)
- Management and monitoring traffic center (associated partners for requirement analysis only)

The main goals of the project were:

- Improve transport **security** for passengers and dangerous goods in Alpine Space infrastructure networks with **joint disaster management**
- Establishment of the JITES platform (Joint Integrated ICT-Technologies for Emergency and Security -management) – which aims at the integration of all relevant information

The JITES system, in case of a traffic interruption due to a natural calamity (like floods, landslides, mudslides, earthquakes), a technical damage (e.g. at bridges, tunnels) or a car accident, amongst others, should be able to:

- Enable the visualisation of all resources which are involved in the management plan for security and emergency of transports as well as all facilities relevant during
- Enable the localization, classification of the interruption and the visualisation of the area interfered through the disaster
- Allow for routing/rerouting of traffic on other itineraries on the basis of different parameters (shortest, quickest, others)
- Should make this information available for the relevant stakeholders, if possible through the internet also for mobile devices

The authors group, called AGIS, was responsible for the "Geo" part but also involved in other tasks. The main aspects were:

- Definition of the JITES architecture, interfaces to other systems, prototype set-up and test
- State-of-the-Art analysis of openGIS systems
- Evaluation of the usage of standard Geo Web Services (GWS) for project purposes
- Project dissemination at local level, media contacts and end-users active involvement, scientific advice

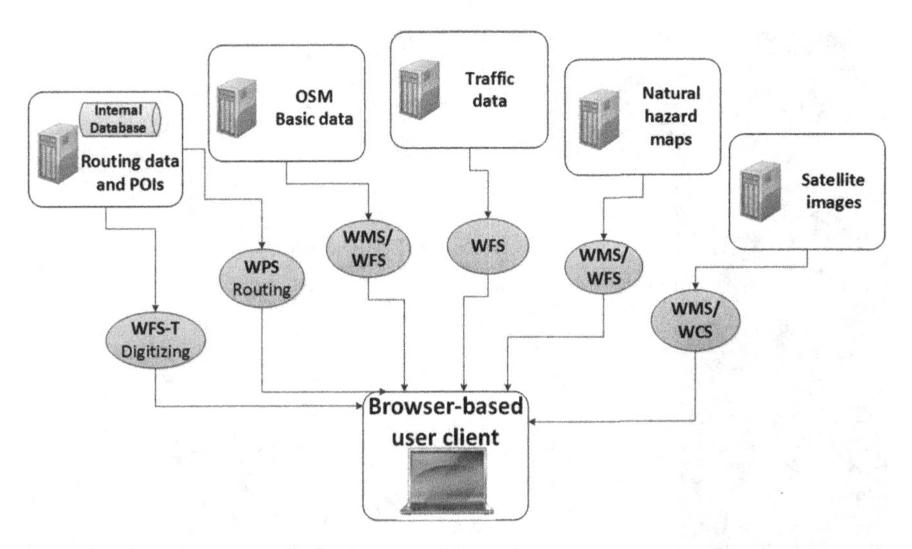

Fig. 14.6 Geo Web Services in the TranSAFE-Alp project

In this paper only a short summary of the concept how geographic information is included in the JITES platform is given.

A service based architecture making use of Geo Web Services was chosen (see Fig. 14.6). Part of the relevant data have been integrated in the JITIS data base but also a number of WMS from other organisation like DLR have been integrated to be able to visualise their relevant data.

A specific client also has been developed to visualise the local data as well as the data provided by the GWS (Fig. 14.7). The client allows for the visualisation of all relevant data and for the results calculated by functions implemented for the project. Parts of this function have been provided through WPS (see Sect. 14.2.2). With respect to length of this paper only a few of the functions are described here:

- Visualise local data and data provided through WMS
- Digitizing the area involved in a disaster from satellite imagery
- Determine the location of the nearest facilities of interest in case of a disaster like hospitals, helicopter sites etc.
- Determine the shortest fastest or safest alternative route in case of the closing of a road. As some of the road segments include vulnerability attributes (assigned to vulnerability classes and related to different natural disasters) a route could be calculated which avoids road segments with high vulnerability attributes which is called "safest" route.

In a test and training session with external experts the feasibility of the developed concepts and tools have been proven. The open source and GWS based approach

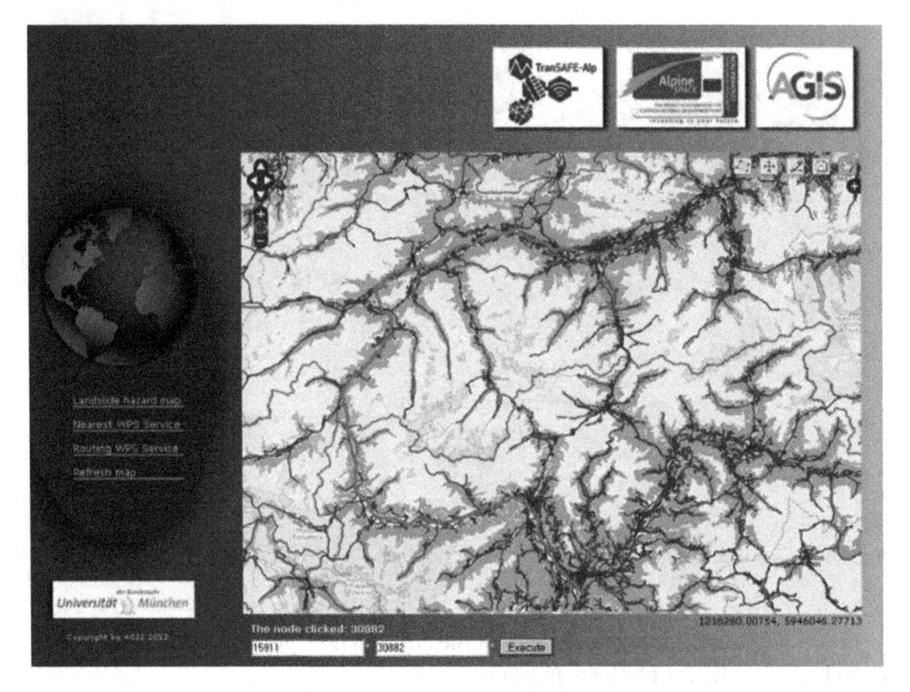

Fig. 14.7 Geo Client developed for the TranSAFE-Alp project

was working well and accepted by the test persons. A more detailed description of the project, the architecture, features, functionalities and the used technology can be found in Reinhardt et al. [42] and Galicz et al. [43].

14.5 Conclusions

Disaster Management is a very broad field where various disciplines are involved and doing research. Geographic Information is one important factor which can contribute to improve activities in all phases of DM. In this paper quite a number of examples from research oriented projects were given which demonstrate the importance of GI (data and methods) for DM. The usage of Geo Web Services for the field is very appropriate, which has been proven in a number of projects, mentioned in the paper. Related to research for GI4DM from the author's perspective there is still a need for further investigations to improve spatial simulation models at least in the field of landslides. Other topics for further research are the integration of chained services in disaster management workflows as well as the usage of social media to improve crowd sourcing based approaches for different phases of disaster management.

References

1. UN (2013) United Nations plan of action on disaster risk reduction for resilience, http://www.preventionweb.net/english/professional/publications/v.php?id=33703. Accessed 19 Dec 2013
2. EU (2013) Proposal for a regulation of the European Parliament and of the council establishing the Copernicus programme and repealing regulation (EU) No 911/2010, http://www.copernicus.eu/pages-principales/library/policy-documents/. Accessed 19 Dec 2013
3. EU (2007) Directive 2007/2/EC of the European parliament and of the council of 14 March 2007 establishing an Infrastructure for Spatial Information in the European Community (INSPIRE), http://eur-lex.europa.eu/LexUriServ/LexUriServ.do?uri=CELEX:32007L0002:EN:NOT/. Accessed 19 Dec 2013
4. Stronk (ed) (2009) Geotechnologien science report, early warning systems in earth management, http://media.gfz-potsdam.de/geotechnologien/doc/Science_reports/SR13.pdf. Accessed 19 Dec 2013
5. National Research Council (2007) Successful response starts with a map: improving geospatial support for disaster management. National Academies Press, Washington, DC
6. Goodchild M, Glennon JA (2010) Crowdsourcing geographic information for disaster response: a research Frontier. Int J Digit Earth 3(3):231–241
7. OSM (2013) http://www.openstreetmap.org. Accessed 19 Dec 2013
8. NCGIA (1990) CGIA Core Curriculum in GIS. University of California, Santa Barbara, http://www.geog.ubc.ca/courses/klink/gis.notes/ncgia/toc.html. Accessed 19 Dec 2013
9. Longley P, Goodchild M, Maguire D, Rhind D (2010) Geographic information systems and science. Wiley, Hoboken
10. OGC (2013) http://www.opengeospatial.org/. Accessed 19 Dec 2013
11. OWS (2013) http://www.opengeospatial.org/standards/common. Accessed 19 Dec 2013
12. Stollberg B, Zipf A (2009) Development of a WPS process chaining tool and application in a disaster management use case for urban areas, UDMS 2009, 27th urban data management symposium, Ljubljana
13. SOA (2013) http://en.wikipedia.org/wiki/Service-oriented_architecture. Accessed 19 Dec 2013
14. UN ISDR (2009) UNISDR terminology on disaster risk reduction, http://www.unisdr.org/files/7817_UNISDRTerminologyEnglish.pdf. Accessed 19 Dec 2013
15. CEDIM (2005) Glossary, http://www.cedim.de/download/glossar-gesamt-20050624.pdf. Accessed 19 Dec 2013
16. Stangl R, Stollenwerk J (2011) Terminologie von Katastrophenmanagement Kreisläufen/-Phasen, Bericht im Rahmen des Kiras (Österreichisches Förderungsprogramm für Sicherheitsforschung) – Projektes SFI@SFU, online: http://www.esci.at/sfi-sfu/sfi_sfu_studie_4_kkm_kreislaeufe.pdf. Accessed 19 Dec 2013
17. Gallus D, Abecker A, Richter D (2008) Classification of landslide hazard in the development of early warning systems, Lecture notes in Cartography and Geoinformatics. Springer ISBN: 978-3-540-78945-1
18. Pradhan B, Oh HJ, Buchroithner MF (2010) Weights-evidence model applied to landslide susceptibility mapping in a tropical hilly area, geomatics. Nat Hazards Risk 1(3):199–223
19. Nittel S (2009) A survey of geosensor networks: advances in dynamic environmental monitoring. Sensors 2009(9):5664–5678
20. SWE (2013) http://www.opengeospatial.org/ogc/markets-technologies/swe. Accessed 19 Dec 2013
21. Kandawasvika A (2009) On interoperable management of multi-sensors in landslide monitoring applications. Dissertation (Ph.D. thesis), University of the Bundeswehr Munich
22. Walter (2012) Untersuchungen räumlicher Dienstestandards zur Umsetzung einer Geodateninfrastruktur eines sensorbasierten Frühwarnsystems. Dissertation (Ph.D. thesis), Universität Rostock, http://dgk.badw.de/fileadmin/docs/c-691.pdf. Accessed 19 Dec 2013
23. ZKI (2013) http://www.zki.dlr.de/de. Accessed 19 Dec 2013

24. Reznik T (2013) Emergency support system management of geographic information for command and control systems, Habilitation thesis, accepted by the Czech University of Defence, Brno
25. ESS (2013) http://www.ess-project.eu/. Accessed 19 Dec 2013
26. Wang and Yuan (2010) Challenges of the sensor web for disaster management. Int J Digit Earth 3(3):260–279
27. Konecny M, Reinhardt W (Guest editors) (2010a). Special issue on early warning and disaster management: the importance of geographic information (Part A). Int J Digit Earth 3(3)
28. Konecny M, Reinhardt W (Guest editors) (2010b) Special issue on early warning and disaster management: the importance of geographic information (Part B). Int J Digit Earth 3(4)
29. O'Sullivan D, Perry GLW (2013) Spatial simulation: exploring pattern and process. Wiley-Blackwell, West Sussex
30. Dransch D, Rotzoll H, Poser K (2010) The contribution of maps to the challenges of risk communication to the public. Int J Digit Earth 3(3):292–311
31. Nuhn E, Reinhardt W (2011) Coupling geoinformation and simulation systems for the early warning of landslides. Appl Geomat 3(2):101–107
32. Nuhn E, Kropat E, Reinhardt W, Pickl S (2012) Preparation of complex landslide simulation results with clustering approaches for decision support and early warning, hicss. In: 45th Hawaii international conference on system sciences, pp 1089–1096
33. Trauner FX, Ortlieb E, Boley C (2009) A coupled geoinformation and simulation system for landslide early warning systems. In: 6th EUREGEO congress Munich 2009, proceedings volume 1
34. Breunig M, Schilberg B, Kuper PV, Jahn M, Reinhardt W, Nuhn E, Mäs S, Boley C, Trauner F-X, Wiesel J, Richter D, Abecker A, Gallus D, Kazakos W, Bartels A (2009) EGIFF – developing advanced GI methods for early warning in mass movement scenarios. In: Stronk (2009) Geotechnologien Science Report, ISSN 1619-7399
35. Merz B, Disse M, Günther K, Schumann A (eds) (2010) Risk management of extreme flood events, NHESS – special issue, http://www.nat-hazards-earth-syst-sci.net/special_issue82.html. Accessed 19 Dec 2013
36. Fischer K, Riedel W, Häring I, Nieuwenhuijs A, Crabbe S, Trojaborg S, Hynes W, Müller I (2012) Vulnerability identification and resilience enhancements of urban environments. In: EURA conference 2012, Vienna, http://link.springer.com/chapter/10.1007/978-3-642-33161-9_24#page-1. Accessed 19 Dec 2013
37. Howe J (2008) Crowdsourcing: why the power of the crowd is driving the future of business. Crown Business, New York
38. Ushahidi (2013) http://www.ushahidi.com/. Accessed 19 Dec 2013
39. Morrow N, Mock N, Papendieck A, Kocmich N (2011) Independent evaluation of the ushahidi project, http://ggs684.pbworks.com/w/file/fetch/60819963/1282.pdf. Accessed 19 Dec 2013
40. TranSAFE-Alp (2013) http://www.transafe-alp.eu/. Accessed 19 Dec 2013
41. ASP (2013) www.alpinespace.org. Accessed 19 Dec 2013
42. Reinhardt W, Gálicz E, Hossain I (2013) Klassifizierung von GIS-Funktionalitäten im Bereich des Katastrophenmanagements. 17. Internationale geodätische Woche Obergurgl 2013, Wichmann
43. Gálicz E, Hossain I, Reinhardt W (2013) Geo Web Services for transport crisis management in alpine region. In: Proceedings of international cartography conference 2013, Dresden

Chapter 15
The Potential of Cloud Computing for Analysis and Finding Solutions in Disasters

Mitica Craus and Cristian Butincu

Abstract In this paper we discuss the potential of cloud computing to deliver services for the study of disasters and provide solutions for mitigation of the effects. Our investigation is focused on cloud services based on data mining techniques and nature inspired heuristics. By estimating the evolution of a disaster or a complex phenomenon that can produce disasters, based on history of similar events, data mining can help to find the best decision to mitigate the effects. Nature inspired heuristics could be useful for finding solutions for influencing the evolution of such events and simulation of defense against adverse effects, when the change of the evolution is not possible.

15.1 Introduction

Cloud computing is a new technology with high potential to deliver services for study of disasters' evolution or phenomena, natural, technological or social, which can generate disasters and to provide solutions that can reduce their effects.

Globalization complicates the problems that the humankind has to solve in order to avoid malfunctions. Global systems and global links dominate this period. Modelling such systems is absolutely necessary to study their evolution and the generated effects. Often these effects, if they are not prevented, can be disastrous. Often, globalization produces chains of disasters. For example, melting glaciers can cause massive flooding and extinction of some species; an effect of the huge flooding could be little food and famine.

M. Craus (✉) • C. Butincu
Gheorghe Asachi Technical University of Iasi, D. Mangeron Bd., 700050 Iasi, Romania
e-mail: craus@cs.tuiasi.ro; cbutincu@cs.tuiasi.ro

H.-N. Teodorescu et al. (eds.), *Improving Disaster Resilience and Mitigation - IT Means and Tools*, NATO Science for Peace and Security Series C: Environmental Security,
DOI 10.1007/978-94-017-9136-6_15,

The study of a complex system requires very refined models. Such models contain many variables, making it impossible to study their evolution without adequate computing environments and without tools for modification in a dynamic manner of model parameters. Most of the time, simulation applications require intervention over the used resources.

The pressure of the complex problems derived from technological progress and globalization, determines the ICT world to build appropriate hardware and software. Parallel and distributed computing systems could be a solution. Multicomputer systems are easily built from the existing hardware resources while the multiprocessor systems are often expensive, but the performance is higher. Building software for these types of systems is not simple and adaptation requires an extensive effort from specialists. The parallel and distributed programming is fundamentally different from the sequentially one. Programmers need to do something more so that the specific technologies of the parallel and distributed computing can become more accessible to those who need this computing power, potentially unlimited.

Nowadays, more and more researchers choose Grid and Cloud as the platform for modeling and simulation of complex processes they confront. The focus on resource concept makes grids deficient in the services area. Cloud technologies tend to correct this deficiency of grids through orientation mainly towards services.

Often, real-time intervention to change the route of the execution of an application or its allocated resources is necessary. An application for simulating the evolution of a complex phenomenon must provide tools for driving the model to situations that may cause unknown effects. The study of these effects is extremely important to prevent unwanted developments and to build mitigation mechanisms. Steering the computation is of great interest due to globalization of the natural, technological and social phenomena.

15.2 Related Work

Several efforts were made in order improve the efficiency of acquiring and sending data (sensor data and/or command data) over wireless networks. One algorithm that is based on dynamic ant colonies is presented in [1]. According to the authors, this new algorithm, SAMP-DSR, supports two operating modes (local and global). Depending on the wireless network topology change events and their magnitude, the algorithm switches between these two modes, ensuring a quick adaptation to the dynamics of the network environment. This kind of algorithms that efficiently route data through wireless networks are extremely important to maintain a constant dataflow in case of disaster, allowing the authorities to take immediate actions.

Disaster recovery as a service (DRaaS) represents a set of processes that are offered by third-party vendors. These processes come to the aid of anyone (business, authority etc.) that needs a disaster recovery plan (DRP). Such a plan is also called a business continuity plan (BCP) or business process contingency plan (BPCP). The plan consists of a set of measures intended to mitigate the effects of disasters and to maintain and resume vital functions of the organizations as quickly as possible.

Although hardware failures and human errors are still the leading causes of unplanned outages, other events (such as storms, earthquakes etc.) are also forcing the industry to take serious steps in implementing a disaster recovery plan [2].

Several methods of flood disaster assessment were developed in the past years. An improved method, presented in [3], is based on cloud model and fuzzy certainty degree. According to these authors, this improved cloud model method is confirmed to be a reliable method for rapid disaster assessment.

When coming to modelling natural disasters, another approach uses social media networks in order to acquire sensor data into the cloud. Combined with other sensor data networks, the data acquired from social network can provide critical information for build a global view of the disaster evolution in real time. All these information can be used to control and minimize the effects of disasters. The paper [4] presents such an approach and as a use case scenario the authors analyze the Deepwater Horizon oil spill disaster.

Based on the facts that the frequency of natural disasters is expected to increase in the future, the paper [5] analyzes the use of cloud computing as a possible integration component in the emergency management. The target is to provide the needed support for business continuity in case a disaster strikes.

A system design that combines devices, autonomous sensors through XMPP and cloud services is presented in [6]. The authors recognize the demand for monitoring and communication systems to support post-disaster management and the fact that such systems can be effectively developed by merging various device classes and wireless communication strategies.

Price [7] proposes a method to observe and track distant thunderstorms using ground networks of sensors. These sensors detect the radio waves emitted by each lighting discharge and can also provide a wide variety of metrics such as polarity of the discharge, peak currents, charge etc. The author observed that changes in some of these metrics are often related to changes in the severity of the storms.

In an effort to accurately monitor and predict weather changes in China, an operational model for weather modification that predicts the microphysical features of cloud and precipitation and helps locate possible seeding areas was proposed in [8].

The paper [9] presents the development of an integrated computer support for emergency care processes by evolving and cross-linking institutional healthcare systems. The authors developed an integrated EMS (Emergency Medical Service) cloud-based architecture that allows authorized users to access emergency case information in standardized document form.

The paper [10] presents an approach based on the use of geo-information and remote sensing in order to increase the awareness on natural hazard prevention and for supporting research and operational activities devoted to disaster reduction.

By using cloud computing and application of fuzzy logic models, the paper [11] proposes an integrated approach for risk assessment of natural disasters. The authors designed a hierarchical fuzzy logic system with several inputs and one output. They defined the risk assessment problem as a multicriterial task that evaluates input variables (natural disaster indicators) and takes into account available information sources and the expert knowledge.

A new cloud computing innovation is being tested by IBM and Marist College [12] and is targeted to assist organizations in preventing disruptions in their data services when a hurricane or other natural disaster occurs. This innovation uses software-defined networking (SDN) technology that allows data center operators to efficiently control data flows across physical and virtual networks. By using this technology, the time needed to move data services to a safe location (that usually takes several days) is measured in minutes.

In 2013, EVault announced a Cloud Disaster Recovery Service, a managed service intended to help organizations to quickly recover their critical systems after a disaster [13].

15.3 Cloud Service for Disaster Management

Cloud computing offers three types of services:

- Application access – SaaS (Software-as-a-Service): Google Apps, Salesforce.com, WebEx;
- Application development platforms – PaaS (Platform-as-a-Service): Coghead, Google App Engine;
- Virtual machines (computing and data storage infrastructures) – IaaS (Infrastructure-as-a-Service): Amazon AWS, Microsoft Azurem, Google Compute Engine.

In our opinion, a cloud service for disaster management must include:

- Big data-bases;
- Dynamic allocation of resources;
- Libraries for data analysis, optimization, simulation;
- Interfaces for resource access, computational steering;
- Tools for visualization.

We consider that the most important advantages of cloud services are the following:

- They aggregate different resources as services.
- Cloud services are accessible when they are necessary.
- They provide fault tolerance because of virtualization and replication.
- They are safe and secure.
- No previous costs are needed.

15.3.1 A Possible Architecture for Cloud Services Used in Disasters

Figure 15.1 illustrates our proposal for a cloud architecture that can be used to implement a series of cloud services needed to mitigate the effects of disasters.

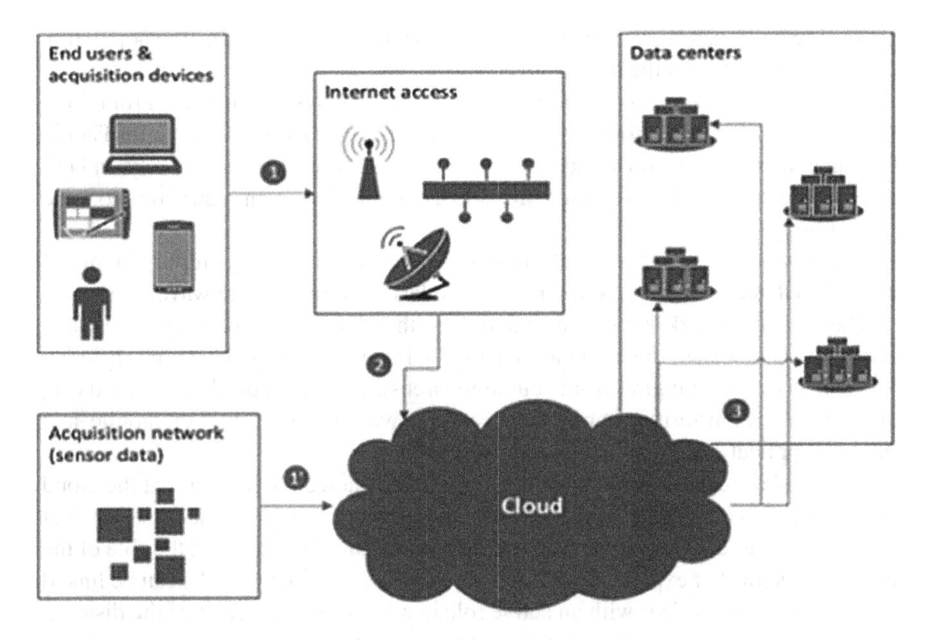

Fig. 15.1 Cloud service architecture that can be used in case of disasters

In case of disasters, end-users can access cloud services through the internet (see arrows 1, 2 in Fig. 15.1) to find an immediate solution to their current situation, but more importantly they can also provide valuable data to the cloud to be analyzed in real-time and to help other users that are also affected.

The cloud can optionally be connected to an acquisition network (1') in order to acquire as much data as possible for the current situation. The acquisition network is composed of a series of hardware sensors (pressure, temperature etc.) and video camera feeds. Moreover, it can also provide information from existing social networks (Facebook, Twitter, etc.) in real time. Based on the data acquired from end-users and from the acquisition network, the cloud software can make informed decisions about how to help its users.

Even if some network paths to the cloud become unavailable, routing methodologies such as anycast can be deployed in order to ensure full availability of cluster services to the end-users, which will be directed to the closest available datacenter (3).

15.3.2 Case Study – Tsunami Earthquake

In case of an earthquake that generates a tsunami, the cloud can provide the means to analyze and make decisions in real-time in order to mitigate the impact of the tsunami over the boats along the coastline and over the population living near the shore. The boats can be directed to a secure zone in the sea/ocean and the population can be evacuated from the danger area.

Several international and regional tsunami warning systems operate around the world (United States National Oceanic and Atmospheric Administration for Pacific Ocean, United Nations for Indian Ocean etc.). These warning systems rely on both a network of sensors to detect tsunamis and a communication infrastructure to issue the warnings.

The network of sensors can vary from case to case, but usually employs a mix of sensors to detect earthquakes, to measure sea level and to measure wave energies.

The sensor network sends real-time data to the cloud. The data is analyzed, and if a threat is detected, the cloud automatically issues warnings. There are different types of warnings, ranging from sounding sirens across the coastline, to notifying authorities over priority channels, to sending warning messages to mobile and desktop applications.

This kind of reaction from the cloud can be considered passive, in that the cloud does not take active measures to mitigate the effects of the potential disaster, that is, it only issues a series of warnings on different channels. However, the role of the cloud can be further expanded and more sensors and control networks can be linked to it in order to provide it with an active role in mitigating the effects of the disaster.

One possible active component can be a control network over the transportation system (traffic lights, highway signals, public transportation time-tables etc.). The main role of this active component is to streamline the traffic and to avoid bottlenecks by constantly monitoring the traffic and by taking informed decisions based on real-time traffic information acquired from a supplementary sensor network (real-time satellite images, local area traffic sensors, public transportation data etc.). The target of this active control component is to evacuate as many people as possible from the danger area by maintaining a steady flow of transportation.

The aforementioned active component can be used in a wide range of disaster mitigation systems such the ones employed in volcano eruptions, terrorist attacks, nuclear power plant failures and any other event that requires mass evacuation of population from a designated danger area in a timely manner.

15.4 Cloud Services Based on Data Mining

Cloud services are ideal for offering users the facilities to use data mining techniques according to their needs. In case of disaster, analyzing the history of similar events is essential for offering solutions to mitigate the effects. Data mining can help to find the best decision to react in case of disasters. Only in a cloud it is possible to accumulate the world experience on complex problems such as preventing or reacting to a disaster.

Data mining consists of:

- data pre-processing through interactive exploration, data cleansing and consolidation, as well as formatting for specific data analysis algorithms;

- applying data analysis algorithms and performing various simulations;
- information validation and presentation to final users.

The main methods for data analysis are association rule discovery, classification, data clustering, fuzzy data mining and interactive data analysis

Association rule discovery implies two stages: finding frequent itemsets and testing frequent itemsets in order to discover strong associations. The most important algorithms are Apriori [14] and Frequent Pattern Growth [15, 27]. The corresponding parallel algorithms (Hash Partitioned Apriori, Multiple Local Frequent Pattern Tree) and, our algorithm Parallel Fast Itemset Miner [16, 17] are very adequate in a cloud service because of the potential of a cloud to run parallel applications, which could dramatically improve the response time, very important in disasters. For example, analyzing the most frequent actions of the population and the association between these actions in case of different past disasters (earthquake, flood, landslides, hurricane, and tsunami), can provide new information for mitigation of the bad effects.

Classification means the process of assigning an element of a set to a class defined in advance, based on known criteria. The most popular technique used in classification is based on the decision trees and the main algorithms in this respect are: ID3 [18], C4.5, C5.0 (improvements of ID3) and SLIQ [19]. A disaster can be included in a class for which exist strategies for mitigation. Then, the decisions can be taken in real time.

Clustering can be defined as follows: a given data set is partitioned in a set of classes that contain elements with common characteristics according to a specified metric function. The most popular algorithms are K-Means [20], CLARANS, PAM, CLARA and FGKA In the case of a war, the current conflict and the past conflict evolutions stored in a big database can be clustered. Depending on the cluster on which the current conflict belongs, decisions can be found that can help to settle the conflict.

Fuzzy data mining operates with fuzzy attributes. Many variables used in analysis of the phenomena, which can generate disasters, have fuzzy features, being suited for modeling employing fuzzy methods. Various algorithms in this field are highly complex, employing a high computational effort. Parallel computing is suitable for such algorithms and the cloud is ideal to offer the necessary virtual resources [21].

15.5 Cloud Services Based on Nature Inspired Heuristics

Cloud services based on nature inspired heuristics could be useful for determination of solutions for influencing the evolution of a complex phenomenon that can produce disasters. When a disaster is ongoing, the simulation of resilience and mitigation could be done by the help of such services. Cloud computing is ideal for implementing nature inspired heuristics because such techniques need large amount of computing and storage resources.

The nature inspired researchers in their efforts to find new methods to solve complex problems for which traditional approaches proved not to be adequate. The most important heuristics inspired from nature are:

- Ant Colony Optimization (ACO): inspired by ant colonies behavior.
- Particle Swarm Optimization (PSO): a class of heuristics inspired by flocks of birds or schools of fish.

15.5.1 Ant Colony Optimization

Ant algorithms were inspired by the observation of real ant colonies, as social insects: they are able to find shortest paths between food and their nest. Starting from ant behavior observations, the researcher Marco Dorigo [22, 23] proposed a method of solving optimization problems, called Ant Colony Optimization (ACO). ACO applied to various problems of optimal paths can be useful to solve the evacuation of people and moving significant resources in a disaster.

ACO's major idea is best illustrated by applying Ant System algorithm to Travelling Salesman Problem (TSP) [22, 24]. Given a set of n cities and a set of distances between them, the Traveling Salesman Problem (*TSP*) is the problem of finding a minimum length closed path (a *tour*), which visits every city exactly once. Thus, we have to minimize:

$$COST\,(i_1, i_2, \ldots, i_n) = \sum_{j=1}^{n-1} d\left(C_{i_j}, C_{i_{j+1}}\right) + d\,(C_{i\,n}, C_{i_1})$$

where $d(C_x, C_y)$ is the distance between cities C_x and C_y.

15.5.1.1 Ant Actions

Each ant moves from city to city until it completes a Hamiltonian tour, trailing their way

$b_i(t)$ = the number of ants in city i at time t, $i = 1,n$.

$m = \sum_{i=1}^{n} b_i(t)$ = the total number of ants

If $\tau_{ij}(t) =$ the intensity of the trail on the connection (i,j) then $\tau_{ij}(t+n) = (1-\rho)\tau_{ij}(t) + \Delta_{ij}(t, t+n)$, ρ $(0 < \rho < 1)$ is a coefficient representing pheromone evaporation.

15.5.1.2 Trail Modification

$$\Delta_{ij}\left(t, t+n\right) = \sum_{k=1}^{m} \Delta_{ij}^{k}\left(t, t+n\right)$$

where $\Delta_{ij}^{k}(t, t+n)$ is the quantity per unit of length of pheromone laid by ant k on connection *(i,j)* at time $t+n$ and is given by:

$$\Delta_{ij}{}^{k}\left(t, t+n\right) = \begin{cases} \frac{Q}{L_k}, \ if\ ant\ k\ uses\ edge\ \ (i, j)\ \ in\ \ its\ tour \\ 0, otherwise \end{cases}$$

where Q is a constant and L_k is the tour length found by the kth ant

15.5.1.3 The Probability that an Ant k in City i Moves to City j

$$P_{ij}{}^{k}(t) = \begin{cases} \dfrac{\left[\tau_{ij}(t)\right]^{\alpha}\left[\eta_{ij}\right]^{\beta}}{\sum\limits_{h \in allowed_k(t)} \left[\tau_{ih}(t)\right]^{\alpha}\left[\eta_{ih}\right]^{\beta}}, \ if\ j\ \in allowed_k(t) \\ 0, otherwise \end{cases}$$

where

allowed$_k$(t) is the set of cities not visited by ant k at time t,
η_{ij} is a local heuristic; for TSP it's called visibility: $\eta_{ij} = 1/d(C_i, C_j)$
α, β control the relative importance of pheromone trail versus visibility.

15.5.2 Particle Swarm Optimization

Particle Swarm Optimization (PSO) is an optimization heuristic that finds solutions by using multiple iterations. At each iteration the candidate solution is improved.

PSO has proven its efficiency in a large category of optimization problems, and hundreds of papers were written to report successful applications of PSO [25].

Initially, the algorithm generates a population of candidate solutions. In PSO terms, each candidate solution is called a "particle". The algorithm moves the particles into the search space according to their position and velocity. The movement of a particle is influenced by both its local best-known position and by the best- known positions found into the search space. These best-known positions are continuously updated as they are discovered by the algorithm. However, the original PSO algorithm can prematurely converge into a local solution. To overcome this drawback, some improvements were proposed to the original algorithm, like employing the idea behind a simulated annealing algorithm [26], or using simplified variations like in [28].

PSO evaluates candidate solutions (particles) the same way genetic algorithms do. That is, a cost function $f : \Re^n \rightarrow \Re$ is provided that takes a candidate solution (particle) as argument and computes the effective cost of that candidate solution. Depending on the solution search type (minimization or maximization), the goal is to find a solution for which the cost is less than or equal (or greater than or equal) to the cost of all other particles in the search space.

The algorithm generates an initial set S of particles using random values, according to some random distribution function. These random values are used to set, for each particle "i" from the set S, a position $x_i \in \Re^n$ into the search space and an initial velocity $v_i \in \Re^n$. In the algorithm, each particle holds its best known position "p_i" found so far. Also, as the algorithm runs, it keeps the best known position "g" of the entire swarm.

Similar to genetic algorithms, the PSO algorithm runs until a certain criterion is met, like the maximum number of iterations was reached, or a solution with adequate cost was found, or the quality of the best swarm solution did not improve over a certain number of iterations.

During each iteration, the PSO algorithm changes the velocity of each particle "i" according to following formula:

$$v_{i,d} = wv_{i,d} + \varphi_p r_p (p_{i,d} - x_{i,d}) + \varphi_g r_g (g_d - x_{i,d})$$

where w is an inertia factor, d is a particular dimension, φ_p and φ_g are accelerating coefficients and r_p and r_g are random values between 0 and 1. Thus, the new velocity is computed based on the previous particle velocity, the particle's best position and the swarm's best position. After this step, the new position for the particle is computed:

$$x_i = x_i + v_i$$

After the new positions for all particles in the current iteration are computed, the PSO algorithm evaluates the particles and, if it is the case, updates the best particle position and the best swarm position. At this point, if the termination criterion is met the algorithm terminates, otherwise, a new iteration is started.

15.5.3 *Case Studies*

15.5.3.1 Quick Inspection of Critical Locations

Let us suppose that authorities want to visit the critical points in an affected area: emergency dispatch centers, fire stations, police agencies and military barracks, water supply stations, hospitals, schools, electrical plants, thermal plants, supermarkets. Several mobile devices can access a cloud service to determine the shortest Hamiltonian tour, by means of ACO technique. So, they can obtain a map of displacement and through GPS devices they can follow this route. If some streets are blocked, they can ask cloud services to build another Hamiltonian shortest tour that avoids blocked areas. All these computations can be made in real time, because of the cloud power and the efficiency of the ACO method.

15.5.3.2 Supply Efficiency

Similarly to the case presented before, in order to efficiently supply, the map of the affected area can be partitioned and every area can be the object of conversion into a graph for which an ACO-TSP algorithm is applied. Having a Hamiltonian tour, a convoy of cars can assure the supply of food, water, medicines and other essential things.

15.5.3.3 Simulations for Mass Evacuation in Case of Disaster

Software simulations can be used to predict and analyze how the population of an entire area (mall, city, state etc.) can be evacuated in a short period of time. The results of such simulations provide an efficient evacuation plan so that as much people as possible can be evacuated successfully in a short period of time.

A possible example consists of the evacuation of an entire city due to a terrorist attack with a dirty bomb. The erratic behavior of people during such events can cause high losses in terms of people lives. Large and chaotic clusters of people usually form and hinder each other while trying to reach safe areas.

The target of simulating and analyzing such an event is to create a sequence of carefully planned actions, synchronized with crowd movements in order to avoid bottlenecks and to direct as much people as possible out of the danger zone. Some of these actions can be controlled automatically directly from the cloud and without human intervention (traffic lights, delayed warning signals in different zones of the evacuation area etc.). Other actions require human intervention, usually from the authorities, and range from setting up road blocks (in order to avoid two or more clusters of people to hinder each other) and clearing paths for crowds by using explosives or other equipment in order to provide shorter routes to safe areas.

These simulations use software agents to simulate people, and a mathematical model based on swarm behavior is used to predict crowd movements. Using the map of the area needed to be evacuated, the agents are placed on the map according to the statistical distribution of the population. Several random variables control the simulation, and for each scenario, a sequence of actions is being computed.

Using this approach, the cloud software generates a sequence of actions to be used ahead of time for each possible scenario. When the actual disaster occurs, the cloud automatically monitors crowd movements based on the data it acquires from a sensor network and it triggers the execution of the correct sequence of actions (computed ahead of time). All these actions are taken in real time. However, the chosen scenario may not fit exactly to the actual crowd movements, so the cloud continuously monitors the entire process of evacuation; if it detects a mismatch between the current and the predicted crowd movements, it simply switches to another scenario that matches the current situation. Thus, a new sequence of actions is chosen (also computed ahead of time) and the process continues until the entire population is evacuated.

Given the fact that the solution space for these scenarios is usually extremely large, an exhaustive search over the entire space is not feasible. Thus the cloud software usually employs different techniques to compute the shortest and/or safest evacuation paths on the map and will try to enforce these paths into the crowd by using a series of control networks (traffic control, road blocks etc.). Ant colony optimization algorithms (ACO) are good candidates that use a probabilistic approach for finding paths close to optimum paths. Another approach can be the use of particle swarm optimization algorithms (PSO). Given an initial set of paths, the PSO algorithm tries to improve the results by moving the swarm of solutions across the solution space iteratively, toward better solutions. These techniques can also be easily parallelized to take full advantage of cloud computing power.

15.6 Conclusions

Cloud services could help authorities and people to improve the resilience to disasters, but also to prepare for mitigating disaster effects. Data mining is a very powerful set of techniques that can be used to analyze the humankind experience during similar disaster occurrences and so to find better solutions for the current events. Nature inspired heuristics can help simulate disasters in advance and can be used to obtain efficient solutions for mitigation and resilience to these events. The cloud provides the best hosting environment for implementing and accessing such tools. High availability services combined with two of the most predominant features found in clouds, virtualization of computing resources and parallel computing potential, make the cloud an ICT leader for disaster prevention, disaster mitigation and improvement of resilience to disaster.

References

1. Ehsan KA, Majid N, Atieh SP (2011) A dynamic ant colony based routing algorithm for mobile Ad-hoc networks. J Info Sci Eng 27:1581–1596
2. Geng S (2013) DaaS, MaaS & DRaaS: the next phase of cloud computing. http://readwrite.com/2013/03/29/the-next-phase-of-cloud-computing-daas-maas-draas

3. Deng W, Zhou J, Zou Q, Zhang Y, Hua W (2013) Improved flood disaster assessment method based on cloud model and fuzzy certainty degree. Info Technol J 12(10):2064–2068
4. Aulov O, Halem M (2012) Human sensor networks for improved modeling of natural disasters. Proc IEEE 100(10):2812–2823
5. Velev D, Zlateva P (2012) A feasibility analysis of emergency management with cloud computing integration. Int J Innov Manag Technol 3(2):188
6. Klauck R, Kirsche M (2013) Combining mobile XMPP entities and cloud services for collaborative post-disaster management in hybrid network environments. Mobile Netw Appl 18(2):253–270
7. Price C (2008) Lightning sensors for observing, tracking and nowcasting severe weather. J Sens 8(1):157–170
8. Lou XF, Shi YQ, Sun J, Xue LL, Hu ZJ, Fang W, Liu WG (2012) Cloud-resolving model for weather modification in China. Chin Sci Bull 57(9):1055–1061
9. Poulymenopoulou M, Malamateniou F, Vassilacopoulos G (2012) Emergency healthcare process automation using mobile computing and cloud services. J Med Syst 36(5):3233–3241
10. Giardino M, Perotti L, Lanfranco M, Perrone G (2012) GIS and geomatics for disaster management and emergency relief: a proactive response to natural hazards. Appl Geomat 4(1):33–46
11. Zlateva P, Hirokawa Y, Velev D (2013) An integrated approach for risk assessment of natural disasters using cloud computing. Int J Trade Econ Finance 4(3):134–138
12. http://www-03.ibm.com/press/us/en/pressrelease/42523.wss (2013) Made in IBM labs: testing cloud invention to prevent natural disaster outages: IBM and Marist College Innovation Could Avert Costly Business Disruptions
13. http://www.businesswire.com/news/home/20130416005055/en/EVault-Launches-Industry-Pioneering-Cloud-Disaster-Recovery-Service#.UsGQE7RkwSo (2013) EVault launches industry-pioneering Cloud Disaster Recovery Service in EMEA
14. Agrawal R, Srikant R (1994) Fast algorithms for mining association rules in large databases. In: Proceedings of the 20th international conference on very large data bases, Santiago, September 1994, pp 487–499
15. Agarwal R, Agarwal C, Prasad VVV (2000) A tree projection algorithm for generation of frequent itemsets. J Parallel Distrib Comput (Special issue on High Performance Data Mining)
16. Craus M (2008) A new parallel algorithm for the frequent itemset mining problem. In: Proceedings of the 7th international symposium on parallel and distributed computing (ISPDC 2008), IEEE Computer Society, Krakow, 1–5 July 2008, pp 165–170
17. Craus M, Archip A (2008) A generalized parallel algorithm for frequent itemset mining. In: Recent advances in computer engineering, New Aspects of Computers
18. Quinlan JR (1986) Induction of decision trees. Mach Learn 1:81–106
19. Mehta M, Agrawal R, Rissanen J (1996) SLIQ: a fast scalable classifier for data mining. In: Advances in database technology – EDBT'96. LNCS, vol. 1057, pp 18–32
20. MacQueen JB (1967) Some methods for classification and analysis of multivariate observations. In: Proceedings of 5-th Berkeley symposium on mathematical statistics and probability, vol 1. University of California Press, Berkeley, pp 281–297
21. Kandell A, Tamir D, Rishe N (2014) Fuzzy logic and data mining in disaster mitigation, improving disaster resilience and mitigation – new means and tools, trends, ARW 984631, Iaşi, 6–8 Nov 2013
22. Dorigo M, Gambardella LM (1997) Ant colony system: a cooperative learning approach to the traveling salesman problem. IEEE Trans Evol Comput 1(1):53–66
23. Dorigo M, Caro GD (1999) Ant algorithms for discrete optimization. Artif Life 5:137–172
24. Craus M, Rudeanu L (2005) Parallel framework for cooperative processes. Sci Program 13(3):205–217, IOS Press
25. Poli R (2008) Analysis of the publications on the applications of particle swarm optimisation. J Artif Evol Appl, 10 pages

26. Mu A, Cao D, Wang X (2009) A modified particle swarm optimization algorithm. Nat Sci 1:151–155
27. Han J, Pei J, Yin Y, Mao R (2004) Mining frequent patterns without candidate generation: a frequent-pattern tree approach. Data Min Knowl Discov 8:53–87
28. Pedersen MEH, Chipperfield AJ (2010) Simplifying particle swarm optimization. Appl Soft Comput 10:618–628

Chapter 16
How to Improve the Reactiveness and Efficiency of Embedded Multi-core Systems by Use of Probabilistic Simulation and Optimization Techniques

Jürgen Mottok, Martin Alfranseder, Stefan Schmidhuber, Matthias Mucha, and Andreas Sailer

Abstract Safe and reliable multi-core technology becomes more and more important in the field of embedded systems. Today's and future embedded systems require increasing performance while being more energy efficient. Moreover, the functional safety for these embedded systems has to be improved or developed completely new. In this chapter, we first address the challenges of embedded multi-core real-time systems. To raise the resilience of such systems we use the deadlock-free synchronization model of Block et al. (A flexible real-time locking protocol for multiprocessors. 2012 IEEE international conference on embedded and real-time computing systems and applications, vol 0, pp 47–56, 2007). The metric mean Normalized Blocking Time (mNBT) is hereby used to measure the timing effects of the blocking behavior of strongly interacting tasks. In a second step, we present a model-based approach to map the tasks of an embedded real-time system to the cores of a multi-core processor. Moreover, we derive an execution time model from runtime measurements of software functions. This information is then used to perform precise probabilistic simulations of different task-to-core mappings and evaluate them with regard to task response times, inter-task blocking overhead and load distribution. Subsequently, we integrate the probabilistic simulation within an optimization technique to systematically improve the task-to-core mapping. We conclude with a case-study, where we demonstrate the effectiveness of the presented approach by optimizing the task-to-core mapping of a practical automotive power-train system.

J. Mottok (✉) • M. Alfranseder • S. Schmidhuber • M. Mucha • A. Sailer
Laboratory for Safe and Secure Systems LaS³ (www.las3.de), Faculty Electro- and Information-Technology, Ostbayerische Technische Hochschule Regensburg, Regensburg, Germany
e-mail: juergen.mottok@oth-regensburg.de; martin.alfranseder@oth-regensburg.de; stefan.schmidhuber@oth-regensburg.de; matthias.mucha@oth-regensburg.de; andreas.sailer@oth-regensburg.de

H.-N. Teodorescu et al. (eds.), *Improving Disaster Resilience and Mitigation - IT Means and Tools*, NATO Science for Peace and Security Series C: Environmental Security, DOI 10.1007/978-94-017-9136-6_16,

16.1 Introduction

16.1.1 Laboratory for Safe and Secure Systems (LaS³)

As a laboratory for software engineering, the Laboratory for Safe and Secure Systems (LaS^3) conducts research in the fields of safety and software intensive systems. It focuses on new concepts for data security and software-based functional safety as well as to optimize architectures in embedded systems to raise their performance, functional safety, security and availability.

16.1.2 Software Engineering

Software Engineering describes the complete software production process. This includes all steps from the system specification to the maintenance (support) of the developed software product. It provides the production process with tools to achieve a required quality. In the case of resilience and mitigation software engineering has influence on the safety and security of real-time systems.[1]

A systematic software production process is needed to produce fail-safe systems in a fast way and to reduce production costs in long term development processes [1, pp. 7–10].

Software products can be characterized by their attributes in the following way as [1, p. 8] showed:

- Maintainability,
- Dependability and security,
- Efficiency and
- Acceptability.

Systematic software production needs *process models* (software paradigms) to get an architectural view of what happens during the product phases, to control them and to react if something fails during a phase. Such process models represent the software production life cycle. Three well-known process models are the waterfall model introduced by Royce [2], the spiral model introduced by Boehm [3] and the V-Model [4, pp. 25–29]. Functional Safety is the software engineering part for safety technical systems and discussed in the next section.

16.1.3 Functional Safety

Serious consequences can arise if a hazardous failure occurs that a technical system produced. Such consequences range from accidents over deaths up to large-scaled

[1] See Sect. 16.1.4

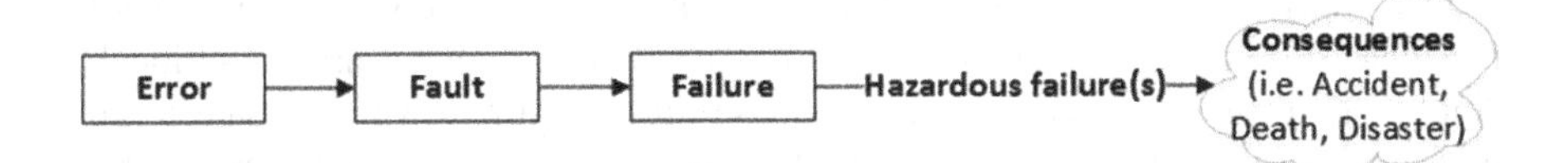

Fig. 16.1 Error-fault-failure chain for technical systems

disaster. But a failure is not a failure at its initial state – it's an error. An *error* can be described as a mistake which occurs during a development phase and may change to a fault. A *fault* is the reason why the system fails. So a *failure* can (but not necessarily) follow out of a fault. This is the state where the system does not execute its specified functions. Their relationships are shown in Fig. 16.1. Functional Safety identifies these failures and categorizes them into different safety levels. A safety level is a maximum tolerable range to measure the appearance of a risk (failure). This means each system that leads to failures which in sum are hazardous, is called *safety-related system*. Moreover, a *safety-critical* system is a system, where one failure alone makes a system hazardous. Goals that can be derived out of safety-related, respectively safety-critical systems are to minimize the appearance of failures, or to determine the frequency of the appearance of failures. Such systems are categorized into safety levels [5, p. 7].

A big influence has the *ISO 26262* standard for functional safety in the automotive field. It was first introduced in 2011 and is a variant for road vehicles of the IEC 61508. The ISO 26262 addresses the safety of electrical, electronic, and software parts of road vehicles. It contains risk analyses and safety goals to classify safety-related/ safety-critical systems into so called *Automotive Safety Integrity Levels* (ASILs). ASIL owns the four categories A, B, C and D, with A as the lowest and D as the highest level [6, p. 1], [28].

Safely Embedded Software (SES) as another functional safety part is placed in the application software level where it guarantees the safety of the complete system. With this approach hardware redundancy is reduced due to a diverse redundancy in the software. SES especially is used to achieve for example ASIL D in technical systems [7, p. 1]. Safety-related/safety-critical systems and with it ASIL A to D are crucial for real-time systems. The next section discusses real-time system in more detail.

16.1.4 Real-Time Systems

A *real-time system* depends on its logical results and on the temporal generation frequency of these results. Kopetz classified system requirements in functional, temporal and dependability requirements, where safety is integrated into dependability. Referring to real-time systems one distinguishes between *hard and soft real-time systems*. In hard real-time systems normally one or more safe states exist. If it has high error-detection coverage in terms of how fast a failure will be detected to reach a safe state, it is called fail-safe. In this case, the safety criterion of hard real-time systems is critical. Soft real-time systems in contrast are non-critical. Another comparable criterion is the response time constraint which is said to be hard for hard real-time systems and mild for soft real-time systems [8, pp. 1–16]. A variant of real-time systems are so called embedded real-time systems.

Embedded real-time systems are intelligent devices inside self-contained systems. They take user information and environmental conditions as input to process them for a desired output. Embedded real time systems are very limited related to the processing power, random access memory (RAM), programming ability and Input-/Output (I/O) devices. Operating systems for (embedded) real-time systems are called *Real-Time Operating Systems* (RTOS). Real-Time Operating Systems have to be distinguished between other operating systems. Two fundamental RTOS properties described by [8, p. 216] are

- deterministic which means to have predictable response times and
- fault-tolerant.

The four most used real-time operating systems are POSIX as the general standard, OSEK (automotive), APEX (avionics) and μITRON (for small systems) [9, p. 419], [8, pp. 18–19], [10, pp. 3–7].

Today embedded real-time systems in fields of automotive or avionics lack of computational capacity due to their increasing amount of information processing. For this reason a single-core processor cannot calculate all the needed information (in real-time) any longer for a given point in time. The change to *multi-core* processors overcomes this problem by the use of two or more cores in one embedded system. Like single-core embedded systems, multi-core embedded systems consist of a task set with several executable tasks. Hence a *scheduling policy* is needed to execute the tasks in their correct order at any point in time. In scheduling theory, task sets are distinguished between *static* and *dynamic*. Static task sets contain a fixed set of tasks with known task properties (i.e. deadline or execution time, execution order). Such task sets are non-preemptable. Dynamic task sets allow task *preemption* at runtime. Tasks can be interrupted (preempted) and later activated at runtime. This concept is useful in the view of schedulability if preemption is limited. In hard-real-time systems tasks have to meet their deadline. The *deadline* of a task is defined as the right uppermost time in which a task must finish its execution to avoid a false system behavior. If the deadline is missed, the system behavior can lead to disaster [9, pp. 23–24].

16.2 Case Study

16.2.1 Concept of the Case Study

Embedded real-time systems are one of the most important research fields of LaS[3]. Real-time, as described above, means that there are strict constraints so that all the functionality is done correctly and within a certain amount of time. If this timing constraint is not fulfilled, we say that the system is suffering from deadline violations. Nowadays the increasing functionality in many fields of embedded real-time systems is the reason to increase the frequency of processors more and more. But due to problems like power dissipation and heat generation, increasing the CPU frequency is not sufficient any more. A change to multi-core processors is inevitable in those cases.

With the help of a discrete, event-based simulator, which will be described in the following part of this chapter, we are able to simulate the runtime behavior of embedded multi-core real-time systems.

In particular, we want to apply our case study based on real existing approaches of real-time systems. As the research partners of our laboratory (LaS^3) are mainly from automotive industries, we are able to use information and create an abstract task model of an automotive powertrain system. Note that we are also able to simulate the timing behavior of any other type of real-time systems, too. Our task model is described within the next sections. In general, those automotive powertrain systems are utilized highly, so it is hard to find a scheduling solution that all deadlines can be met. In our case study we use a task-fix priority-, partitioned scheduling algorithm (OSEK-Scheduling) like it is mostly used in automotive systems. We assume that we can use a processor containing three symmetric cores.

A further point is that in practical embedded real-time systems tasks are dependent from each other (e.g. data-flow, event chains). That means that they have to share global memory resources. Synchronization mechanisms are mandatory to avoid simultaneous access to shared data. In single-core systems, this problem can be solved by a resource allocation protocol. But if multi-cores are used, this synchronization between the different cores leads to blocking overhead ("busy wait"). This overhead extends the general execution time of several tasks and may lead to deadline violations.

16.2.2 Goal of the Case Study

In our case study we want to optimize the most important issues of real-time systems. At first, the reactiveness of the systems has to be optimized. That means that all tasks should be terminated as early as possible. Followed by this, all deadlines of all tasks should be kept in every point in time (real-time requirement). The second optimization criterion is the efficiency, which is denoted as the minimal synchronization overhead (blocking time) for the system. The third and weakest criterion is that we want to achieve well-balanced core utilization in order to prolong the life-time of the multi-core processor.

16.2.3 Discrete Event Simulation

In contrast to desktop or web-based systems, a malfunction of software in safety-critical embedded systems can lead to severe damage or even physical harm. Consequently, faulty behavior is not tolerable. Errors in such embedded systems however cannot only arise from functional incorrectness but also from temporal inaccuracy. This is due to the fact that additional requirements, the so called timing constraints, have to be met. That means that software functions and their interactions must guarantee a result before a strict deadline. Hence, embedded systems in general have to meet a higher quality standard.

To design or to verify the timing behavior of such an embedded system is not only one of the most challenging works during the development but also the most crucial one at the same time. Therefore it is not surprising that "analysis of real-time properties such as response-times, jitter, and precedence relations are commonly performed in development of the examined applications" [11, p. 4]. Over the years the industry brought forth a variety of different methods, technologies and tools to specify and reason about timing properties. According to [12], the field of timing-related analyses can be divided into three different approaches: execution time analysis, timing analysis with model checking, and simulation-based timing analysis.

The former one determines the worst-case execution time (WCET) for each task. Based on those values, a schedulability or feasibility analysis is executed. That way it is possible to make a statement about the timing behavior of the system. However, due to simplifications in the system model and the assumptions that have to be made, this approach leads to very pessimistic results and is thus inapplicable [13].

The second approach instead uses model checking to verify a model against formally specified requirements. Therefore the system behavior has to be described in a model whose elements also have formally well-defined semantics, like a finite-state automaton. The state space explosion problem of model checking however leads to the fact that complex embedded systems, as used in the industry nowadays, can hardly be analyzed due to the unrealistic amount of memory and run-time required [14].

The last approach for a timing-related analysis finally is to use a simulation-based timing analysis. There the system behavior is simulated by a discrete event simulation [15]. The system behavior and thus the input of the simulation thereby are represented by a model, the simulation model. Depending on the granularity of the model, the system can be analyzed at different levels of abstraction. The simulation results are then used to determine the necessary real-time properties of the system. Although the simulation model cannot describe the real system in every detail, this approach allows exploration of different scenarios, which are otherwise hard to evoke on the real system. This fact also makes it possible to counteract the circumstance that the results of the simulation greatly depend on the quality of the simulation model. It cannot guarantee, however, that the obtained results are valid for all system executions. Therefore a simulation-based timing analysis is not

Table 16.1 Properties of the artificial task set

			Recurrence [ms]			Net execution time [ms]		
Task	Priority	Deadline [ms]	Min	Avg	Max	Min	Avg	Max
1	12	1	1			0.248	0.331	0.457
2	11	2	2			0.697	0.828	1.311
3	10	4.93	4.93	12.95	29.92	2.00	2.28	2.86
4	9	4.94	4.94	12.96	29.92	1.96	2.25	2.87
5	8	5	5			0.87	0.925	1.07
6	7	10	10			2.42	2.49	2.63
7	6	20	20			2.32	2.37	2.48
8	5	50	50			0.749	0.809	1.04
9	4	100	100			3.38	3.52	3.94
10	3	200	200			2.77	3.32	3.68
11	2	500	500			2.75	3.04	3.45
12	1	1,000	1,000			1.12	1.37	1.50

suitable for the determination of WCETs [16]. Instead it provides an ordinary and thus a more realistic view on the system behavior. Another great advantage of this approach is that a simulation can be much faster than an execution of the real system, since it can be performed on workstations and is thus not restricted by the limited system resources available on embedded systems. Based on these advantages, we chose the simulation-based timing analysis approach for our research.

16.2.4 Data Basis

Since the simulation results, as already mentioned in the section above, greatly depend on the quality of the simulation model, this section addresses the data basis of the presented case study. Although the data basis for our case study was created artificially, it is comparable in complexity to a practical automotive engine management system.

It consists of a task set with 12 real-time tasks and 2,500 functions, also called runnables. In total 28,000 data accesses to 20,000 shared global variables are performed by these runnables. The information which runnable accesses what data element was thereby created randomly following a uniform distribution. In addition, 1,600 semaphores are used to protect those data accesses. They are used to protect the write accesses to global data, while all tasks can read data simultaneously. The bit size of each global data element was again determined randomly. Therefore a value of the interval from 8 to 64 with a step size of 8 was assigned following a uniform distribution. Further it is also assumed that each data element is transferred in 16Bit blocks with a transfer time of 160 ns each, in order to consider the data transfer times for read and write accesses in the simulation. An overview of the properties of the described task set is given in Table 16.1.

All tasks are periodically activated except for Task 3 and 4. Those two are engine triggered, that means that their recurrence is based on the current engine speed. The engine speed follows in this use case the form of a saw-tooth wave with the provided minimum, average and maximum. This emulates the acceleration of the engine rotation speed of a vehicle with a subsequent deceleration phase. In addition, the computational power demand of the tasks is stated as a Weibull distribution defined by the minimum, average and maximum net execution time.

The used hardware model consists of a processor with three symmetric cores and a clock frequency of 160 MHz each. It further is assumed that exactly one instruction can be processed by each core in each clock cycle. Furthermore, each core is managed by an independent OSEK scheduler (partitioned scheduling).

Although the data basis for our case study was created artificially, we want to present a way how information on the model can be gathered in the following. This step has already been presented in detail in [17]. There a hybrid approach is described, which consists of a static as well as a dynamic analysis.

The former means that artifacts, like source code or descriptions of domain specific languages (DSL), are analyzed to get the pieces of information needed. A DSL in the automotive industry for example is the Automotive Open System Architecture (AUTOSAR). That way all parts of the system and thus the most information needed for the model can be acquired. This includes among others the number of tasks, their temporal activations, as well as the functions mapped to those tasks. In addition to that, information about data accesses and the use of semaphores in the individual functions can be determined easily.

What is missing, however, is an estimate about the actual behavior of the system during runtime. This part is covered by the dynamic analysis. There the system is executed while performing run-time measurements on the individual software parts. Run-time measurement in this context means that the relevant events of the system, like the activation, start, preemption, or termination of tasks, are detected and stored for later off-line analysis. This process is called trace recording. This footprint of the system can then be used to determine significant properties of the system, like for example the execution time of functions or sporadic stimulations of tasks. Both are examples for properties that cannot be determined by a static analysis since the system behavior depends on the chosen set of inputs.

This variety of the system behavior is the reason why certain properties are described by a statistical distribution in our model. An example for this can be found in Table 16.1, where the execution times of the tasks are described by a Weibull distribution. The random variable represented by the statistical distribution thus represents the probability that a possible outcome was measured. Whether the execution time of a function varies following a certain statistical distribution can be determined with the help of a goodness-of-fit test, like the Pearson's chi-squared test [18].

Another part of the dynamic system behavior that is hard to describe in a model in a precise way, is the occurrence of sporadic stimulations. However, we found the concept of arrival curves [19] to be an appropriate way to handle this information. In general, arrival curves are a pair of curves that represent the upper and lower

Table 16.2 Comparison of the WFD, WFI, BFD and BFI bin-packing heuristics

Bin-packing heuristic	Task sorting scheme	Task allocation scheme
Worst-Fit Decreasing (WFD)	Decreasing	Least loaded core
Worst-Fit Increasing (WFI)	Increasing	Least loaded core
Best-Fit Decreasing (BFD)	Decreasing	Most loaded core
Best-Fit Increasing (BFI)	Increasing	Most loaded core

bound of stimulations that may occur within any time interval. Those boundaries are determined by sliding a time window over the events stored in the trace recording and registering the maximum and minimum number of stimulation events within this time window. This procedure is then repeated with a size of the time window that gets successively increased until all information is covered by the arrival curves. By doing so, the timing characteristics of the system can be modeled in a compact and also precise way.

16.2.5 Optimization Approach

In order to achieve the aforementioned goals with partitioned scheduling, we have to optimize the task-to-core mapping. This task allocation problem is equal to the bin-packing problem, which is known to be NP-hard [20]. For the task set presented in Sect. 16.2.4 for example, there are $3^{12} = 531,441$ possible solutions. Due to the fact that reactiveness, efficiency and core load distribution should be optimized simultaneously, suitable multi-objective approaches have to be considered.

Genetic algorithms are often used to solve that kind of problems [21]. Moreover, they are particularly well-suited for the problem at hand [22, 23]. Since we do not possess allocation rules for the presented task set, which would render a full-scale design space exploration unnecessary, we use a genetic algorithm to find multiple trade-off solutions for task allocation. These are compared to the WFD, WFI, BFD and BFI bin-packing heuristics. Table 16.2 depicts the differences between those heuristics. When bin-packing is performed, the task set is sorted regarding the load they will inflict on the cores. Following the initial sorting, each task is assigned to a core in a specific way. When WFD bin-packing is used for example, the tasks are first ordered in decreasing order of their load. After that, beginning with the task that has the highest load, that specific core which has the most available load left is selected for allocation (considering the previously allocated tasks).

16.2.6 Genetic Algorithm

Genetic algorithms simulate natural selection and natural evolution in an iterative process [23]. Figure 16.2 depicts the different steps of a genetic algorithm. First, the

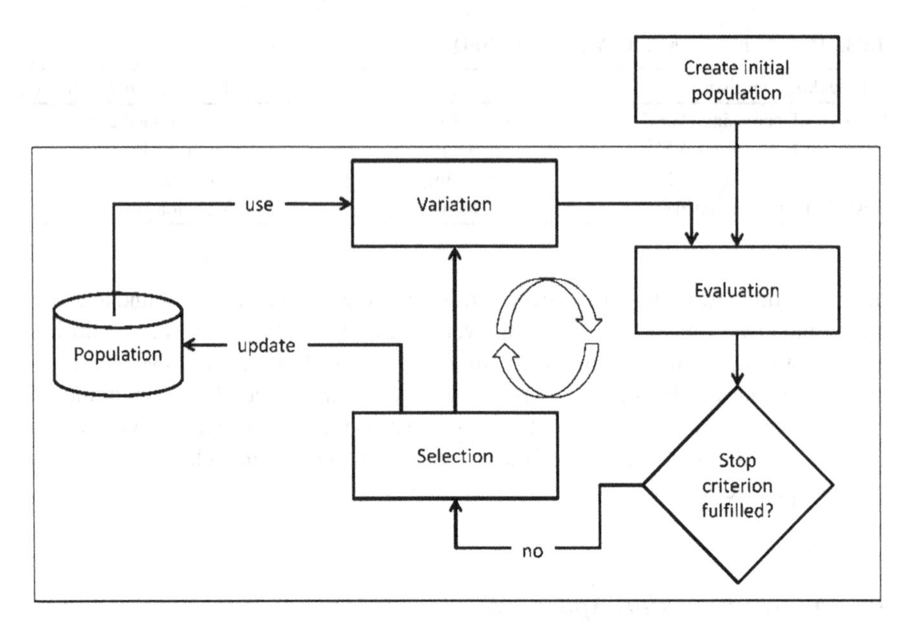

Fig. 16.2 Outline of a genetic algorithm

initial population, a subset of the population, is created. The population represents working set of solutions of the algorithm. Solutions in the initial population are typically randomly created. After the initial population has been created, the solutions are evaluated and hereby assigned a real number called the fitness. The next step is to check the stop criterion. After a pre-defined number of iterations or when a stagnation criterion is met, the algorithm stops. When the stop criterion is not fulfilled, the selection step is conducted. The selection step determines a subset of the population, which is called parent solutions. According to the survival-of-the-fittest idea, the best parent solutions are selected to become the origin of newly created solutions. These new solutions are created during the variation step by using mutation or crossover operators. While mutation modifies a parent solution to create a new one, crossover combines the properties of two or more solutions to create new solutions. Subsequently the newly created solutions are again evaluated and assigned a fitness value. This process continues until the stop criterion is fulfilled.

In our work, we have decided to adapt the *Nondominated Sorting Genetic Algorithm II* (NSGA-II) [24] to our task allocation problem. In literature it is considered well tested and efficient [23] and requires only the size of the population as parameter.

For fitness assignment, NSGA-II uses the crowded-comparison operator [24] to define a total order between the solutions. First, the non-domination level for each solution is determined. This refers to the Pareto-criterion for dominance [22]. Consider a multi-objective optimization problem with $n \in \mathbb{N}^+$ objectives: one solution (weakly) dominates another solution (i.e. is a better trade-off), if it performs better in one objective and at least as good in every other $n - 1$ objectives. The subset of non-dominated solutions depicts the first front of non-dominated solutions. Every solution in that subset is assigned the lowest (best) rank 1. Subsequently, the subset of non-dominated solutions is determined when the previously ranked solutions are no longer considered. The procedure is continued until a non-domination level has been assigned to every solution. As the non-domination level only describes a partial order between solutions, multiple solutions can have the same non-domination level. Therefore the so called crowding distance is used as a tie-breaker. The crowding distance denotes the proximity of other solutions in objective space for a certain solution. When two solutions have been assigned the same non-domination level, the solution with the greater crowding distance is considered better.

In order to adapt NSGA-II to our problem domain, we specified how to create the initial population, how to perform variation and defined a stop criterion. The initial population is created by using a uniform distribution to create solutions with random task-to-core allocations. In order to create one new solution for each parent solution, we use the bit-flip mutation technique. A uniform distribution is used to randomly select a task and to randomly choose a new core it should be allocated to. The stop criterion is defined by a fixed number of evaluated solutions. When this threshold is reached, the algorithm stops. Note that in order to avoid duplicate solutions, we implemented a uniqueness check when creating new solutions. Moreover, we implemented a feasibility check that prevents cores from being loaded beyond 100 %.

16.2.7 Optimization Metrics

In order to quantify our optimization goal, we chose the following metrics for reactiveness, efficiency and equal distribution of core loads.

The *maximum Normalized Response Time (mNRT)* [17] is used to quantify to what extent deadlines have been met on task set level. Consider a task set $\tau = \{T_1, T_2, \ldots, T_N\}$ with n tasks and m instances (or jobs) $J_{i,j} = \{J_{i,1}, J_{i,2}, \ldots, J_{i,M}\}$ of each task T_i. The response time of a job $J_{i,j}$ is calculated as the difference of its finishing time $FT_{i,j}$ and its activation time $AT_{i,j}$. The mNRT is calculated as the maximum normalized response time over all tasks and over all jobs of a task. Normalization is done by dividing the maximum job response time by the relative deadline D_i of the respective task. When the mNRT is less or equal to 1.0, the deadlines of all task and all jobs of each task have been met.

$$RT_{i,j} = FT_{i,j} - AT_{i,j}$$

$$\mathrm{m}NRT(\tau) = \max_{T_i \in \tau} \left(\frac{\max_{J_{i,j} \in T_i} (RT_{i,j})}{D_i} \right)$$

To quantify the efficiency of a solution, we use the mean Normalized Blocking Time metric (mNBT) [25]. First, the *mean Normalized Task Blocking Time* $mNTBT(T_i)$ of each task T_i is calculated. This is done by the determination of the arithmetic mean over all jobs of the blocking time $BT_{i,j}$ to net execution time $NET_{i,j}$ ratio. The mNBT is then calculated as the arithmetic mean of all mean Normalized Task Blocking Times. It denotes the mean share of net execution time of all tasks in the systems which are blocked by synchronization [25].

$$mNTBT\left(T_i\right) = \frac{\sum_{J_{i,j} \in T_i} \frac{BT_{i,j}}{NET_{i,j}}}{m}$$

$$mNBT\left(\tau\right) = \frac{\sum_{T_i \in \tau} mNTBT\left(T_i\right)}{n}$$

In order to quantify to what extent the cores are equally loaded, we calculate the standard deviation over all differences of each core load and the average core load. The average core load equals the sum of all core loads divided by the number of cores.

16.2.8 Experimental Setup

The NSGA-II optimizer has been configured in the following way: We chose a population size N = 360 and configured the optimizer to stop after the evaluation of 18,000 unique solutions. We used version 13.11.0 of the simulator TA Toolsuite [26]. Hereby, a simulation time of 20 s of system runtime has been chosen for each solution. As a result of prior experimentation, this simulation time has been identified as good trade-off between accuracy and simulator runtime. The optimization was performed on a PC with an Intel Core i7-3930 K processor, a MSI X79A-GD45 mainboard and 16 GB DDR3-1600 RAM. A 90 GB Corsair Force GT SSD hard drive was used for the operating system while a 320 GB Seagate Momentus 7200.4 hard drive was used for the experiment. On this PC, one simulation of 20 s system runtime takes around 10 s of real time to complete. In the experiment, 12 simulations were performed in parallel to harness the maximum thread concurrency of the processor. The total runtime of the experiment was 64 h.

16.2.9 Results

Generally, the results of the case study show that it is very hard to allocate the applied task-set on the three cores without any deadline violations. Figure 16.3 shows only those solutions among the best trade-offs, which have met all their

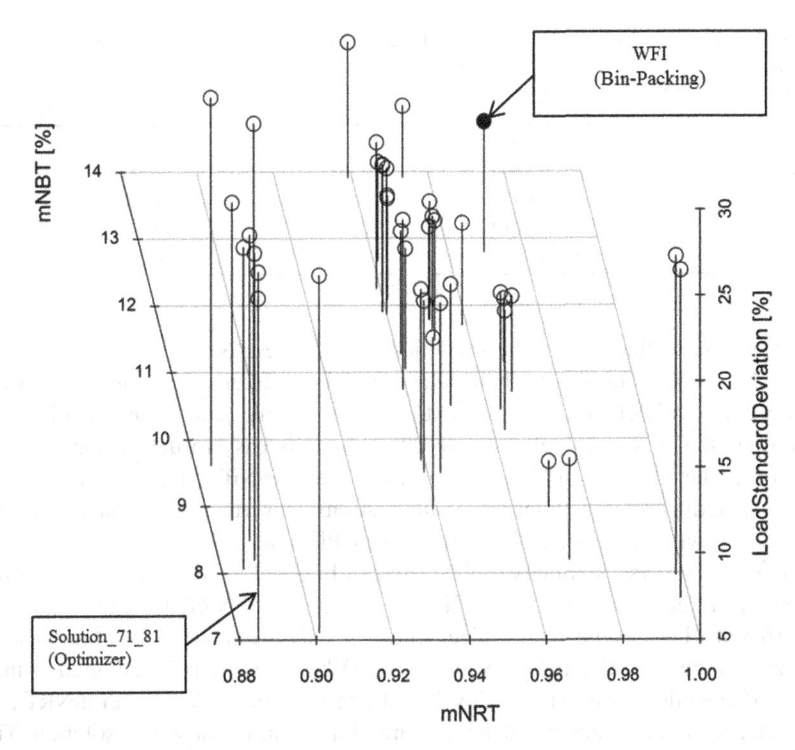

Fig. 16.3 Comparison of the results of the genetic optimization of the applied task set without deadline violations. The depicted solutions are only the ones which did not have any deadline violations during the simulation duration

deadlines during the simulation duration (mNRT < 1.0). As you can see there, from the family of bin-packing heuristics, only the WFI (filled point in Fig. 16.3) produced a solution for the task set in which all deadlines were met. All other types of bin-packing have failed in this approach. In contrast to this, the genetic optimization gave us 40 different solutions that met all their deadlines. These are really few compared to the 18,000 created solutions of the optimization approach, but the trade-offs of its best results outperform the bin-packing solutions.

In detail, Fig. 16.3 shows us the following: The x-axis describes the most important metric in our view, the mNRT. It is the most important one because the main requirement for real-time systems (beside the correctness) is to meet the deadlines. The smaller the mNRT, the better is the solution. The next important metric is depicted on the y-axis, the mNBT which describes the blocking of the system overhead (share of time at which the system is blocked due to synchronization). For the mNBT, the best solution is also the minimum value. Finally, the z-axis visualizes the standard deviation of the CPU load over the 3 assumed cores. Above we have written that we want to achieve a well-balanced CPU load which means that here the minimum is also the best.

Table 16.3 Results of one selected solution of the genetic optimization compared to the standard bin-packing heuristics

Allocation	mNRT	mNBT [%]	LoadStdDeviation [%]
Solution_71_81	**0.8850**	**7.01**	**26.35**
WFI	**0.9694**	**12.80**	**12.58**
WFD	1.0436	12.15	14.17
BFD	44.7912	7.69	49.77
BFI	1483.42	7.69	49.77

If we want to find out the best solution of these results, we first take that one where the mNRT is minimal. There can be found 4 different solutions nearly at the same value of mNRT in Fig. 16.3. Next, we take that one (out of these 4 solutions) which has the minimum value of mNBT. Now it is easy to detect that solution 71_81 has the best trade-offs regarding mNRT and mNBT and followed by this the best reactiveness and efficiency of all solutions. Nevertheless, we have to accept a standard deviation of 26.35 % regarding the CPU load.

Table 16.3 shows a summary of the results of bin-packing heuristics and solution 71_81 of the genetic optimization. If we compare the mNRT, we can observe in Fig. 16.3 that besides the 40 optimizer solutions only the value of WFI bin-packing is smaller than one. The mNRT value of WFD bin-packing is barely greater than 1.0 (1.044) while the BFD and BFI values are much greater. A value of mNRT that is greater than one means that deadlines are violated in the observed solution. The very high value for the BFI (1483.42) is caused by the fact that there was one task at minimum which was activated, but never executed. This happens when the task has a low priority and other tasks with higher priorities utilize the corresponding core to 100 % by executing or waiting for resources. The selected genetic optimization solution has also a better mNBT (7.01 %) than all bin-packing heuristics. Only the BFD and BFI are nearly at the same level, but they contain deadline violations (see the mNRT). Only for the standard deviation of CPU load, the WFD bin-packing provides a better solution. Nevertheless, because the mNRT and mNBT are the most important metrics according to the real-time behavior of the system, the optimization approach provides better results than standard bin-packing heuristics for our applied task set.

16.2.10 Conclusions

In this case study we showed that the genetic optimization algorithm NSGA-II leads to better results than standard bin-packing heuristic for our approach. In future work we also want to apply the optimization approach on task sets from different fields of real-time systems, not only automotive powertrain systems. By means of that we want to make a contribution that multi-core real-time systems become safer regarding meeting their deadlines and thus avoiding possible disasters. Further we want to compare and improve different genetic optimization algorithms. In order to fulfill the constraints of safety certification regarding ISO 26262, we started statistical investigations in which we want to prove that after a certain simulation duration or number of simulations the maximum of response time is reached with a diagnostic coverage >99 %.

Acknowledgments Individual parts of this work were supported by Bayerische Forschungsstiftung under grant agreement S^3Core (FKZ: AZ-980-11) and by ITEA 2 Call 4 under grant agreement AMALTHEA (FKZ: 09013) respectively.

References

1. Sommerville I (2011) Software engineering 9. Pearson Education
2. Royce W (1970) Managing the development of large software systems: concepts and techniques. In: 1970 IEEE WESTCON, Los Angeles, pp 1–9
3. Boehm B (1986) A spiral model of software development and enhancement. SIGSOFT Softw Eng Notes J 11:14–24
4. Vogel-Heuser B (2003) Systems software engineering. Angewandte Methoden des Systementwurfs für Ingenieure. Oldenbourg
5. Smith D, Kenneth G (2004) Functional safety, second edition: a straightforward guide to applying IEC 61508 and related standards, 2nd edn. Butterworth-Heinemann
6. Ward D, Crozier S (2012) The uses and abuses of ASIL decomposition in ISO 26262. In: 7th IET international conference on system safety, incorporating the cyber security conference 2012, pp 1–6
7. Braun J, Mottok J (2013) Fail-safe and fail-operational systems safeguarded with coded processing. In: IEEE EUROCON 2013, pp 1878–1885
8. Kopetz H (2011) Real-time systems: design principles for distributed embedded applications. Real-time systems series, LLC. Springer, New York
9. Buttazzo G (2011) Hard real-time computing systems: predictable scheduling algorithms and applications. Real-time systems series, LLC. Springer, New York
10. Iyer S, Pankaj G (2003) Embedded realtime systems programming. McGraw-Hill Education (New Delhi) Pvt Limited
11. Hanninen K, Maki-Turjia J, Nolin M (2006) Present and future requirements in developing industrial embedded real-time systems-interviews with designers in the vehicle domain. In: 13th annual IEEE international symposium and workshop on engineering of computer based systems 2006, p 9
12. Kienle H, Kraft J, Mueller H (2011) Software reverse engineering in the domain of complex embedded systems. InTech
13. Bernat G, Colin A, Petters S (2002) WCET analysis of probabilistic hard real-time systems. In: 23rd IEEE real-time systems symposium
14. Clarke E, Grumberg O, Jha S, Lu Y, Veith H (2001) Progress on the state explosion problem in model checking. In: Informatics. Springer Berlin Heidelberg, pp 176–194
15. Deubzer M (2011) Robust scheduling of real-time applications on efficient em-bedded multicore systems. Dissertation, Technische Universität München
16. Wilhelm R, Engblom J, Ermedahl A. Holsti N (2008) The worst-case execution-time problem – overview of methods and survey of tools. In: ACM transactions on embedded computing systems (TECS), vol 7
17. Sailer A, Schmidhuber S, Deubzer M, Alfranseder M, Mucha M, Mottok J (2013) Optimizing the task allocation step for multi-core processors within AUTOSAR. In: IEEE international conference on applied electronics (AE), pp 1–6

18. D'Agostino R (1986) Goodness-of-fit-techniques. Statistics: a series of textbooks and monographs. Taylor & Francis, New York, USA
19. Chakraborty S, Kuenzli S, Thiele L (2003) A general framework for analysing system properties in platform-based embedded system designs. DATE
20. Coffman E, Garey M, Johnson D (1996) Approximation algorithms for bin packing: a survey. Approximation algorithms for NP-hard problems, pp 46–93
21. Jones D, Mirrazavi S, Tamiz M (2002) Multi-objective meta-heuristics: an overview of the current state-of-the-art. Eur J Oper Res, vol 137, pp 1–9
22. Deb K (2005) Multi-objective optimization. In: Search methodologies, Vol 10. Springer, New York, pp 273–316
23. Konak A, Coit D, Smith A (2006) Multi-objective optimization using genetic algorithms: a tutorial. J Reliab Eng Syst Saf, vol. 91, 992–1007
24. Deb K (2002) A fast and elitist multiobjective genetic algorithm: NSGA-II. In: 2nd IEEE transactions on evolutionary computation, pp 182–197
25. Alfranseder M, Mucha M, Schmidhuber S, Sailer A, Mottok J (2013) A modified synchronization model for dead-lock free concurrent execution of strongly interacting task sets in embedded systems. In: IEEE international conference on applied electronics (AE), pp 1–6
26. Timing-Architects (2013) TA Toolsuite version 13.11.0. In: TA Academic & Research License Program
27. Block A, Leontyev H, Brandenburg BB, Anderson JH (2007) A flexible real-time locking protocol for multiprocessors. In: 2012 IEEE international conference on embedded and real-time computing systems and applications, vol 0, pp 47–56
28. Van Eikema Hommes Q (2012) Review and assessment of the ISO 26262 draft road vehicle – functional safety. SAE technical paper 2012-01-0025

Chapter 17
Design of Safe Systems

Peter Hlousek and Ivan Konecny

Abstract This chapter deals with brief overview of Railway Signalling and Interlocking (RSI) systems and its characteristics, safety definition, safety integrity levels, methods of safety achievement, problems of safety data transmission and safety approval process. The main purpose of use of these systems is assurance of railway traffic safety that means prevention against railway accidents. There is rich experience with their design gained during more over than 150 years of their evolution. We believe that it can be good inspiration for other systems, for increasing their resilience in disaster conditions.

Keywords Safety-related systems • Safety integrity • Resilience to faults

17.1 Introduction

The focus of this chapter is targeted on Railway Signalling and Interlocking (RSI) systems, which are so called safety-related controlling systems in the field of railway transportation. Their main task is a provision of safety by prevention of railway accidents.

High dynamics of train movement means potentially disastrous consequences in case of accident – many fatalities, high material or environmental damages. Main causes of accidents are human errors and failures of technical equipment. The human operators controlled railway traffic at the beginning and they were the most perfect controlling system for long time. Still, they were not perfect and made mistakes, which led to several catastrophic accidents. Defence is realized by

P. Hlousek (✉) • I. Konecny
Department of Applied Electronics and Telecommunications, Technical University of Pilsen, University of West Bohemia, Univerzitni 8, Pilsen 306 14, Czech Republic
e-mail: hlousek@kae.zcu.cz

H.-N. Teodorescu et al. (eds.), *Improving Disaster Resilience and Mitigation - IT Means and Tools*, NATO Science for Peace and Security Series C: Environmental Security,
DOI 10.1007/978-94-017-9136-6_17,

technical system supervision of people decisions to assure correct control of railway traffic, which is the obvious reason why the very strict safety requirements have to be applied to RSI systems. It leads to more sophisticated and complex systems and therefore more complicated process of system design.

The basic safety-related presumption for railway transportation is that the stop of train movement is equal to highest safe state. This is the big advantage of railway transportation, compared to road or air transportation. The safe reaction to failures of RSI device has to lead to limitation of traffic due to safety protection.

The standardization of methodology of design process of these systems is the content of many European standards, see more details in Sect. 17.2. This collected knowledge and experience can be inspirational and useful for other branches.

17.2 Historical Overview of RSI Systems

History of development of railway signalling and interlocking systems is more than 150 years long now. This rich history and collected experience can be an inspiration for other systems.

The first systems were developed soon after the railway transportation started to be popular and frequent. At that time (1860) the level of technical means was low from current point of view so the principles and functions had to be very simple. Naturally, these first systems were based on simple mechanical principles. The mechanical lock-ups creating logical functionality and mechanical robustness to prevent faults were used. Electromechanical systems started to be used around 1900 still used of mechanical lock-ups, but also electrical signals to create logical functionality over long distances. When the electric started to be used frequently, evolution of electrical systems (1930) began. They were based on use of special safety relays and their contacts creating required logical functionality. Present modern systems use of programmable electronics and their evolution started much later than in industry (1980) because of harsh operational environment.

The different types of RSI systems can be distinguished from operational function purpose point of view. First, interlocking systems are used ensure safe movement of trains in railway stations. Line signaling systems are used on open lines to prevent train collisions between stations. Level crossing systems ensure safety on crossings of road and rail track, where the train has to have right of way. Automatic train protection systems have fixed and mobile part and they supervise the engine driver to correctly obey orders from line signaling and interlocking systems. Modern technology enables to create complex systems, when dispatcher in one controlling central office supervises railway traffic over large area, e.g. long line with several stations. For more details, see [1].

Development of electronic complex systems is very demanding and complicated process today. Several standards were created as helping guides through that process to ensure defined safety level of the final product. The target of these European standards is to unify criteria for design, assessment and comparison of all newly created systems. The first versions of these standards were released about 1995 with some updates until today.

The standard EN 61508 [2] is targeted for general industrial safety-related systems and forms base for other standards valid for specific branches. EN 50126 [3], EN 50128 [4], EN 50129 [5] and EN 50159 [6] are such standards for railway transportation sector. They deal with problems of development and operation of railway signalling and interlocking systems. These standards are based on philosophy that the system safety can be threatened in every single phase of system life-cycle. The requirements for every particular phase, what should be done to keep the safety, are defined in the standards.

17.3 Characteristic Features of RSI Systems

17.3.1 Safety Requirements

Formally, the system safety can be divided into two parts, functional and technical one, but, of course, both of them must be fulfilled in safety-related system.

1. Functional safety

This term means that the system performs correctly its functions in faultless conditions. The system behavior, which is designed in system definition phase of the life-cycle, must not threaten the safety of railway traffic. The system functions should be processed as checked sequence of partial actions with independent supervision of action results leading to quick detection of incorrect behaviour.

2. Technical safety

This term means that the safe behavior, after any fault occurrence in the system, is assured. The system must enter in predefined safe state and stay in until predefined corrective action is carried on.

17.3.2 Principles of Technical Safety Assurance

First presumption, which is taken in the approach to keep wieldy, is the occurrence of only one independent fault in system at one particular time moment. Justification of the presumption is based on the basic postulate of reliability theory. Next, there are six axioms that describe the basic approach:

1. Any fault must not threaten the safety of the system.
2. Quick and proper detection of any fault, according to failure rate, is necessary.

3. If the second axiom is not fulfilled, any other fault has to be expected to appear. Both faults together must not still threaten safety of the system.
4. If occurrence of next dependent faults is possible due to the first one, it is necessary to consider all fault combinations as one fault. The first axiom still must be fulfilled.
5. Automatic negation of system outputs after detection of fault is required, so the system enters the safe state.
6. Safe state preservation against any next fault occurrence is required until restoration of correct function.

17.3.3 Relation and Difference Between Safety and Reliability

It is very important to correctly understand the difference between reliable and safe system, although it is quite often considered as the same. It is impossible to absolutely eliminate possibility of fault occurrence, either random or systematic, in complex real systems. That is why the system with arbitrarily high reliability cannot be automatically considered to be safe. As a safe system, it can be considered only that one with well-defined safe mode of behavior after occurrence of any possible fault. This safe reaction mostly consists in some variant of graceful degradation of function, which leads to traffic limitation (fluency violation, operational anomalies). Such a system need not to be highly reliable from technical point of view, because any fault cannot immediately threaten the safety. The railway traffic cannot be naturally stopped for whole period required for restoration, typically several hours, though. During this period the traffic is controlled in emergency mode, which is stress situation prone to mistakes and there is increased risk of potential accidents due to RSI system function limitation or even turn off. So, the high reliability (dependability) of the system is also necessary from the whole railway transportation process point of view.

17.3.4 Types of Failures

Resilience to failures is the key topic in design of safety-related systems. It is important to distinguish among different types of failures, because the protection measures have to be also different. There are three main different sources of failures: HW faults, human errors (mistakes) and physical influences. Nature of failures is either random or systematic.

17.3.5 Random Failures

This type of failures occurs randomly in time and the failures of HW components are the best-known source. Random failures are quantifiable and can be associated with a failure rate using the theory of reliability methods, especially means of probabilistic calculations. These are based on known data for component failure rates and detection times of random failures. For inherently fail-safe components the hazard rate is considered as negligible (zero), but the correct justification has to be carried out.

17.3.6 Systematic Failures

Systematic failures are major threat to safety for present complex electronic systems. In contrast to random failures, they are non-quantifiable and cannot currently be associated with a failure rate. They are caused by explicit reason and occur each time under the same conditions. The systematic faults have different effects in the different life-cycle phases and counter-measures are application dependent. It is therefore not possible to list all individual causes of systematic faults during the life-cycle phases and carry on a quantitative analysis for the avoidance of them. In spite of this, they have to be completely assessed and extensively mitigated by the well-managed development and operation process and technical measures. Main causes of systematic failures are:

- Human errors – made in specification, design, project, manufacture, maintenance process, but also SW errors,
- HW faults – common for all items of same type or production runs; higher probability of complicated errors in complex components (integrated circuits),
- Tools faults (compiler, development tools),
- Physical influences (EMI, ESD, climatic, power supply, etc.).

17.3.7 Common Cause Failures (CCF)

This special type of failures is relevant for composite fail-safety systems realized by redundant structures, see details in Sect. 17.4.3. These are faults common to items that are intended to be independent. The main threat to safety is the same fault occurrence in all channels of redundant system, which cannot be naturally detected by comparison.

Main causes of common cause failures are internal and external physical influences – EMI, faults of common power supply, lack of galvanic insulation, extreme temperatures, mechanical perturbations, etc.

Another important sorting of failure types is on credible and incredible failures. The incredible failure modes are considered as they cannot happen.

17.4 Methods of Achievement of Fail-Safety

The three basic principles, described below, how to achieve fail-safety of the controlling systems, were recognized until now. The design of modern complex systems mostly takes advantage of the combination of these principles.

17.4.1 Inherent Fail-Safety Systems

These systems are based on items with suitable properties, resilient to faults and their proper system implementation. This principle was used in all non-electronic generations of the RSI systems from the beginning of historical evolution. This principle means that none fault can cause hazardous state of the system, so it can be realized as one channel device, both HW and SW. Analysis verifying that all the credible failure modes of the items used in the system are non-hazardous (e.g. have inherent physical properties) is necessary to carry on. The catalogue of failure modes, which is included in Appendix C of the EN 50129 standard, is the helping guide for such analysis. The Failure Modes and Effects analysis (FMEA) is mostly used for this purpose. This method involves the study of the all possible fault modes which can occur in every component and the determination of its effects on the required functions of the system. Skilled state of the art design and validation process performed under strict quality control have to be used in development process of these systems. Use of this principle in electronic systems is very problematic, because not many components have suitable properties. Especially complex electronic systems using integrated circuits cannot be based on this principle.

The next problem of this principle is not quick enough fault detection and negation, because the behavior of failed system is the same as non-failed one when some condition for function execution is not fulfilled. In addition, the system information, why some function is not performed, is not usually available. This problem was not serious in older systems, which were operated by experienced human operators, who were able to identify the problem. It is obviously not the case of modern automated systems.

17.4.2 Reactive Fail-Safety Systems

Reactive fail-safe systems are based on the quick detection of all safety-related faults in the main channel (Function A) by independent checking device. In case of fault detection, negation of the outputs must be done to preserve safe state, see figure B.2 in [5]. These systems are realized by implementation of combined HW and SW diversification in most cases. The means of fault detection can be continual testing or coding of processed data. This architecture by itself mitigates some threats, especially common cause faults, taken into account in the composite fail-safety architecture. The reaction time and fault coverage completeness has to be specifically evaluated. This principle is rarely used in systems developed in Europe today, because of bad experience with safety proving in the past. But, there is a chance of these systems revival in future, due to advances in technology.

17.4.3 Composite Fail-Safety Systems

This is the third principle, the most often used one today, which can be used for design of modern electronic systems, when components have no suitable fail-safety properties.

Composite fail-safety principle is implemented by redundant structures – multi-channel processing. The 2*oo*2 (two out of two) and 2*oo*3 (two out of three) structures are normally used in railway applications. The example of basic 2*oo*2 structure is shown on figure B.2 in [5]. The different types of redundancy, typically their combination, can be used to realize multi-channel processing, e.g. multiple HW, multiple SW or data redundancy. Important part of these structures is common device (fail-safe comparator) ensuring the channel comparison and the negation of failed channel outputs. Comparison can be realized as HW device or SW. The HW comparator has to be an inherent fail-safe device and has to be designed and evaluated accordingly. As alternative solution, the voters in every channel may be used.

Another characteristic feature is that an information processing (provided system functions) and protection against faults are two separated functions in these systems. Realization of the particular channels needs not to be realized to fulfil safety requirements by itself.

This principle is based on presumption of fault occurrence only in one channel at one moment, therefore the channel independency is required to assure integrity against common cause failures (CCF). Another safety requirement, which is absolutely necessary, is the quick detection and negation of faults and enforcement of unchangeable safe state of outputs. This is necessary feature as a protection against second possible fault, which is potentially dangerous, see figure B.2 in [5]. There is the basic recommendation, see details in [5], for the length of detection plus negation time:

$$T_{sf} \leq 1/(1000 \cdot \lambda), \qquad (17.1)$$

where λ is the sum of failure rates of the items, whose simultaneous malfunctioning could be hazardous. This figure basically says that the detection plus negation time should be at least three orders shorter than MTBF value of the system.

Diversification is the next safety measure used very often, at least partially. It shall be implemented, as far as possible, to ensure protection against systematic and common mode failures. Again, there are many forms possible, e.g. different processors used in particular channels or different algorithms created by different programming teams etc. On the other side, diversification generally causes problems with the comparison of channel results and therefore some compromise solution has to be chosen.

At the end it should be stressed that implementation of these redundant structures is totally different from backing-up, which is implemented to achieve higher availability, but not safety.

17.5 Safety Assessment of RSI Systems

This section deals with brief description of safety assessment philosophy and approach used for safety-related RSI systems. Nowadays, two points of view on this problem can be distinguished:

1. Qualitative viewpoint

This is the older approach, which was applied from the very beginning of historical development of these systems, but it is still valid today. Characteristics "Qualitative" means that system behavior can be evaluated as safe or not safe. It has to be reminded that safe behavior means the system ability to limit consequences of its faults by preserving safe state of its outputs (faults negation). If we take for granted that the probability of incredible faults is zero, we can call inherent fail-safety systems as "absolutely safe". Of course, the zero probability cannot be guaranteed in real world. Similarly, the composite fail-safety systems could be considered as "absolutely safe" if none second fault can happen in the detection plus negation time interval T_{sf}, i.e. if $T_{sf} = 0$. Again, in real world this interval can only be more or less close to zero. This little bit simplified approach was successfully used for non-electronic RSI systems, because they were functionally relatively simple and were constructed as state of art technical systems of its time. But the different safety requirements were, at least partially, respected by the different device categories, targeted on main/ regional lines, station/hump yard etc.

2. Quantitative viewpoint

This point of view started to be stressed, when advances in technology enabled to construct very complex electronic systems (e.g. processor based) mainly using the composite fail-safety principle. The question was raised, if all created RSI systems have to be state of art, which means also very expensive.

This approach admits that some faults may occur when the system will not react correctly. It will still be considered as safe, if the probability of these faults is lower than some required limit. It means that the safety can be quantified as the probability of non-presence of hazard in system for given time and given operational and maintenance conditions. This also means that there is a connection between safety and reliability of the system. With raising complexity of the systems, the existence of relation among safety, reliability and cost of the system started to be clear. It is necessary to search for reasonable compromise, but even though the safety should be the highest priority. It has to be stressed that to minimise the risk, the combination of both approaches has to be applied correctly.

17.5.1 Methodology of Safety Quantification

The methodology of safety quantification is based on *risk concept*, which is taken as a combination of two factors:

- the probability of hazard events occurrence or rate of these occurrences,
- the hazard consequences.

The *hazard* is a condition that could lead to an accident. The risk severity can differ and it obviously depends on how often some hazard can occur but also how catastrophic are the results. For example, the same risk severity could be assigned to the hazard with frequent occurrence, but no casualties and the other one with several casualties, but extremely rare occurrence.

The process of *risk analysis* has to be carried on at the very beginning of the new system development. The first part of the analysis consists of the identification of all possible hazards. The classification of occurrence rate and severity of these hazards is done in the next one. The determination of *Tolerable Hazard Rate* (THR) for each hazard is the last part of the risk analysis, but it is the most problematic one. The risk tolerability is not technical problem, but social and legislative one. In addition, there is no European consensus on this issue and also several different methods to calculate THR exist.

The result of risk analysis, which is the list of identified hazards and their associated THRs, is the basis for system hazard & safety risk analysis. This analysis has to identify possible causes of each hazard in system, measures to prevent hazards from occurring or reduce their frequency/probability/consequences and defines the required safety integrity.

Safety Requirements The *System Requirements Specifications* can be divided into requirements, which are not related to safety (including operational functional requirements) and requirements related to safety. The safety requirements are usually contained in a separate *Safety Requirements Specification* and may be considered in two parts. First, *Safety Functional Requirements* which are the actual safety-related functions which the system is required to carry out. Second, *Safety Integrity Requirements* which define the level of Safety Integrity required for each safety-related function.

Safety integrity Safety integrity relates to the system ability to provide required safety-related functions under given operational conditions. The higher the safety integrity, the lower the probability that the system will fail to perform its safety-related functions. The safety integrity is comprised of two components (also see fig. A.8 in [5]):

- *Systematic failure integrity*;
- *Random failure integrity*.

Both the Systematic and the Random failure integrity requirements have to be satisfied, if adequate safety integrity is to be achieved.

Systematic failure integrity is the non-quantifiable part of the safety integrity and relates to hazardous systematic faults (HW or SW). Systematic faults were characterized in detail in Sect. 17.3. Systematic failure integrity is achieved by means of the processes of quality and safety management, plus technical measures; details are described later in this section.

Random failure integrity is that part of the safety integrity, which relates to hazardous random hardware faults, which were characterized in detail in Sect. 17.3.

Safety Integrity Levels The safety relies both on adequate measures to prevent systematic faults and on adequate measures to control random failures. Measures against both causes of failure should be balanced in order to achieve the optimum safety performance of a system. The concept of *Safety Integrity Levels* (SIL) is used to create balance between measures to prevent systematic and random failures. Safety integrity is specified as one of four discrete levels. Level 4 means the highest level of Safety Integrity; level 1 the lowest. Level 0 is used only for the purpose to indicate that there are no safety requirements.

System functions shall have a qualitative safety target and/or a quantitative target attached to them. The qualitative target shall be in the form of a SIL covering systematic failure integrity. The quantitative target shall be in the form of a numerical failure rate covering random failure integrity.

The requirements (see figure A.8 in [5]), which have to be fulfilled to achieve specified SIL, are:

- Quality Management conditions
- Safety Management conditions
- Functional and Technical safety conditions
- Quantified safety targets

It has to be stressed that all of these factors in need to be fulfilled in order to achieve the specified SIL. For example, fulfilment of only quantified safety target does not, by itself, mean that the corresponding SIL has been achieved. As it can be seen from this figure, SILs are used as a criteria to group methods, tools and techniques which, when used effectively, are considered to provide an appropriate level of confidence to Systematic failure integrity.

Techniques and measures to avoid systematic faults An example of techniques and measures for safety-related systems for the avoidance of systematic faults and the control of random and systematic faults is shown below in this section. This is the method how to fulfil qualitative requirements necessary to achieve particular SIL. The guidance is provided in EN 50129, annex E [5] in the form of several tables, which show differentiation of the requirements and measures by the particular Safety Integrity Level.

Selection of suitable system architecture

Table E.4 in [5] shows differentiation of the requirements and measures by the Safety Integrity Level for selection of suitable system architecture. It can be seen, for example, that Single electronic structure with only own check and supervision should not be used for SIL 3 or 4 systems, because it is recommended only for lower SIL 1 or 2.

Relationship between SIL and Safety Targets The table A.1 in [5] shows the relationship between SILs and THRs. In the absence of numerical safety targets this table has to be applied to each function performed by the system, in order to derive the necessary quantitative targets required. The table identifies the required SIL for the system from the quantitative requirement, when the safety targets exist.

The actual Safety Integrity Level and quantified safety target for each particular railway application is the responsibility of the relevant railway or safety authority. There is the effort to unify these safety targets across the Europe, but it is very difficult long-term process.

17.5.2 Evidence of Safety Integrity Level Achievement

The evidence of all the conditions necessary to achieve some particular SIL has to constitute global document called *Safety Case*, which is submitted to the relevant Safety authority in order to obtain safety approval for the system. Some details of this documentation content is given below.

Quality Management This documentation has to provide evidence of Quality Management process throughout the entire life-cycle of the system. Briefly said, the requirements are similar to those defined in the quality standards like ISO 9001 [7] and others.

The purpose of the Quality Management process is to minimize the incidence of human errors at any given phase of the life cycle, in order to reduce the risk of systematic faults in the system and the approach is described in the *Quality Plan*. This document describes the planning carried out by the organization for project development, production management and the system installation. It also provides necessary details regarding the Organization's aspects, specific quality procedures, resources employed, activities, schedules, etc.

Safety Management This documentation has to provide evidence of Safety Management throughout the entire life-cycle of the system. The purpose of the safety management process is to guarantee the implementation of the safety measures, specified by current standards, during the entire project lifecycle, in order to reduce the incidence of human errors related to safety and, therefore, to minimise the residual risk of systematic faults, which results in assurance of Systematic failure integrity of the system.

The *Safety Management Report* is used to demonstrate that the project has been defined, developed and produced in accordance with a safety management system consistent with the EN 50126 requirements and that the organizational structure complies with the EN 50129 standard. It also describes the activities carried out to guarantee the project safety in a manner consistent with the requirements of EN 50129.

Functional and Technical safety This part consists of technical evidence for the safety of the design, which shall be documented in the *Technical Safety Report*. The Technical Safety Report shall explain the technical principles, which assure the safety of the design, including all supporting evidence (design principles and calculations, test specifications and results, safety analyses etc.). It has also to provide assurance of correct functional operation, description of effects of faults, assurance of operation with external influences, description of safety-related application conditions and safety qualification tests. For more details, see [5].

17.6 Safety-Related Data Transmission

Data transmission is very important part of the system functions in modern RSI systems. Generally, the RSI systems have its parts distributed over quite large area, e.g. railway station, which have to communicate together. The processing of correct and actual information in all parts of the system is obviously required.

The older systems transmitted information in form of very simple signal, one bit (active/passive state) per one wire. The modern electronic systems of course use modern digital communication standardized means (buses, protocols) and even commercial-of-the-shelf communication devices (switches, routers), which are available today. The information is sent typically in form of data packets. The problem is that these means do not offer and guarantee the required safety level of communication in safety-related systems. Generally, transmitted safety-related information must not be incorrect (incorrect sender, message type or data content) or obsolete (too much time-delayed or received in incorrect order).

The basic transmission problems, which have to be solved, are:

- message falsification in transmission channel,
- the incorrect message creation or decoding during broadcasting or receiving (faults of terminal equipment).

Naturally, the quick detection and negation of transmission faults is required.

Solution to these problems, which is mainly used today, consists in use of specially designed safety-related functions in application layer in addition to the general data transmission system. If the correct implementation is realized, the result is the safe data transmission system as a complex. It means that the received data will not be dangerously changed neither by interference during transmission nor by faults of transmission or terminal equipment.

Again, the helping guide on how to create communication system with the safety level required by RSI system is content of the standard EN 50159 [6].

When the specification and design of safety-related communication system is carried out, the consideration of relevant possible transmission faults has to be made and corresponding defense measures have to be set.

Let's give one example of protection measures combination for closed transmission system, which is often sufficient. One measure can be the coding as a defense against interferences, the next can be the source and destination identifier as a defense against crosstalk and the time stamping as a defense against excessive time delay.

Fault overview and possible effective defence measures can be found in table 1 in [6].

17.7 Safety Approval and Acceptance of RSI Systems

An operation of RSI system must not adversely affect traffic safety and therefore the system must be approved by the safety authority, as safe enough, for the particular application. The objective of assessment is to arrive at a judgment that the product or the system (subsystem, equipment) fulfils the conditions required for the defined safety integrity level based on credible evidence. The complex verification and confirmation of appropriate SIL, availability, compatibility and interoperability of the RSI system in accordance with European and national standards is necessary to carry on. It must be stressed that the process of assessment and approval cannot be only purely administrative act, including just only verification of existence and positive conclusions of the documents created by the producer. It is highly recommended to start and carry out the assessment concurrent to the development process and beyond system acceptance as appropriate.

Necessary condition for safety approval is the independent safety assessment of the system by competent and certified assessor and also verification of the system functions and long-term dependability in operational conditions (*Safety Qualification Tests*). List of fulfilled conditions, which are required by the safety assessor to approve new system, follows:

- system and safety requirements specifications proving correctly defined requirements for the new system development,
- well-organized and methodical process of the new system development,
- documentation proving that required level of functional and technical safety of designed system was reached,
- completion of appropriate system tests proving its suitability for proposed application,
- documentation proving that the conditions necessary for assurance of required safety for the whole life cycle were created.

The key role of documentation is stressed by European standards, which require the strict structure (*Safety Case*) and systematic processing of documents to provide clear and understandable evidence for safety assessment.

17.8 Summary

As a summary of this chapter it can be said that there is no easy way to design present complex electronic safety-related systems. Systematic approach must be applied and close cooperation with end user of system and safety assessor from initial phases of system life-cycle in the best case. Safety approval process with independent safety assessment is the key activity, which provides confidence that the appropriate level of safety was achieved. There is the helping guidance of EU standards, as was mentioned in the chapter, available nowadays. The correct application of them can assure that the required safety integrity level was achieved. They also serve as a help for comparison of different systems from different producers to support interoperability.

References

1. Composite authors: Railway signalling & interlocking international compendium, ISBN 978-3-7771-0394-5, 2009
2. EN 61508 Functional safety of electrical/electronic/programmable electronic safety-related systems, ed. 2, 2010
3. EN 50126 Railway applications – the specification and demonstration of reliability, availability, maintainability and safety (RAMS), 1999
4. EN 50128 Railway applications – communication, signalling and processing systems: software for railway control and protection systems, ed. 2, 2011
5. EN 50129 Railway applications – communication, signalling and processing systems: safety related electronic systems for signalling, 2003
6. EN 50159 Railway applications – communication, signalling and processing systems – Safety-related communication in transmission systems, ed. 2, 2010
7. ISO 9001 Quality management systems – requirements, ed. 2, 2002

Chapter 18
Emergency-SonaRes: A System for Ultrasound Diagnostics Support in Extreme Cases

Constantin Gaindric, Svetlana Cojocaru, Olga Popcova, Serghei Puiu, and Iulian Secrieru

Abstract Statistics show that in cases of natural disasters, catastrophes and accidents about 70 % of affected persons need specific healthcare approach limited in time. Ultrasound diagnostics at the site of disaster is aimed at determining the level of urgency to save lives and to prevent any complications for people at risk. The Emergency-SonaRes system provides help to physicians from emergency crews in determination of adequate diagnosis in proper time for saving lives of patients. It will offer on-the-fly instructions to obtain qualitative images of affected organs and recommendations for victims sorting: to provide medical assistance in the site or after quick evaluation of the patient state to send him immediately to clinic specialized in organ's pathology or injured organs' system. Two versions of the system are presented: the first one, which can be used at the accident site, as well as in most departments of medical clinics, involved in emergency care – Critical and Intensive Care Service, Emergency Department, Imaging Department. The second version is recommended for utilization in Emergency Medical Services.

18.1 Introduction

We live in a world where the everyday citizens' life is more and more frequently influenced by accidents, cataclysms, natural calamities etc. For example, territories of Romania and the Republic of Moldova are subjected to seismic danger of very

C. Gaindric (✉) • S. Cojocaru • O. Popcova • I. Secrieru
Institute of Mathematics and Computer Science of the Academy of Sciences of Moldova, Academiei, No. 5, Chisinau MD 2028, Republic of Moldova
e-mail: gaindric@math.md

S. Puiu
Moldova State University of Medicine and Pharmacy "N. Testemitsanu", Chisinau, Republic of Moldova
e-mail: puiusv@yahoo.com

H.-N. Teodorescu et al. (eds.), *Improving Disaster Resilience and Mitigation - IT Means and Tools*, NATO Science for Peace and Security Series C: Environmental Security,
DOI 10.1007/978-94-017-9136-6_18, © Springer Science+Business Media Dordrecht 2014

high level. In some regions the seismic intensity may reach 9° (according to seismic 12° MSK scale). Experience of the latest earthquakes demonstrated that also in relatively stable conditions such cataclysms can cause considerable damage.

Along with earthquakes we can mention also other factors of damage, which can hit our country, and namely: landslides, inundations, ecocatastrophes. Concerning the last subject we can remind grave cases of the river Dniester pollution with toxic waste, spilt by Ukrainian chemical industrial enterprises. There persists the danger of catastrophes, which can be provoked by the cause of arsenals and storehouses of rocket propellant situated at the left side of Dniester.

Lack of building plot areas, especially in Chisinau, is compensated, naturally, by the growth of the number of many-storied buildings which are vulnerable at cases of both fires and earthquakes.

In many countries the acts of terrorism become more and more frequent. The dynamics of road accidents becomes alarming as well. In all these cases, the people life-saving depends on the actions of emergency services. Net of the centers of urgent medical help, assures the assistance which is efficient enough in the cases when the number of victims is not large. But if the number of victims is large (mass casualty and disaster setting) the new methods are necessary which will determine quickly the lesion location, first aid measures at the place of accident and the help that should be provided for victims to be transported to specialized clinics. To assure adequate and efficient measures of providing of urgent medical help to victims the efficient on-the-fly diagnostics methods are necessary.

18.2 Ultrasound in Emergency Diagnostics

In hospital conditions, when the possibility of qualified assistance exists, the examination time is not so critical.

When making diagnostics in extreme cases, the strategy of physician is reduced to sorting victims into several groups [1]:

- Priority 1 (Red) Serious but salvageable life threatening injury: the highest priority for transportation and treatment.
- Priority 2 (Yellow) Moderate to serious injury (not immediately life-threatening).
- Priority 3 (Green) "Walking-wounded" Victims who are not seriously injured. Generally, the walking wounded are escorted to a staging area out of the "hot zone" to await fatal injuries.
- Priority 5 (Black) Victims who are found to be clearly deceased at the scene with no vital signs.

There already exist experiences of practical application of ultrasound utilized in an emergent diagnostic imaging capacity.

The sonography at present follows the tendency to be at the top of diagnostic imaging methods and is named "stethoscope of the future". Being a noninvasive method, easily applicable and prevalent, the urgent sonographic examination has the possibility to meet a lack of quick diagnostics in medicine for the catastrophes and calamities, this fact being due to the appearance of portable sonographic scanners.

There are several scenarios for application of ultrasound diagnostics in emergency cases out of hospital [2]: (helicopter) Emergency Medical Services (EMS), mass casualty and disaster setting, austere or remote area, peacekeeping and tactical field, scarce resource health setting.

The goal of emergency ultrasound is to answer a focused clinical question in a timely manner [3] when the physicians sometimes have only minutes to find the cause and treat it correctly before the patient decompensates.

The following arguments speak to ultrasound diagnostics for emergency cases (especially out of hospital) [4]:

- Many trauma patients have injuries that are not apparent on the initial physical exam;
- Significant bleeding into the peritoneal, pleural, or pericardial spaces may occur without obvious warning signs.

In [5] a case study of 202 patients made in six medical centers is presented, in order to determine whether prehospital ultrasound imaging improves management of abdominal trauma. The accuracy of physical examination at the scene was compared with the accuracy obtained using ultrasound. Later examinations in the emergency department (ultrasound and/or computed tomography) were used as the reference standard. The sensitivity, specificity and accuracy of ultrasound examinations were 93 %, 99 % and 99 %, respectively, compared with 93 %, 52 % and 57 % for physical examination at the scene.

The idea of Point-of-care ultrasound, already accepted today in most countries, has its own history of affirmation. The first successful attempts were made in Japan and Germany in 1970s, followed by the implementation as the initial imaging test of choice for trauma care in the United States [4]. In recent years, there have been developed and accepted a number of ultrasound examination protocols for critical situations targeted to specific organs or pathological states: the UHP ultrasound protocol for the empiric evaluation of the undifferentiated hypotensive patient [6], the RUSH protocol (Rapid Ultrasound in Shock) in the evaluation of the critically ill patient [7], International evidence-based recommendations for point-of-care lung ultrasound adopted by the International Consensus Conference on Lung Ultrasound in 2011 [8] etc. The last document was elaborated to guide implementation, development, and training on use of lung ultrasound in all relevant settings; it will also serve as the basis for further research and to influence and enhance the associated standards of care.

However, there are not so many widely acknowledged international guidelines, and only a few scattered national examples exist that address ultrasound training for physicians, nurses, and paramedics performing ultrasound examinations in emergency and critical care settings [2].

The ultrasound examination method of blunt trauma, called FAST – Focused Assessment with Sonography in Trauma [4], has been accepted rapidly enough worldwide. Initially intended for trauma of abdominal organs, in a short time it has been extended for cases of identifying free fluid (usually blood) generally in the pericardial, pleural or intraperitoneal spaces. Based on the assumption that many life-threatening injuries cause bleeding, FAST exam focuses on limited scanning planes, when physicians are only trying to find free fluid and not do a comprehensive survey of the involved organs.

On the other hand, existing guidelines and protocols propose only algorithms, which specify the examination sequence (Right Upper Quadrant Left Upper Quadrant, etc.), containing indications like: "The heart will be surrounded by a rim of echogenic pericardium. Any discrete blackness between this rim and the heart wall represents fluid in the pericardial sac" [9].

Given the fact that in critical situations rapid and qualitative decision is an important factor, there is the need for development of decision support systems that meet these requirements.

18.3 SonaRes: A Decision Support Systems in Ultrasound Diagnostics

At present there exist several decision support systems in sonographic diagnostics, one of them, named SonaRes, being elaborated in the Institute of Mathematics and Computer Science of the Academy of Sciences of Moldova in collaboration with Moldova State University of Medicine and Pharmacy "N. Testemitsanu" [10–12]. The system is oriented to support sonographic examinations of hepato-pancreato-biliary pathologies. It is founded on both the images analysis and analysis made on the basis of rules – the majority of known world-wide systems being based only on one of these approaches.

The SonaRes system is designed to be used in imaging departments of medical centers and clinics. In addition, it can be used as well for training of novices or students by accessing the detailed diagnostics mode and explanation module. In cases when physicians are pressed by time in their decision making process, the SonaRes system provides the possibility for conclusion deduction based on a minimal set of necessary facts – "quick examination".

In conditions of out-of-hospital examination at the site another approach is required than the "quick" one. In most cases, it is to establish if free fluid (blood) is present and the cause (from which organ appears).

In the process of the system SonaRes elaboration some problems had been already solved (medical description of hepato-pancreato-biliary region organs, structuring of pathologies, anomalies; formalization of the data obtained from experts group; development of unified DB; development of methods of image collection and storage; elaboration of diagnostics validation tool; development of user interface).

Methodology, technology and medical knowledge representation form, used in framework of the development of the SonaRes decision support system, have allowed the creation of the Emergency-SonaRes system, which will help physician-sonographist to obtain quickly a correct diagnosis directly in the places of accidents, catastrophes, and acts of terrorism. It will offer on-the-fly instructions to obtain qualitative images of affected organs and recommendations for sorting: to provide medical assistance in the site or after quick evaluation of the patient state to send him immediately to clinic specialized in organ's pathology or injured organs' system.

18.4 Emergency-SonaRes

One can distinguish two versions of the Emergency-SonaRes: an extension of the SonaRes system and a special application Emergency-SonaRes (Table 18.1).

In the first case the algorithm of its use is the following: (1) examine the status of patient under the scenario of the Emergency-SonaRes incorporated in the SonaRes system; (2) if the status of patient allows, the user (physician-sonographist) can select "quick examination" in the SonaRes system in order to perform a detailed examination; (3) if the patient is in serious or critical condition, physician makes decision using the conclusion, offered by the system based on scenario of the Emergency-SonaRes, as the second opinion.

The second version is a standalone application, developed for emergency situations. This application contains only modules and knowledge base necessary for determination of the injury severity and danger of the patient's life.

The first version can be used at the accident site, as well as in most departments of medical clinics, involved in emergency care – Critical and Intensive Care Service, Emergency Department, Imaging Department.

The second version is recommended for utilization in EMS. This version can be incorporated into medical equipment or the portable electronic means.

Let follow the changes in the knowledge base on the example of the liver ultrasound examination (Table 18.2).

One can observe that to describe any pathology and/or abnormality of the liver the SonaRes system uses 167 facts, mostly based on sonographic signs, when the standalone application for emergency cases operates just with 67.

Table 18.1 SonaRes system versions and their application domains

SonaRes system		Extended SonaRes system with Emergency-SonaRes scenario		Emergency-SonaRes system	
In hospital	Out of hospital	In hospital	Out of hospital	In hospital	Out of hospital
Imaging department	No	Critical and intensive care service Emergency department Imaging department	EMS	No	EMS

Table 18.2 Changes in knowledge base and user interface depending on utilization of the system

SonaRes system		Extended SonaRes system with Emergency-SonaRes scenario		Emergency-SonaRes system	
Number of facts and decision rules	Examination mode	Number of facts and decision rules	Examination mode	Number of facts and decision rules	Examination mode
Liver – 167 facts and 31 rules	3 (quick, normal, deep)	Liver – 67 (167) facts and 11 (31) rules	2 (emergency – by default, quick – additional)	Liver – 67 facts and 11 rules	1 (emergency)
Gallbladder – 335 facts and 54 rules		Gallbladder – 52 (335) facts and 5 (54) rules		Gallbladder – 52 facts and 5 rules	
Pancreas – 231 facts and 52 rules		Pancreas – 45 (231) facts and 4 (52) rules		Pancreas – 45 facts and 4 rules	
Bile ducts – 257 facts and 15 rules		Bile ducts – 31 (257) facts and 3 (15) rules		Bile ducts – 31 facts and 3 rules	

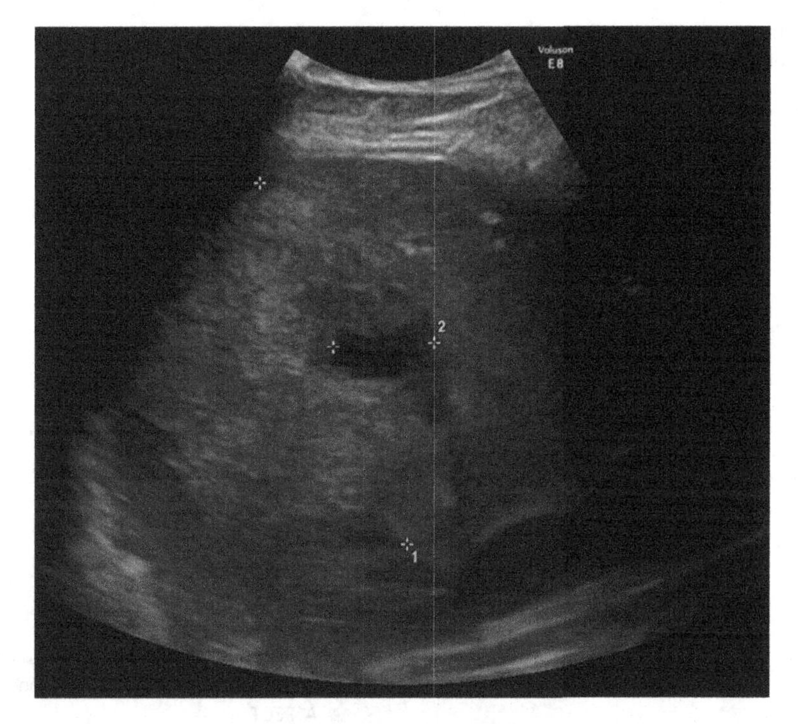

Fig. 18.1 Post-traumatic lesion of liver. Recent (acute) hepatic hematoma

If the liver injury is suspected (Fig. 18.1) the ordinary examination will produce the following conclusion: A heterogeneous lesion of considerable size with not clear defined boundary is viewed in the right lobe of the liver. The heterogeneity is caused by the inhomogeneous zones with increased echogenity, which represent the parenchyma contusions and by the avascular anechoic and hypoechoic areas, which mean small hematomas, representing consequences of recent closed abdominal trauma.

For the same case the Emergency-SonaRes system needs the answers only the following questions: Is liver contour discontinued? Are modifications in the liver structure (mostly circumscribed) observed? Are collection(s) in the liver surrounding areas detected? Has the patient severe pains with acute onset (post-traumatic)? Of course, the last question should not be answered if the patient fainted and the process stops on the third point.

The physician can conclude that the liver contour is not affected, a circumscribed (focal) lesion is detected in the liver parenchyma, and no fluid (blood) collection in perihepatic areas or in abdomen. The patient complains pain. The final decision for victims sorting in the case of examination at disaster scene corresponds to yellow code, which doesn't mean a life-threatening situation and patient should be directed to a medical clinic for surveillance, to prevent any complication.

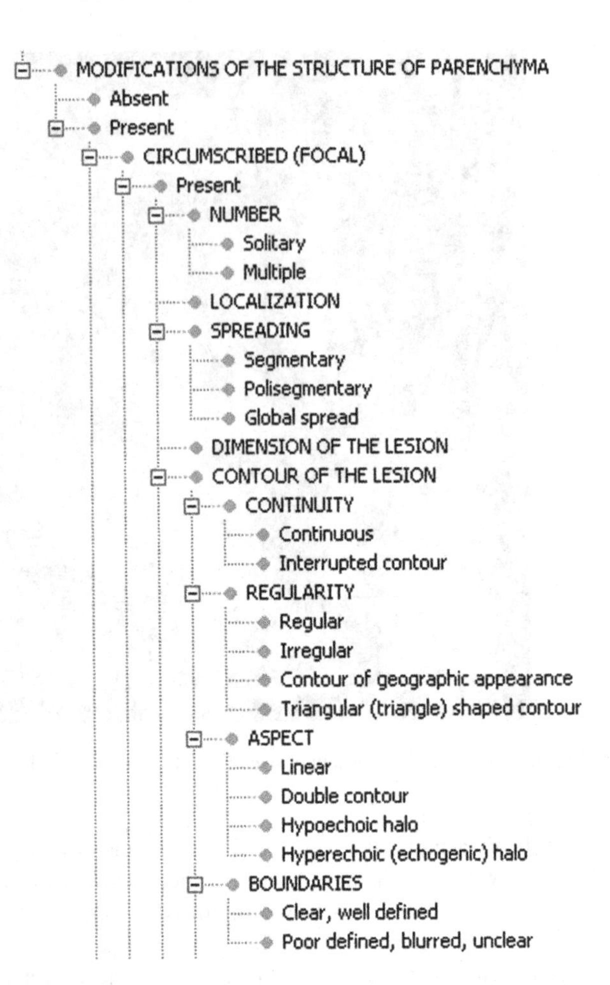

Fig. 18.2 A fragment of the base of facts used for description of modifications of the structure of parenchyma

To confirm these facts the Emergency-SonaRes system goes to description of the structure modifications and the detected collections. Toward this end, a minimal set of 68 facts is used. To confirm or reject them the physician answers eight questions on average (the maximum being equal to 14), if the patient's status is not critical. In critical cases only four questions are enough.

Figure 18.2 represents a fragment of the base of facts used for description of modifications of the structure of parenchyma. Based on it the decision rule of the liver injury is created.

The evaluation of SonaRes system which has been carried out by 149 physicians [13] shows that an experienced specialist needs 5–7 min for hand-free obtaining of an ultrasound examination protocol. Duration of data input into the system at the beginning of the work with SonaRes was 8–10 min and 3–5 min after habituation to the system and its frequent using.

According to the average of questions one have to answer in the process of examination using the Extended SonaRes we expect its duration no more 2–3 min while for the Emergency-SonaRes we expect its duration of diagnostics to be no more than 1–2 min.

18.5 Conclusions

Human life is the most precious and the most fragile thing that we must take care in disaster situations. Each emergency case when people suffer requires much effort from the rescue teams that can be facilitated and accelerated by applying technologies capable of saving lives and health of victims.

The Emergency-SonaRes system, which is now under development, will be useful both in on-the-fly diagnostics and training in the critical care domain. It allows for qualitative and fast primary sorting of victims in order to provide the necessary, but differentiated assistance.

Collection and systematization of data, obtained using the Emergency-SonaRes system, will allow analyzing actions of the special services in emergency cases and, if needed, making necessary changes to better coordination of EMS and In Hospital departments.

The experience obtained in knowledge formalization and definition of the main decision-making rules for ultrasound diagnostics of hepato-pancreato-biliary region in emergency situations can be used to extend the knowledge base of the Emergency-SonaRes system with information and data about other organs.

Acknowledgments This work has been supported by the Supreme Council for Science and Technological Development under projects 11.817.08.03A “Advanced technologies of intelligent systems development for Information Society” and 13.824.18.170T “Clinical decision support system in domain of ultrasound examination of the hepato-pancreato-biliary zone (SonaRes 13)”.

References

1. Amick RD (1998) Mock disaster operations. Emergency medical response, Jefferson county open space. http://bcn.boulder.co.us/community/explorer/ep493d4c.htm.Accessed19Dec2013
2. Neri L, Storti E, Lichtenstein D (2007) Toward an ultrasound curriculum for critical care medicine. Crit Care Med 35(5),(Suppl):290–304
3. Ashar T (2005) Introduction to emergency ultrasound. Israeli J Emer Med 5(4):299–302
4. Reardon R (2005) Ultrasound in trauma – the FAST exam. Focused assessment with sonography in trauma. http://www.sonoguide.com/FAST.html.Accessed19Dec2013
5. Walcher F, Weinlich M, Conrad G, Schweigkofler U, Breitkreutz R, Kirschning T, Marzi I (2006) Prehospital ultrasound imaging improves management of abdominal trauma. Br J Surg 93(2):238–242
6. Rose JS, Bair AE, Mandavia D, Kinser DJ (2001) The UHP ultrasound protocol: a novel ultrasound approach to the empiric evaluation of the undifferentiated hypotensive patient. Am J Emerg Med 19(4):299–302

7. Perera P, Mailhot T, Riley D, Mandavia D (2010) The RUSH exam: rapid ultrasound in SHock in the evaluation of the critically ill patient. Emerg Med Clin North Am 28(1):29–56, vii
8. Volpicelli G et al (2012) International evidence-based recommendations for point-of-care lung ultrasound. Intensive Care Med 38:577–591
9. Logan P, Lewis D (2004) Focused Assessment with Sonography for Trauma (FAST). Emergency Ultrasound UK
10. Rîbac E, Cojocaru S, Gaindric C, Puiu S, Ţurcanu V (2005) The process of designing and implementing the examination support in sonographic investigations. In: Proceedings of the international conference on advanced information and telemedicine technologies for health (AITTH'2005), Minsk, 8–10 Nov 2005
11. Burtseva L, Cojocaru S, Gaindric C, Popcova O, Secrieru S (2011a) Ultrasound diagnostics system SonaRes: structure and investigation process. In: Proceedings of the second international conference on modelling and development of intelligent systems (MDIS 2011), Lucian Blaga University Press, Sibiu, 29 Sept–02 Oct 2011
12. Burtseva L, Gaindric C, Cojocaru S (2011b) Images and rules based decision support system in ultrasound examination of abdominal zone. Memoirs of the Scientific Sections of the Romanian Academy. Computer Science, SERIES IV, Tome XXXIV:173–184
13. Cojocaru S, Gaindric C, Puiu S (2013) The evaluation of DSS SonaRes – the initial results. In: Proceedings of the 4th IEEE international conference on e-health and bioengineering (EHB 2013) "Improving quality of life through research and innovation", Iasi, 21–23 Nov 2013

Chapter 19
FireFight: A Decision Support System for Forest Fire Containment

Jaume Figueras Jové, Pau Fonseca i Casas, Antoni Guasch Petit, and Josep Casanovas

Abstract The FireFight project is being developed in collaboration with the GRAF wildland firefighting department (Generalitat de Catalunya, Spain). The main objective is the development of a web-accessible decision support system based on an integrated simulation and optimization framework for optimal wildfire containment. FireFight uses the tooPath (www.toopath.com) web server infrastructure to acquire the broadcasted real-time GPS position of approximately 1,650 land and aerial firefighting resources deployed across the territory. The short-term goal of the project is to help managers in making decisions about the number of extinguishing teams that should be deployed, the design of the water supply chain to bring water and other supplies to the firefighting teams, and the design of the change-of-shift transportation problem.

19.1 Introduction

Forest fires have become increasingly devastating and uncontrollable phenomena and their management is a major issue in Mediterranean countries, where they are an annual occurrence. This situation is certain to worsen with the anticipated impact of climate change and the continuous decline in agricultural land use, which allows the extension and continuity of forested areas to increase. Land abandonment and the cessation of traditional management practices is favoring scrubland and forest expansion throughout Europe, reducing the extent of many semi-natural open habitats of a high ecological value [1]. Figure 19.1 shows a typical map of forest fire risk during the summer period in the Catalonia region (north east of Spain); most of the land has a medium to extreme risk of forest fire.

J. Figueras Jové • P. Fonseca i Casas • A. Guasch Petit (✉) • J. Casanovas
Universitat Politècnica de Catalunya, Edifici B5-S102, C/Jordi Girona, 31,
Barcelona 08025, Spain
e-mail: Jaume.figueras@upc.edu; pau@fib.upc.edu; toni.guasch@upc.edi; josepk@fib.upc.edu

H.-N. Teodorescu et al. (eds.), *Improving Disaster Resilience and Mitigation - IT Means and Tools*, NATO Science for Peace and Security Series C: Environmental Security,
DOI 10.1007/978-94-017-9136-6_19,

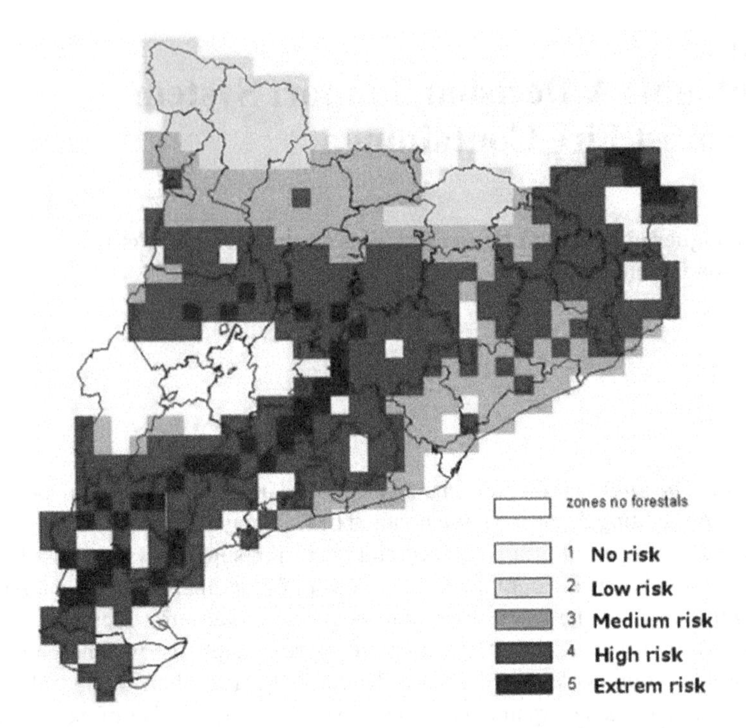

Fig. 19.1 Map of forest fire risk, August 2012 (Source: Generalitat de Catalunya)

The management of these fires is a serious concern for many governments around the world, and responsibility for decisions regarding containment falls to specialist wildfire managers; in the event of a reported wildfire, they select which of the firefighting resources at each base station should be dispatched to the fire site, establish a plan of action, and determine the firefighting tactics that should be employed. Taking such decisions is difficult due to the time constraints, the dynamic and uncertain behavior of fires, and the limited availability of firefighting resources. Correct decision-making is crucial for saving lives, property and natural resources.

In Catalonia, forest fire decisions are taken by the GRAF team lead by Marc Castellnou. The GRAF team is a group within the firefighting department of the Government of Catalonia and is responsible for preventive actions as well as resource deployment and attack. A typical preventive action is the burning of specific areas in winter to prevent excessive connectivity of large forest areas.

Fuel treatment of the wildland-urban interface is an approach that can mitigate the risk of fire propagation through urban or residential areas [2]. In 2003, Law 5/2003 was passed in Catalonia laying down mandatory measures for residential and rural buildings designed to prevent forest fires. A summary of these regulatory measures is given below:

- Buildings must be surrounded by a protective perimeter at least 25 m wide, cleared of shrubs and with reduced tree fuel.

- Plots within the protective perimeter must be cleared of dry vegetation.
- Residential areas must have a network of approved fire hydrants.
- An approved self-protection plan must be in place in these areas.

Unfortunately, not all wildland-urban interfaces are protected by a cleared perimeter, and the problem is worsened by the reduction of resources devoted to wildfire protection and the continued expansion of the wildland-urban interface. In large wildfire scenarios affecting urban areas or isolated farms, the decision-making process is a critical task with potential repercussions on human life and property. The ability to predict fire spread behavior and to evaluate the impact of alternative tactics is crucial to decision-making for wildfire containment [3].

FireFight is an ongoing project concerned primarily with the development of a decision support system (DSS) based on an integrated simulation and optimization framework for optimal wildfire containment.

19.2 State of the Art

Modeling wildfire and the related containment processes is not an easy task due to the inherent complexity of the different processes involved. As such, various efforts to model wildfires have focused on one or more specific aspects of the phenomenon. However, the most obvious element to be modeled is fire propagation across the landscape. Some theoretical approximations involve the use of operations research techniques to model fire behavior [4–6]. In these approaches, the representation of wildfire behavior is generated taking into account landscape, wind and many other factors that can affect a fire's movement. The modelling of individual wildfire factors is often challenging due to their specific complexities, as is the case of wind, which has been the focus of several studies, in particular [7, 8], which present local descriptions of wind behavior for risk areas in Wales and the Netherlands. Other factors have also been the focus of specific studies. Thus, [9] presents a model for determining the probability of fire occurrence in forest stands of Catalonia, and [10] proposes the Dynamic Fire Risk Index (DFRI), which establishes the risk of fire occurrence on the basis of different static and dynamic factors.

All of these models rely heavily on the availability of accurate data, so it is necessary to verify the correctness of the data fed into the model as well as ensuring the early detection of the phenomenon. In this sense, notable studies include [11], on early detection, and [12], on the mapping of possible and probable fires and the retrospective modelling of past forest fire scenarios. An interesting study that considers the problems deriving from low visibility due to cloud cover was published by [13].

However, these theoretical models do not take into account the containment process. If a complete description of the phenomenon is needed and several alternatives for the containment process are considered, a holistic approach is required. In this case, it is necessary to have not only a robust model for representing the fire behavior but also a model for representing resource allocation and containment strategy.

Existing infrastructures for representing wildfire behavior and containment include FARSITE [14], a fire behavior and growth simulator used by the USDA FS, USDI NPS, USDI BLM, and USDI BIA Fire Behavior Analysts. It is designed for use by planners and managers familiar with fuels, weather, topography, wildfire situations, and the associated concepts and terminology. FARSITE and other simulation tools such as HFire [15], Prometheus [16] and SiroFire [17] focus heavily on the representation of fire evolution, although they also offer some capabilities for representing fire containment.

Systems that focus more specifically on containment include the proposal of [18], who present a unified framework based on a mathematical programming formulation that integrates various decisional problems arising in the management of different kinds of natural hazards. Donovan and Rideout [19] analyzes how the resources needed for the containment process can be optimized using integer programing. Although this approach is powerful, certain limitations have led to the examination of other alternatives, such as the use of stochastic simulation [20]. SIADEX [21] comprises four main components: a web server, which centralizes the flow of information between the system and the user; the planning and monitoring servers, which are offered as intelligent services through the web server; and the ontology server, for sharing and exchanging knowledge between all components. Hu and Ntaimo [3] presents a stochastic mixed-integer programming model for initial attack to generate firefighting resource dispatch plans using as input fire spread scenario results from a standard wildfire behavior simulator. The same study also presents an agent-based discrete event simulation model for fire suppression, used to simulate fire suppression based on dispatch plans from the stochastic optimization model.

The use of Multi-Agent Simulation (MAS) to represent containment has also been explored by [22], who propose a model for coordinating teams of computational agents that can be used for different domains but is specifically applicable to forest firefighting. The model is based on Pyrosym [23]. MAS has been also been used to represent evacuation under fire conditions [24].

The use of formal languages to model wildfire behavior is a promising recent development. A model to represent wildfire propagation, using a formal representation of a cellular automaton based on DEVS, is presented in [5]. Ntaimo [25] presents DEVS-FIRE, which also uses DEVS as a formal language to represent the phenomenon. DEVS-FIRE can be integrated with a stochastic optimization model to determine the optimal firefighting resources to dispatch to guarantee wildfire containment in the shortest possible time and with minimal cost. Other studies of DEVS for representing the wildfire phenomenon include [26], which presents an interesting combination of Cell-DEVS with CD++ [27] to reduce the algorithmic complexity for the modeler, enabling the use of complex cellular timing behaviors and different Cell-DEVS quantization techniques to decrease execution time. Similar approaches to reduce the inherent complexity of this kind of simulation can be reviewed in [28]. Other formal languages have been used to represent wildfire phenomenon, like Specification and Description Language [29]. One of the main issues that must be addressed in this type of approach is how to represent the environment with the selected formalism, which often entails the representation of cellular automaton structures. Fonseca i Casas et al. [30] presents a method for using SDL to formally represent cellular automaton structures and specifically a wildfire propagation model. Gronewold and Sonnenschein [31] explains how to model cellular automaton structures using Petri nets.

Finally, it is interesting to note studies that focus on validation and verification. Niazi et al. [32] presents the verification and validation of an agent-based model of forest fires using a combination of a Virtual Overlay Multi-Agent System (VOMAS) validation scheme with Fire Weather Index (FWI) to validate the forest fire simulation.

19.3 FireFight Architecture

The proposed experimental environment project architecture is shown in Fig. 19.2. Its principal component is an optimizer that uses fire spread and contention simulations to determine the best fire attack strategy for a particular scenario. The main processes and outputs of the FireFight system are as follows:

Web GUI and tracking server: Designed to facilitate user interaction with the FireFight system. Two different operational modes are provided: a training or evaluation mode that uses an on-line fire spread simulator to generate different evaluation scenarios; and a real-time mode intended to support the management of ongoing wildfires.

Fire & logistics scenario: A simulated wildfire scenario consists of the information defined by [33]. This information is read by the GUI, which tells the controller to generate the scenario from the geospatial database. An ongoing wildfire scenario is monitored via data and other relevant information obtained from field sensors (i.e. GPS data) and from firefighting personnel.

Controller: Synchronizes the different application processes.

On-line fire spread simulator: Simulates a real fire, applying the best tactics proposed by the optimizers. This tool is intended for training and evaluation using models of real fires or hypothetical fire scenarios.

Optimizer: Searches for the best feasible solutions from a set of fire suppression tactics and decisions using the resources available in a particular scenario.

Feasible plans: The set of feasible solutions found by the optimizer which must be simulated in order to evaluate their potential outcomes.

Best plan: The best of the feasible solutions proposed by the optimizer, which must be implemented by the firefighters. This solution is used by the reality simulator.

Fire spread and containment logistics simulator (proposed plans): Evaluates the impact of the proposed containment logistics decisions (feasible plan) proposed by the optimizer.

Results: The output of the fire spread and containment logistics simulator for the proposed plans. These results are used by the optimizer to search for a good solution.

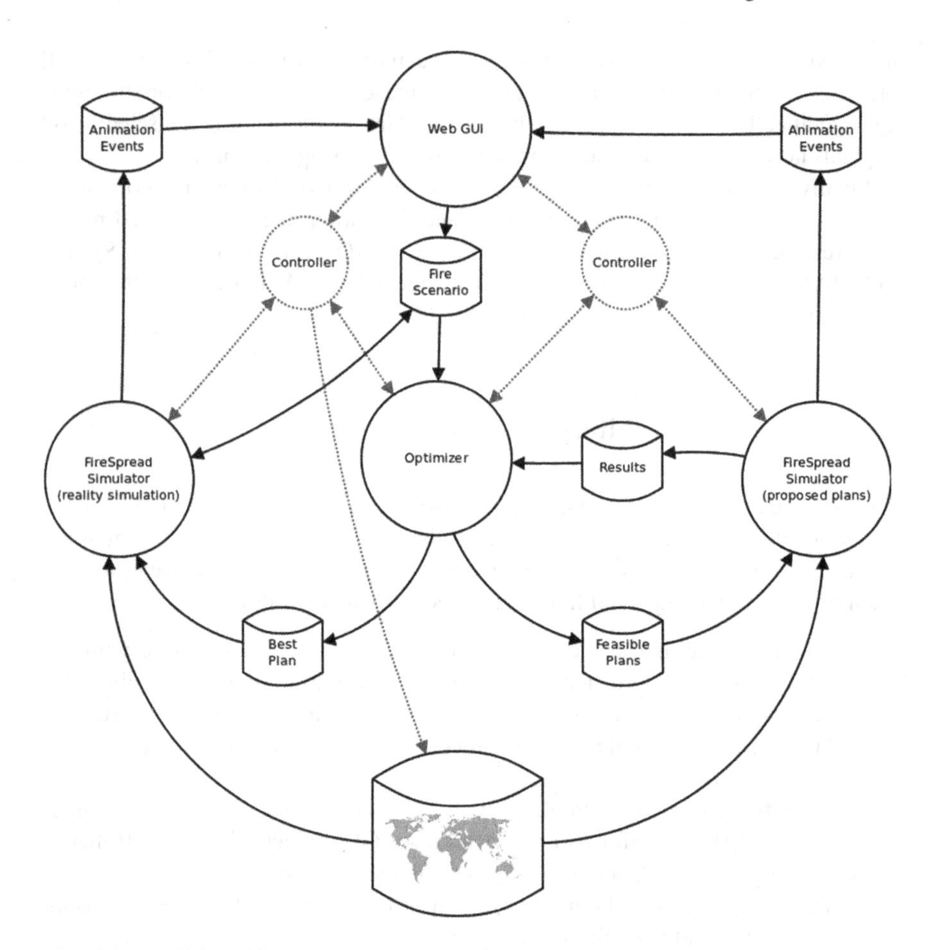

Fig. 19.2 FireFight architecture

The FireSpread simulator is currently being used offline by the GRAF team to predict fire evolution. Analysis of the simulation identifies an anticipated active fire line that is fed into the optimizer to determine the best plan for the deployment of containment resources.

19.3.1 Web GUI and Tracking Server

Personnel portable GPS devices became mandatory in Catalonia for all firefighting teams following recommendations issued after the Horta de Sant Joan fire of July 2009, during which five GRAF elite firefighters died when trapped by fire, despite the deployment of a last resort fire shelter, consisting of an aluminum blanket that protects against flames and heat.

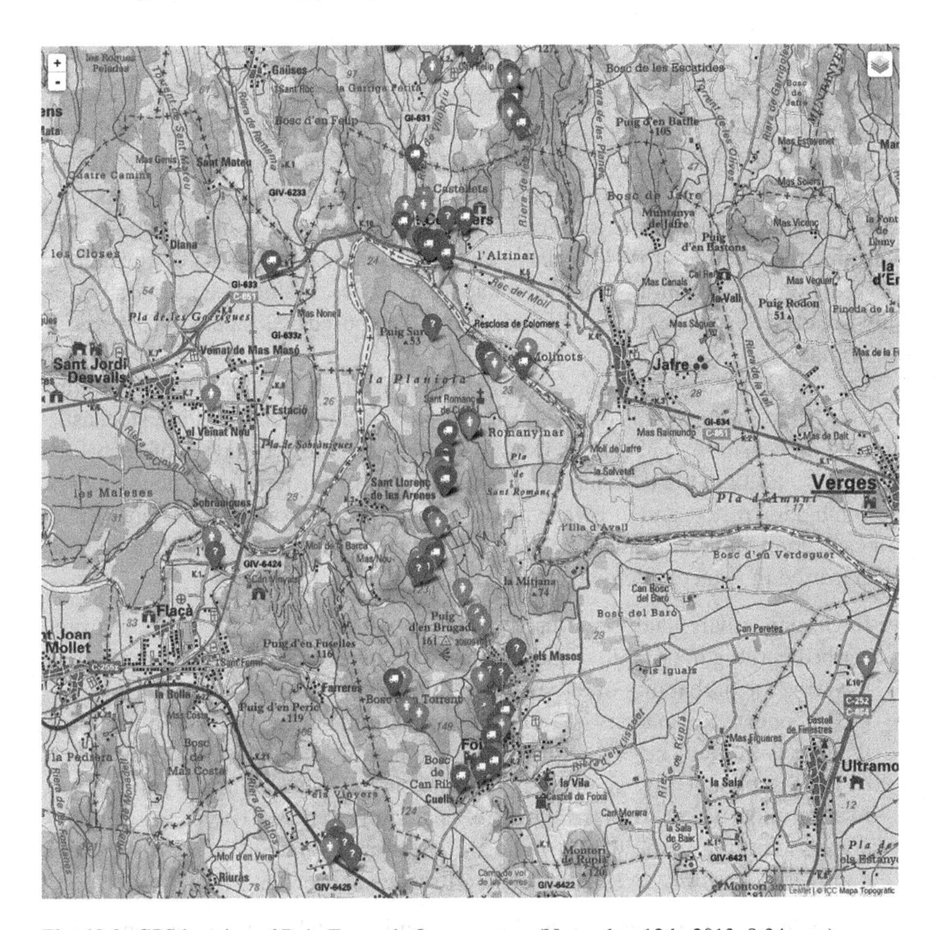

Fig. 19.3 GPS location of Baix Emporda fire resources (November 12th, 2013, 8:34 a.m.)

In Catalonia, all aerial and ground firefighting resources are now located and tracked by GPS. According to the latest figures, 1,650 GPS devices currently send information to the location servers. Figure 19.3 shows the position of all resources deployed for a wildfire on November 12th, 2013. This information is also broadcast in real-time to our tooPath (www.toopath.com) tracking server.

GPS data is a valuable source of information for real-time decision making and post-fire analysis. The effective use of this information is a key focus for future research and development in the area of DSS for wildfire containment logistics.

The containment effectiveness of each ground firefighting team depends on natural factors such as vegetation density, vegetation type and topography, and logistic factors such as the availability of aerial resources. One of our first goals is the systematic analysis of the GPS data in order to obtain real measures of the effectiveness of the attack force. A containment table for initial attack effectiveness has already been proposed [34], and our analysis will expand on current results with a focus on Mediterranean forest fires.

19.3.2 Fire Spread Simulation

Prescribed burning is commonly used in the management of fire-prone vegetation in order to create fuel breaks which may reduce the occurrence of large-scale wildfires [35, 36]. Therefore, when firefighting resources are scarce it is very important to select the most appropriate locations for fuel breaks. One possible approach for selecting the location of fuel breaks is to use fire spread simulations. For example, [37] propose a simulation-optimization approach where a stochastic simulation of fire spread estimates the fire risk of each solution proposed by a metaheuristic search algorithm.

The GRAF firefighting team of the Government of Catalonia uses Wildfire Analyst (http://tecnosylva.com) software to predict fire spread. Offline stochastic simulation sensitivity analysis is used to predict the future behavior of each fire, taking into account the dominant wind in each area of the territory. The team uses the simulation results and its own expertise to build a Fire Propagation Graph (FPG) for each dominant wind situation in all forest areas across the Catalonia region. The FPG is used in two complementary ways:

- FPG connectivity analysis is used to decide where to execute off-season prescribed burning in order to reduce the fuel load in specific arcs of the graph, thus lowering the probability of large-scale wildfires.
- FPG connectivity analysis is also used for real-time decision making when the wildfire is active. The FPG is used in combination with the expected evolution of the fire obtained by simulation with the current meteorological information to determine where to place the extinguishing resources. The decision is not always to minimize the spread of the fire; the protection of urban areas is usually considered a priority, even at the cost of further fire growth.

Dynamic real-time access to fire information is a vital element for controlling wildfires. Fire managers who respond to wildland fire outbreak need regular updates on the state of the fire. Details of the active fire fronts are therefore crucial because they can be used to estimate how fast the fire is moving and to predict where the fire is likely to spread in the future [38]. The anticipated active fire line is used for tactical-level decision-making.

19.3.3 Containment Optimization

Strategic planning is designed to minimize the loss and suffering associated with wildfires by preventing fire outbreaks. Strategic planning also encompasses decisions related to the fire season as a whole. Once a forest fire has broken out, however, tactical-level decisions are needed to bring the fire under control [25]:

1. Strategic decisions: How much money should be spent on forest fire management each year? How many aerial devices should be used and where should they be

placed? How many firefighting teams should be hired? Where should preventive actions be carried out off-season? How should the firefighting teams and aerial resources be deployed each day? When and where should special preventive measures like restricted fire zones and restricted travel zones be invoked?

2. Tactical decisions: How many firefighting and aerial resources should be allocated to a specific fire? Where should ground resources be placed? How many water trucks are needed to replenish each fire water truck?

19.4 Decision Support System Development

A DSS is a computer-based information system that supports decision-making activities [39]. The FireFight project can be classified as an active DSS [40] providing assistance based on a model-driven DSS [41].

An active DSS is a support system that is capable of providing decision suggestions or solutions to different problems without subsequent feedback or refinement from the advisor in charge of the decision-making process. As described in Sect. 19.3.3, the decision suggestion is processed by an optimizer that provides best solutions, thus if no additional data are provided the optimizer cannot refine its search. To overcome the limitations of an active DSS relative to a cooperative DSS, the solutions are recalculated at regular intervals taking into account newly received data.

A model-driven DSS generates suggestions and solutions on the basis of simulation and optimization models. As described in Sects. 19.3.2 and 19.3.3, FireFight DSS uses a fire spread and containment simulator to simulate fire propagation during a fire event and an optimizer to calculate feasible and optimal solutions.

19.4.1 Optimization in FireFight DSS

There are three problems the DSS must solve. The main problem is optimizing the deployment of extinguishing teams, which are composed of an engine and an engine crew. The second problem is ensuring a supply of water and equipment to each engine. The third problem is relieving each engine crew at the end of every operational period (usually 24 h) with new crews arriving from all over the territory.

19.4.1.1 Deployment Optimization

The deployment optimizer determines the best way to contain the fire by answering the following questions: How many engine crews are needed to contain the fire by a specific date/time? What is the outcome of increasing or decreasing the number of crews? Where can crews be placed, taking into account the road/path/trail network, the fire expected evolution, the topography and the cartography?

The answers are calculated taking into account the fire-spread simulation, the topography, the land-cover vegetation and the road/path/trail network. With this information and the GPS tracking data, the DSS calculates the wet-line containment crew speed, enabling the optimizer to determine the optimal deployment.

19.4.1.2 Supply Chain Design

The DSS system designs the water supply chain to ensure that each fire truck has a continuous supply of water. In configuring the design, the following questions must be answered: How many water supply trucks are needed for each extinguishing team, given a geographical distribution of water depots? How many additional tankers or engines are needed to refill finite capacity water depots?

Exact data on water consumption by firefighter crews are unavailable and must therefore be estimated by interpolating historical data on wet-line containment crew speed.

19.4.1.3 Shift Changes Between Operational Periods

The FireFight DSS has to calculate how each of the different crews are replaced with new crews for another operational period. The change-of-shift problem arises because the firefighting vehicles used to transport the crews are located at the site of the wildfire and cannot move from there, hence other resources are needed. In Catalonia, temporary transport is provided by the police, so the DSS has to answer the following questions: How many police cars are needed to pick up 'fresh' fire crews teams from geographically distributed fire stations and bring them to the fire? Are they enough to return the firefighters finishing their shifts to their base stations? What route and timetable should each car follow in order to deliver fire crews to the site on time?

This problem is modeled as a VRP problem with time windows and capacity constraints. Although it is similar to the school bus problem (SBP), this particular optimization problem does not take into account the return of the crews ending their shifts to their base stations, the time of which must be minimized.

19.5 Conclusions

This article focuses on the architecture and approaches for optimizing the solutions to different problems in the emergency management of a wildland fire and presents the FireFight experimental environment architecture for planning optimal deployment and management for wildfire containment and suppression. The system will use real data from different sources. The optimization and simulation framework uses standardized information to ensure interoperability with different systems. The use of different simulators to represent the current state and forecasted state of the fire makes the system robust to changes and enables the optimizer to adapt to new realities. This study and the improvements it reports are being developed in conjunction with the firefighting department of the Government of Catalonia, with the aim of solving problems present in the wildland firefighting workflow.

Acknowledgments This study was funded by the Ministry of Science and Innovation, Spain. Project reference TIN2011-29494-C03-03.

References

1. Ascoli D, Lonati M, Marzano R, Bovio G, Cavallero A, Lombardi G (2013) Prescribed burning and browsing to control tree encroachment in southern European heathlands. For Ecol Manage 289:69–77. doi:10.1016/j.foreco.2012.09.041
2. Syphard AD, Keeley JE, Brennan TJ (2011) Comparing the role of fuel breaks across southern California national forests. For Ecol Manage 261(11):2038–2048. doi:10.1016/j.foreco.2011.02.030
3. Hu X, Ntaimo L (2009) Integrated simulation and optimization for wildfire containment. ACM Trans Model Comput Simul 19(4):1–29. doi:10.1145/1596519.1596524
4. Andrews P (2007) BehavePlus fire modeling system: past, present, and future. ... of 7th symposium on fire and forest meteorology. Retrieved from http://www.researchgate.net/publication/8400756_Diving_ability_of_Anopheles_gambiae_(Diptera_Culicidae)_larvae/file/9fcfd4ff51c851cff4.doc
5. Innocenti E, Santucci J, Hill DRC, Cnrs IUMR, Cedex A (2004) Active-DEVS: a computational model for the simulation of forest fire propagation *
6. Rothermel RC (1983) How to predict the spread and intensity of forest and range fires. Boise, Idaho, p 166
7. Luo W, Taylor M, Parker S (2008) A comparison of spatial interpolation methods to estimate continuous wind speed surfaces using irregularly distributed data from England and Wales. Int J Climatol 959:947–959. doi:10.1002/joc
8. Luo W, Taylor M, Parker S (2004) Spatial interpolation for wind data in England and Wales, University of York. doi:10.1002/joc.1583
9. González JR, Palahí M, Trasobares A, Pukkala T (2006) A fire probability model for forest stands in Catalonia (north-east Spain). Annals For Sci 63(2):169–176. doi:10.1051/forest:2005109
10. Hernández-Leal PA, González-Calvo A, Arbelo M, Barreto A, Alonso-Benito A (2008) Synergy of GIS and remote sensing data in forest fire danger modeling. IEEE J Sel Top Appl Earth Obs Remote Sens 1(4):240–247, Retrieved from http://ieeexplore.ieee.org/xpls/abs_all.jsp?arnumber=4703188
11. Pennypacker C, Jakubowski M, Kelly M, Lampton M, Schmidt C, Stephens S, Tripp R (2013) FUEGO – fire urgency estimator in geosynchronous orbit – a proposed early-warning fire detection system. Remote Sens 5(10):5173–5192. doi:10.3390/rs5105173
12. Carvalheiro LC, Bernardo SO, Orgaz MDM, Yamazaki Y (2010) Forest fires mapping and monitoring of current and past forest fire activity from meteosat second generation data. Environ Model Software 25(12):1909–1914. doi:10.1016/j.envsoft.2010.06.003
13. Chladil M, Nunez M (1995) Assessing grassland moisture and biomass in Tasmania – the application of remote-sensing and empirical-models for a cloudy environment. Int J Wildland Fire 5(3):165. doi:10.1071/WF9950165

14. Finney M (1994) FARSITE: a fire area simulator for fire managers. In: The proceedings of the Biswell symposium, pp 55–56 . Retrieved from http://www.firemodels.org/downloads/farsite/publications/Finney_1995_PSW-GTR-158_pp55-56.pdf
15. Morais ME (2001) Comparing spatially explicit models of fire spread through chaparral fuels: a new algorithm based upon the Rothermel fire spread equation. Retrieved from http://firecenter.berkeley.edu/hfire/hfire_fire_spread_body.pdf
16. Tymstra C, Flannigan MD, Armitage OB, Logan K (2007) Impact of climate change on area burned in Alberta's boreal forest. Int J Wildland Fire 16(2):153. doi:10.1071/WF06084
17. Chi S, Lim Y, Lee J, Lee J (2003) A simulation-based decision support system for forest fire fighting. AI* IA 2003: advances in ..., 2013. Retrieved from http://link.springer.com/chapter/10.1007/978-3-540-39853-0_40
18. Minciardi R, Sacile R, Trasforini E (2009) Resource allocation in integrated preoperational and operational management of natural hazards. Risk Anal 29(1):62–75. doi:10.1111/j.1539-6924.2008.01154.x
19. Donovan G, Rideout D (2003) An integer programming model to optimize resource allocation for wildfire containment. For Sci 49(2):331–335, Retrieved from http://www.ingentaconnect.com/content/saf/fs/2003/00000049/00000002/art00017
20. Fried JS, Gilless JK, Spero J (2006) Analysing initial attack on wildland fires using stochastic simulation. Int J Wildland Fire 15(1):137. doi:10.1071/WF05027
21. de la Asunción M, Castillo L, Fernámdez-Olivares J, García-Pérez O, González A, Palao F (2005) SIADEX: an interactive knowledge-based planner for decision support in forest fire fighting. AI Commun 18(4):257–268, Retrieved from http://dl.acm.org/citation.cfm?id=1218883.1218887
22. Moura D, Oliveira E (2007) Fighting fire with agents – an agent coordination model for simulated firefighting. In: Proceedings of the 2007 spring simulation multiconference, vol 1, pp 71–78. Retrieved from http://dl.acm.org/citation.cfm?id=1404680.1404691
23. Sarmento LM (2004) An emotion-based agent architecture. Retrieved from http://paginas.fe.up.pt/~niadr/PUBLICATIONS/thesis_Masters/LuisSarmento.pdf
24. Yi S, Shi J (2009) An agent-based simulation model for occupant evacuation under fire conditions. In: 2009 WRI global congress on intelligent systems, IEEE, pp 27–31. doi:10.1109/GCIS.2009.442
25. Ntaimo L (2004) Forest fire spread and suppression in DEVS. Simulation 80(10):479–500. doi:10.1177/0037549704050918
26. Muzy A, Innocenti E, Aiello A, Santucci J-F, Wainer G (2002) Cell-DEVS quantization techniques in a fire spreading application. In: Proceedings of the winter simulation conference, vol 1. IEEE, pp 542–549. doi:10.1109/WSC.2002.1172929
27. Wainer GA (2004) Modeling and simulation of complex systems with Cell-DEVS. In: Ingalls RG, Rossett MD, Smith JS, Peters BA (eds) Proceedings of the 2004 winter simulation conference
28. Yang J, Chen H, Hariri S, Parashar M (2005) Self-optimization of large scale wildfire simulations. Computational Science–ICCS ..., pp 615–622. Retrieved from http://link.springer.com/chapter/10.1007/11428831_76
29. ITU-T. (2011) Specification and description language – overview of SDL-2010, p 68
30. Fonseca i Casas P, Colls M, Casanovas J (2010) Towards a representation of environmental models using specification and description language-from the fibonacci model to a Wildfire Model. In: KEOD. Retrieved from http://upcommons.upc.edu/handle/2117/11032
31. Gronewold A, Sonnenschein M (1998) Event-based modelling of ecological systems with asynchronous cellular automata. Ecological Modelling, 18. Retrieved from http://www.sciencedirect.com/science/article/pii/S0304380098000179
32. Niazi Ma, Siddique Q, Hussain A, Kolberg M (2010) Verification & validation of an agent-based forest fire simulation model. In: Proceedings of the 2010 spring simulation multiconference on – SpringSim'10, 1. doi:10.1145/1878537.1878539

33. Nader B, Filippi J, Bisgambiglia P (2011) An experimental frame for the simulation of forest fire spread. In: Jain S, Creasey RR, Himmelspach J, White KP, Fu M (eds) Proceedings of the 2011 winter simulation conference, pp 1010–1022. Retrieved from http://dl.acm.org/citation.cfm?id=2431637
34. Bratten FW (1978) Containment tables for initial attack on forest fires. Fire Technol 14(4):297–303. doi:10.1007/BF01998389
35. Beaumont KP, Mackay DA, Whalen MA (2012) The effects of prescribed burning on epigaeic ant communities in eucalypt forest of South Australia. For Ecol Manage 271:147–157. doi:10.1016/j.foreco.2012.02.007
36. Stephan K, Kavanagh KL, Koyama A (2012) Effects of spring prescribed burning and wildfires on watershed nitrogen dynamics of central Idaho headwater areas. For Ecol Manage 263:240–252. doi:10.1016/j.foreco.2011.09.013
37. Rytwinski A, Crowe KA (2010) A simulation-optimization model for selecting the location of fuel-breaks to minimize expected losses from forest fires. For Ecol Manage 260(1):1–11. doi:10.1016/j.foreco.2010.03.013
38. Fei Y, Xianlin Q, Bo H, Xi Z, Zengyuan L (2013) 1 . Automatic extraction of active fire line using Landsat imagery. In: Proceedings of dragon 2 final results and dragon 3 kick-off symposium, Noordwijk, p 649
39. Keen PGW (1980) Decision support systems: a research perspective. Cambridge, Massachusetts: Center for Information Systems Research, Afred P. Sloan School of Management. Retrieved from http://hdl.handle.net/1721.1/47172
40. Haettenschwiler P (1999) Neues anwenderfreundliches Konzept der Entscheidungsunterstützung. In: Gutes Entscheiden in Wirtschaft, Politik und Gesellschaft. vdf Hochschulverlag, Zurich, pp 189–208
41. Power DJ (1997) What is a DSS? DSstar, 21 October 1997, 1(3). http://dssresources.com/papers/whatisadss/index.html

Chapter 20
Vulnerability Analysis and GIS Based Seismic Risk Assessment Georgia Case

Nino Tsereteli, Vakhtang Arabidze, Otar Varazanashvili, Tengiz Gugeshashvili, Teimuraz Mukhadze, and Alexander Gventcadze

Abstract The paper presents a framework for the vulnerability analysis to assess seismic risk for the Republic of Georgia. Firstly, detailed inventory map of elements at risk was created. Here elements at risk are comprised of buildings and population. The grid of 0.025° size was created to summarize building and population data into the grid cells for further analysis. The custom programming script was created that summarizes the following information for each grid cell: the total number of buildings, total area of buildings, total population, total number of buildings per each taxonomy class and total area of buildings for each taxonomy class. Secondly, seismic hazard maps were calculated based on modern approach of selecting and ranking global and regional ground motion prediction equation for region. Thirdly, on the bases of empirical data that was collected for some earthquake intensity based vulnerability study were completed for Georgian buildings. Finally, probabilistic seismic risk assessment in terms of structural damage and casualties were calculated for the territory of Georgia for 2.8 km grid cells using obtained results. This methodology gave prediction of damage and casualty for a given probability of recurrence, based on a probabilistic seismic hazard model, population distribution, inventory, and vulnerability of buildings.

N. Tsereteli (✉) • O. Varazanashvili • T. Gugeshashvili • T. Mukhadze • A. Gventcadze
Seismic Hazard and Disaster Risks, M. Nodia Institute of Geophysics of I. Javakishvili Tbilisi State University, 1 Aleksidze str., Tbilisi 0171, Georgia
e-mail: nino_tsereteli@tsu.ge; otarivar@yahoo.com; tengiz.gugeshashvili@yandex.ru; nino66_ts@yahoo.com; aleko.gvencadze@yahoo.com

V. Arabidze
Agricultural University of Georgia, David Agmashenebeli Alley 140, Tbilisi, Georgia
e-mail: vakhara@yahoo.com

H.-N. Teodorescu et al. (eds.), *Improving Disaster Resilience and Mitigation - IT Means and Tools*, NATO Science for Peace and Security Series C: Environmental Security,
DOI 10.1007/978-94-017-9136-6_20,

20.1 Introduction

Risks of natural hazards caused by natural disaster are closely related to the development process of society. Disasters pose hazard to sustainable development of the country. Out of the world populations, 75 % live in places prone of various disasters such as earthquakes, floods, hurricane, drought etc. A recent review of worldwide natural hazard losses during 2012 has identified 905 natural disasters, resulting in 96,000 deaths, $170 billion in economic losses, and $70 billion in insured losses [1].

Upward trend of disaster's intensity and frequency are observed worldwide (e.g. [2]) and are also discussed on the national level (e.g. [3, 4]). It is the result of the urbanization and rapid increase of the density of the population, global climate change and economic globalization, environmental deterioration etc. The natural disasters risk occurs when the hazardous hydro-meteorological, geological and other phenomena interact with physical, social, economic and ecological factors of vulnerable social and environmental infrastructure.

The high level of natural disasters in many countries makes necessary to work out the national programs and strategy. The main goal of these programs is to reduce the natural disasters risk and caused losses. Risk mitigation is the cornerstone of the approach to reduce the nation's vulnerability to disasters from natural hazards. Therefore, proper investigation and assessment of natural hazards and vulnerability of the element at risk to hazards is very important for an effective and proper assessment of risk. This paper issues a call for advance planning and action to reduce natural disaster risks, notably seismic risk through the investigation of vulnerability and seismic hazard for Georgia.

Georgia is prone to multiple natural hazards, the most dangerous and devastating of which are strong earthquakes. By the index of disaster risk, Georgia relates to the countries with medium and high level risk [5]. So, the natural hazard, notably seismic hazard in Georgia have to be considered as a standing negative factor in the development process of the country. Such approach implies the necessity of more active actions by all possible means to reduce the seismic risk at each level and maintain the sustainable economic development of the country. Moreover, the risk level has not been quantified so far in terms of different elements at risk and with respect to vulnerability of corresponding elements at risk.

The importance of arising problems stimulates an active investigation of vulnerability of elements at risk and seismic hazard for seismic risk reduction. The attempt has been approached within the framework of the project "EMME – Earthquake Model for Middle East Region: Hazard, Risk Assessment, Economic and Mitigation".

20.2 Seismotectonic and Seismicity of Georgia

Georgia, the southernmost part of Eastern Europe, occupies the western part of the South Caucasus. The main morphological units of Georgia are the mountain ranges of the Greater and Lesser Caucasus separated by the Black Sea-Rioni (Colkhis) and Kura (Mtkvari)-South Caspian intermountain troughs.

Recent geodynamics of Georgia and adjacent territories of the Black Sea-Caspian Sea region, as a whole, are determined by its position between the still-converging Eurasian and Africa-Arabian plates. The geometry of tectonic deformations in the region is largely determined by the wedge-shaped rigid Arabian block intensively intended into the relatively mobile Middle East-Caucasian region. In the first place, it influenced the configuration of main compressional structures developed to the north of the Arabain wedge (indentor) – from the Periarabicophiolite suture and main structural lines in East Anatolia to the Lesser Caucasus, on the whole, and its constituting tectonic units including the Bayburt-Karabakh and Talysh fold-thrust belts. All these structural morphological lines have clearly expressed arcuate northward-convex configuration reflecting the contours of the Arabian Block.

Related tectonic activities caused moderate seismicity in the region. Three principal directions of active faults compatible with the dominant near N–S compressional stress produced by northward displacement of the Arabian plate can be distinguished in the region—one longitudinal (WNW–ESE or W–E) and two transversal (NE–SW and NW–SE).

All historical and instrumental earthquakes observed ($4.5 < Ms < 7.0$) were located along the fault systems of the Greater and Lesser Caucasus and the intermountain depressions. The maximum magnitude of observed earthquakes is $Ms = 7.0$. The moderate earthquakes reflect the regional tectonics that are largely determined by the position of the Caucasus range between the still-converging Eurasian and Africa-Arabian lithosphere plates. By this classification, the southern slopes of Greater Caucasus are characterized by thrust and thrust strike slips, while the Javakheti upland is mostly characterized by strike-slip faults [6]. Georgia is vulnerable to earthquake hazards. A seismic risk is very high here. A single large earthquake, however, can cause far more damage than the average estimate from other natural hazards. Direct losses for residential houses and infrastructure damages resulting just from the Racha earthquake (29 April 1991) in Georgia (with more than 200 deaths) was assessed in the order of 10 billion rubles (as of 1991, approximately 5.5 billion USD at commercial exchange rate and 16 billion USD at official course) [6]. So, to estimate potential losses is vital for the region. Annualized earthquake loss addresses two components of seismic risk: the probability of ground motion and the consequences of ground motion on different element at risk. Some study carried out to combine seismic hazard with the value

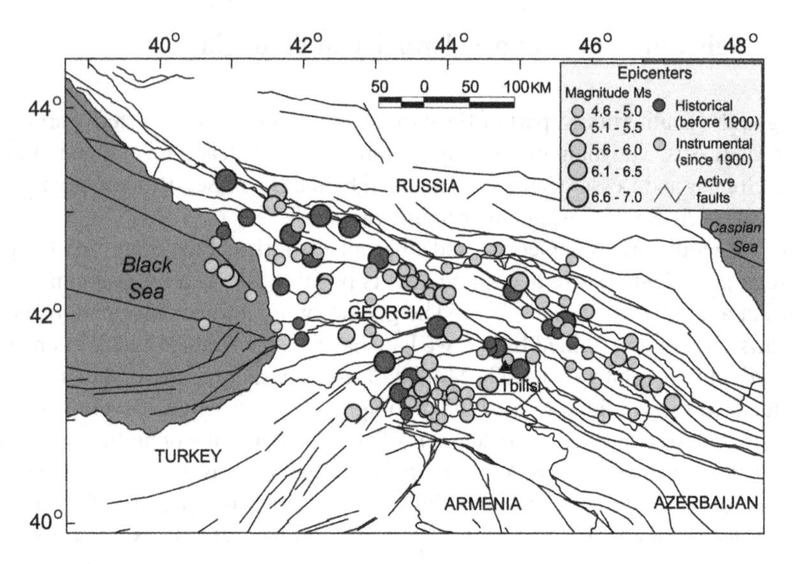

Fig. 20.1 Seismicity and active faults of Georgia

of the building inventory in the country to provide estimations of economic losses and casualty from earthquakes in Georgia. Figure 20.1 presents seismicity with earthquake magnitude Ms more than 4.5 and active faults of Georgia.

20.3 Inventory

In risk assessment, one important parameter is the inventory of elements at risk. In urban area, elements at risk include buildings, lifeline systems, population, socio-economic activities. Extensive and comprehensive collection of element at risk inventories is vital for estimation of losses during the earthquake.

The following GIS layers have been collected from the different sources and united in the centralized geo database: administrative division of country (regions, municipalities, election districts and precincts, cadastral zones, sectors and quarters); base topographic map layers (hydrography road network, elevation); scanned topographical and geological maps; satellite and aerial imagery; settlements and population data, and data on buildings.

The grid of 0.025° (2.8 km) size was created to summarize building and population data into the grid cells for further analysis with the seismic hazard modeling software ELER v3 [7].

Collected buildings GIS layer contains attributive data on building age, material, functionality, and number of floors. This quality and reliability of the data was not sufficient for arranging building into the predefined taxonomy classes, so we conducted additional inventory of the buildings using detailed resolution aerial

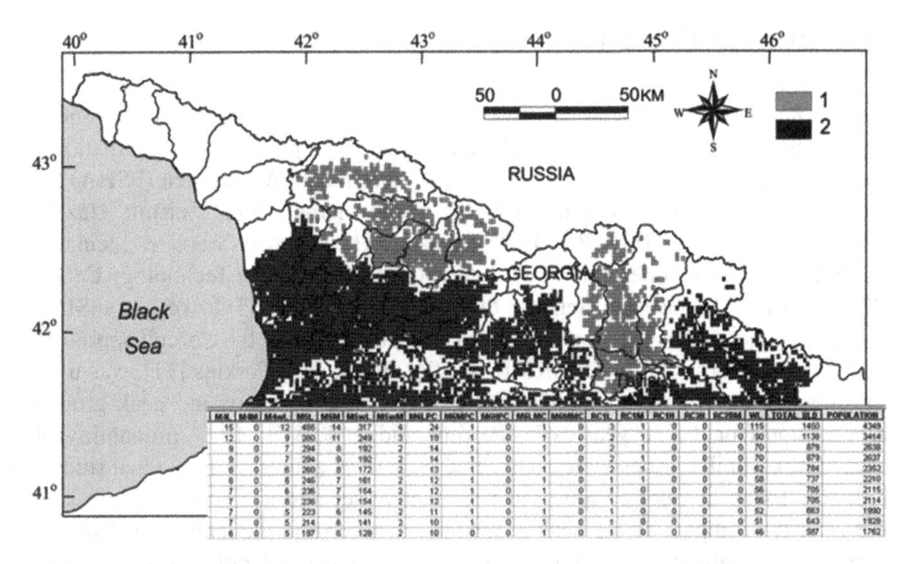

Fig. 20.2 Databases for buildings and population in GIS for territory of Georgia

imagery, building photos (where available) and expert judgment about the known building types in different regions of Georgia. As a result, all buildings of Georgia were assigned to taxonomy classes according to Lagomarsino and Giovinazzi [8] and we obtained a GIS database of classified buildings. Population data for each grid cell was obtained from the basis of the 2007 general population census, GEOSTAT [9]. Population value from the census data was assigned to each settlement point on the GIS map. Summarizing the population into the 2.8 km grid cells using settlement points did not give correct information on population distribution, since in most cases one settlement is located in more than one grid cell, so we needed to distribute the population to the buildings or residential quarters for more exact calculation. For this purpose, the GIS layer of residential quarters (quarter polygons) was used, which models the population distribution fairly well for the whole country. Using GIS analytical tools, we automatically assigned each residential quarter to the nearest settlement point that contained population data (using settlement unique codes). Then the population of each settlement point was distributed among its quarter polygons with the proportion of polygon area (bigger quarters were assigned higher population value). The population for each grid cell was calculated by summarizing populations of each residential quarter falling into the grid cell; if the quarter was located in more than one cell, its population was distributed to the cells by area proportion. Finally, the custom programming script was created that summarizes the following information for each grid cell: the total number of buildings, total area of buildings, total population, total number of buildings per each taxonomy class and total area of buildings for each taxonomy class (Fig. 20.2). This data constituted the input for the ELER software, for further calculation of risk.

20.4 Seismic Hazard Assessment

Seismic hazard assessment for Georgia had been a subject for several investigations in various international and national projects that were published in international and local literature. Namely, Probabilistic Seismic Hazard Assessment (PSHA) for the Caucasus had been undertaken in the frame of the Global Seismic Hazard Assessment Program GSHAP [10]. The Seismic Hazard maps were re-calculated in 2002–2007 during operation of the International Science and Technology Centre (ISTC) project: "Caucasian Seismic Information System" ISTC A651 (CauSIN). The assessment of seismic hazard for the Caucasus, based on the Cornell approach, namely computer program SEISRISK III after Bender and Perkins [11], was used for compilation of the set of maps for macroseismic intensity and peak ground acceleration (for the 50 year exposure time and 2 %, 5 %, 10 % probability of exceeding). Similar issues were considered by several authors in individual studies [12–16]. Today in Georgia are used a normative seismic hazard map expressing the severity of impact by the peak ground acceleration and the intensity (MSK 64) for a 2 % probability of exceedence in 50 years. A generalization of this map to seismic hazard map that was calculated in the frame of the GSHAP project [10] showing considerable differences resulting from generalization. Taking into account that the GSHAP seismic hazard map for the Caucasus region is calculated for a 10 % probability of exceedence in 50 years, these differences became even larger. The Georgian normative seismic hazard map shows a lower susceptibility than the seismic hazard map recently calculated by Slejko et al. [13]. It became evident that there were gaps in seismic hazard assessment and normative seismic hazard maps needed a careful recalculation.

The methodology for the probabilistic assessment of seismic hazard includes the following steps: produce comprehensive catalogue of historical earthquakes (up to 1900) and the period of instrumental observations with uniform scale of magnitudes; produce models of seismic sources zone (SSZ) and their parameterization (estimate recurrence law, the maximum magnitude, the distribution of earthquakes depth and nature of the movement.); develop appropriate GMPE ground motion prediction equation models; develop seismic hazard curves for spectral amplitudes at each period and maps in digital format.

The new seismic catalog of Georgia: Using a complete catalog is the most important to determine seismic hazard analysis in the region. The content of work package WP1 of EMME project involved the compilation of a new earthquake catalog for the Middle East region, including Georgia. In this work, there were two main goals: (1) Drafting a new catalog of historical earthquakes of Georgia and (2) Compilation of new instrumental seismic catalog of Georgia. In the process of compiling the level of the magnitude completeness was limited to: $Ms \geq 2.9$ or $Mw \geq 4.0$.

The main principles for compiling a catalogue of historical earthquakes are considered in Varazanashvili et al. [17]. The study and detailed analysis of these original documents and researches have allowed us to create a new catalogue of historical earthquakes of Georgia from 1250 BC to 1900 AD. The method of the study is based on a multidisciplinary approach, i.e. on the joint use of methods of history and paleoseismology, archeoseismology, seismotectonics, geomorphology, etc. A new parametric catalogue of 44 historic earthquakes of Georgia has been obtained and a full "descriptor" of all the phenomena described in it.

After 1962, the seismic network in the Caucasus and in Georgia improved, but inaccurate velocity models do not give good results. To reveal the best velocity model the hypocenter parameters of the earthquake were calculated, using the HYPO-71 program [18]. More detail this work are described in Tsereteli et al. [16]. During the period 1963–2002, parameters of about 120 earthquakes were re-calculated for Georgia. The difference between obtained and existing locations of individual earthquakes was sometimes more than 20 km. Thus, after the determination or specification of the basic parameters of earthquakes, we present a new catalog of historical earthquakes (pre-1900) of Georgia and the refined instrumental earthquake catalog (1900–2009) of Georgia.

Seismic Sources (SS): The identification of area SSZ was obtained on the bases of structural geology, parameters of seismicity and seismotectonics. In determining the SS, the slope of the appropriate active fault plane, the width of the dynamic influence of the fault, power of seismoactive layer are taken into account. The relationship of strong earthquakes with this fault is set based on data on the position and orientation of the earthquake sources, and the character of motion in them: the position of the hypocenter; data on direction of the first isoseismals; information about the orientation of the scattering area of forshocks and aftershocks; materials on the focal mechanism, the position of the seismogenic landslides and seismodislocations. Each SSZ was defined with the parameters: the geometry, the percentage of focal mechanism, predominant azimuth and dip angle values, activity rates, maximum magnitude, hypocenter depth distribution, lower and upper seismogenic depth values.

Ground motion prediction equation models: Data bases of strong motion are very poor for Georgian region, especially for strong earthquakes with $MS \geq 5$ that is very important from engineering point of view. Quite many works have done in the Caucasus to estimate this equation [19–21], but because of poor data the desired result were not given. Some attempt was to complete the database with the data from the tectonically similar region (Caucasus, north-west Turkey, central Italy and others) to derive more accurate ground motion prediction equation GMPE [12, 13].

To capture epistemic uncertainty in GMPE in different regions and tectonic regimes, a selection of a set of appropriate GMPEs (candidate GMPE) is required. For that, the number of selected GMPEs is kept as small as possible, but enough to capture the uncertainty. The selected GMPEs should be robust enough to cover a wide range of magnitudes, distances and frequencies. The Next Generation Attenuation project [22] developed a series of GMPEs intended for application to geographically diverse regions (including Turkey and Caucasus). Together with NGAs GMPE models by us were further considered European and regional models. From these the candidate GMPEs for shallow active crustal regions are collected and

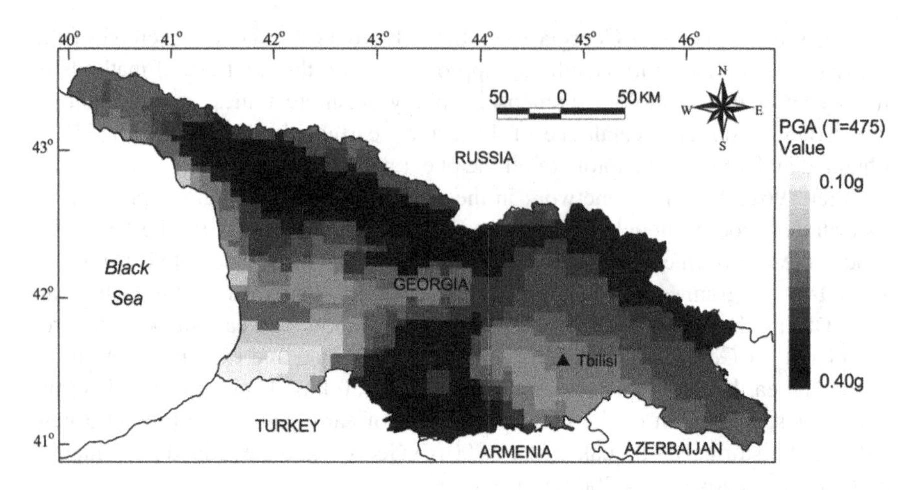

Fig. 20.3 Seismic hazard map in terms of PGA for *10 %* probability of exceedence in 50 years for rock (Vs = 760 m/s)

they are subjected to a further elimination based on the criteria proposed in Cotton et al. [23] [24]. Accelerograms from Turkey, Iran, Georgia, Armenia, Pakistan and Jordan were collected in the EMME databases that were used for ranking and selecting candidate GMPEs using the data-driven testing procedure proposed by Scherbaum et al. [25, 26] and Kale and Akkar [27]. Proceeding from this analysis, GMPEs [28–31] were used with equal weights in the logic tree combination in the seismic hazard calculation [24].

On the basis of obtained area seismic sources, the probabilistic seismic hazard maps were calculated showing peak ground acceleration (PGA) and spectral accelerations (SA) at 0.2, 1 s periods for 10 % probability and 2 % probability in 50 years using selected GMPEs correspondingly for Rock (Vs = 760 m/s) and soil (Vs = 300 m/s). Seismic hazard (SH) calculation has been performed with two software OpenQuake [42] and EZFRISK [43]. They are used the standard methodology for probabilistic seismic hazard analysis. Both of them gave equal value of SH for Georgia. Results some of them are presented in Figs. 20.3 and 20.4.

20.5 Vulnerability

The Vulnerability Module quantifies the vulnerability of the assets subjected to the seismic hazard. Combination of the vulnerability with the value of the affected assets gives possibility to estimate total direct losses in physical terms from earthquake (DL). Collected inventory map of element at risk allow estimation vulnerability just for buildings for the territory of Georgia. So available data just

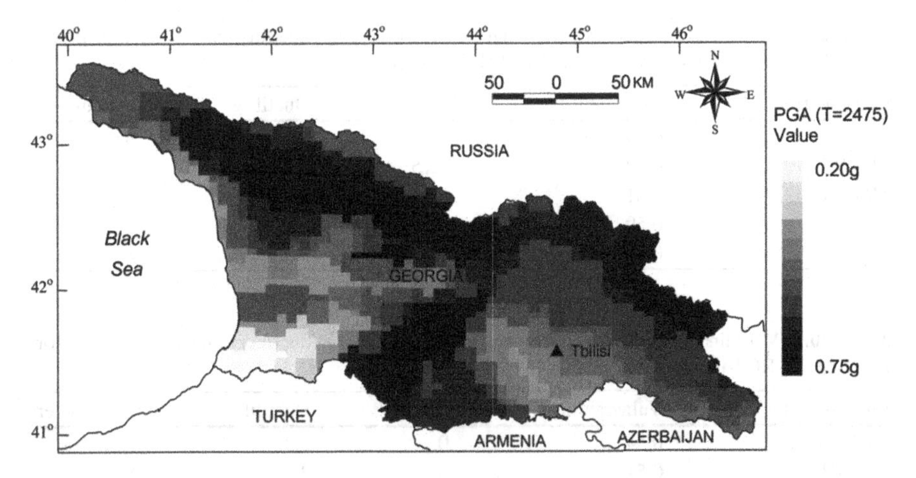

Fig. 20.4 Seismic hazard map in terms of PGA for 2 % probability of exceedence in 50 years for rock (Vs = 300 m/s)

can show a picture of the consequences in terms of direct damages caused mainly on infrastructure. For any developing country, the influence of DL on national economies is important. So, the vulnerability analysis of building trough the country can works as proxy for vulnerability at a large of a country

A review of existing literature in the assessment of risk shows the big gaps in understanding of vulnerability for Georgia. There was not such clear definition of vulnerability as presented by Lagomarsino and Giovinazzi [8]. In particular, there is neither a vulnerability definition with respect to the individual element at risk for the territory of Georgia. Very often in seismic risk assessment, structural engineers use the so-called grade of sensitivity of damage [15, 32, 44, 45] or the vulnerability index [33, 34], which is defined as a relationship between the hazard intensity and the degree of loss. There were attempts to calculate the economic loss expressed in deciles using the technique presented in Dilley et al. [35], Tsereteli [36, 37], Chelidze et al. [38], but the obtained maps were very approximative due to the very limited amount of data available. The first step was done in Varazanashvili et al. [6] for the assessment of the historical and present vulnerabilities for six types of natural hazards (drought, hurricane, hail, frost, flash floods and earthquakes).

In the frame of EMME project was possible to develop intensity based vulnerability for Georgian buildings. The European building taxonomy classification proposed by Lagomarsino et al. [8] were used and taxonomy classification was done for buildings with available data on the bases of building information and aerial photos. On the bases of empirical data that was collected for Racha earthquake (Ms = 6.9) on 29 April of 1991 and Tbilisi earthquake (Ms = 4.6) on 25 April of 2002 some intensity based vulnerability study were completed. These records do not include information about the building height. For this, according to the actual observable inventory and expert opinion the following assumptions are made: simple stone are consider as mid-rise M3M; Pre code masonry reinforced

Table 20.1 V (vulnerability), ΔV_r (regional vulnerability), Q (ductility) and t parameters for M3M, M6LPC, M6LMC, RC2PM and RC2PH buildings

Building name	Vulnerability	ΔV_r	Ductility	t parameter
M3M	0.99	0.25	2.3	8
M6LPC	0.72	0.15	2.3	8
M6LMC	0.61	0.12	2.3	8
RC2PM	0.59	0	2.3	8
RC2PH	0.66	0	2.3	8

Table 20.2 V (vulnerability), ΔV_r (regional vulnerability), Q (ductility) and t parameters for RC2BM and RC2BH buildings

Building name	Vulnerability	ΔV_r	Ductility	t parameter
RC2BM	0.54	0	2.3	8
RC2BH	0.51	0	2.3	8

concrete rc floors are considered as low rise M6LPC; MC masonry rc floors are consider as mid-rise M6LMC. Also industrial types of building like mid-rise large panel buildings RC2PM that do not have analog in European buildings were investigated. Comparison of obtained data with the data proposed by Lagomarsino and Giovinazzi [8] gave us the possibility to develop the regional vulnerability factors for these typologies. For high-rise large panel buildings RC2PH the influence of number of floors based on experts judgment have been taken into account. Obtained results are presented in Table 20.1.

The vulnerability for large block mid-rise RC2BM and high-rise RC2BH buildings were developed on the data bases of the same house in Irkutsk [33] on the bases of experts' judgment. Results are presented in Table 20.2. For building typologies three classes of height have been considered (L = low-rise, M = mid-rise, H = high-rise) differently defined in terms of floor numbers for masonry (L = 1–2, M = 3–5, H $\geq$ 6) and reinforced concrete buildings (L = 1–3, M = 4–7, H $\geq$ 8).

Developed regional vulnerability applied to the vulnerability classification by Lagomarsino and Giovinazzi [8] for masonry and masonry rc floors typology (M3, M4, M6). Other is not changed.

For verifying our results, we calculated the seismic risk in terms of building damage and causalities using software ELER for Racha and Tbilisi earthquakes considering the source as reverse faults using selected GMPEs. There are quite good agreements with calculated and empirical value for damage distribution. D4 + D5. D4 + D5 damage for Racha earthquake was 4,459 from empirical data. Using obtained vulnerability model, calculated D4 + D5 damage is 4,074. Empirical value of D3 damage for Racha earthquake is 11,694 and calculated D3 is 16,013. The same comparison were done for Tbilisi earthquake. The empirical value of D4 + D5 damage is 200 and the calculated value is 402. Empirical value of D3 damage for Tbilisi earthquake is 4,996 and calculated D3 is 3,157. This allows us to conclude that our estimations of vulnerability are good enough.

For estimation of casualty for Racha and Tbilisi earthquakes, we used Coburn and Spence [39], Samardjieva and Badal [40], Bramerini et al. [41]. Calculation values obtained by Coburn and Spence [39] gave the estimations that were close to empirical data. In this methodology, casualty estimations are based on the structural damage levels D4 and D5.

20.6 Seismic Risk

Probabilistic seismic risk assessment in terms of structural damage and casualties were calculated for the territory of Georgia for 2.8 km grid cells using obtained results. This methodology gave prediction of damage and casualty for a given probability of recurrence, based on a probabilistic seismic hazard model, population distribution, inventory and vulnerability of buildings. ELER software [7] was used for this calculation. The approach used in damage estimation is to obtain a normally distributed cumulative damage probability for each building type. The damage probability distribution is a function of each building's vulnerability and ductility parameters [8]. Figures 20.5 and 20.6 illustrate the structural damage distribution for probabilistic seismic map (in terms of PGA) for 10 % probability of exceedence in 50 years for soil and probabilistic seismic map for 2 % probability of exceedence in 50 years for soil correspondingly. Information from this calculation provides quantitative bases for the prioritization of the risk mitigation activities.

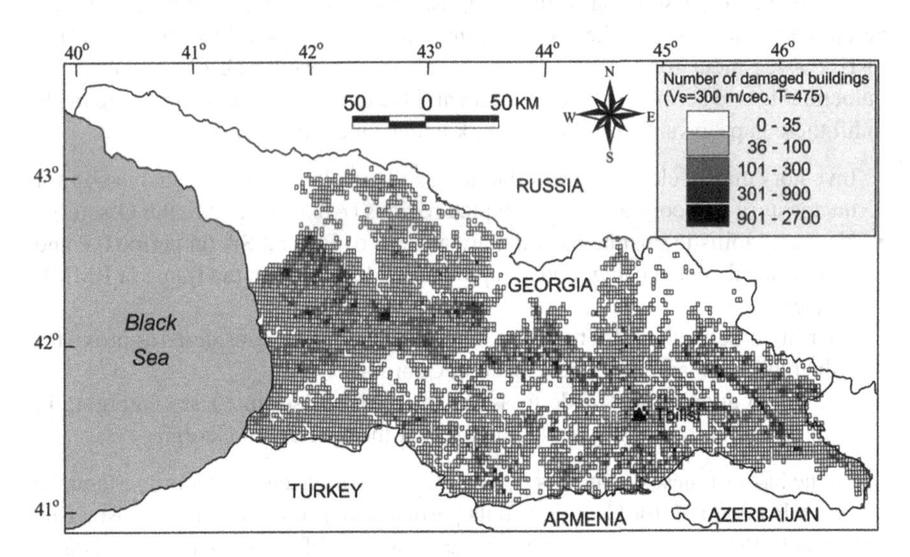

Fig. 20.5 Structural damage distribution for probabilistic seismic map (in terms of PGA) for *10 %* probability of exceedence in 50 years for soil

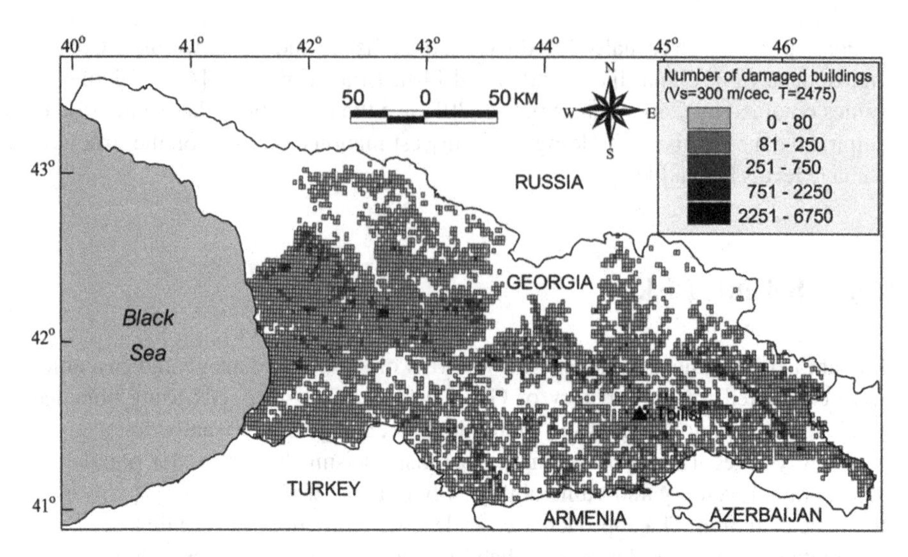

Fig. 20.6 Structural damage distribution for probabilistic seismic map (in terms of PGA) for 2 % probability of exceedence in 50 years for soil

20.7 Conclusions

The compilation of inventory maps, seismic hazard and vulnerability are definitely the pre-requisite for solution of the final task-risk management and reduction vulnerability of the country. The review of international and local literature according to risk assessment for Georgia, however, clearly showed lack of knowledge in vulnerability and seismic hazard assessment. Due to this, some attempts were made to fill these gaps towards an enhanced risk management in Georgia:

- Inventory map of element at risk (building and population) was created in GIS for the territory of Georgia except occupied regions (Abkhazia and South Ossetia).
- New probabilistic seismic hazard map in terms of PG and SA (at period 0.2 and 1 s) were calculated using modern approaches developed in the frame of EMME project.
- Intensity based vulnerability for the buildings that can be considered as proxy for vulnerability at large were developed in Georgia.
- Finally, an attempt was made to assess probabilistic seismic risks emerging in terms of structural damage and casualties for the territory of Georgia.

On the basis of these results it is possible to create a system indicating earthquake vulnerability and risk for Georgia and its particular regions/communities also. This will enable the managing agencies to carry out the effective policy of advance planning and action to reduce seismic risk through the development of resilient cities. The major obstacle in disaster management in Georgia is the absence of active and powerful national agencies that will work in systematic manner for assessing archiving, mapping, monitoring, predicting managing of all catastrophic events. GIS Mapping of seismic risk in Georgia can be considered as the early warning tool for long-term disaster preparedness of national and local authorities The presented work can contribute to a reduction of disaster losses in Georgia and will foster future efforts of harmonization of risk management strategies in the country.

Acknowledgments The authors would like to acknowledge the valuable advice and consultation provided by the Steering Committee and Work Packages leaders, moreover, the authors kindly acknowledge the insightful comments of project managers and all team of EMME project.

References

1. Munich Re (2012) Topics Geo natural catastrophes 2011: analyses, assessments, positions. Munich Reinsurance Company, Munich. http://preventionweb.net/go/17345
2. Swiss Re (2012) Sigma 2/2012: natural catastrophes and man-made disasters in 2011. Swiss Re, Zurich. http://media.swissre.com/documents/sigma2_2012_en.pdf
3. Downton M, Miller J, Pielke R (2005) Reanalysis of US national weather service flood loss database. Nat Hazards Rev 6:13–22
4. Barredo J (2007) Major flood disasters in Europe: 1950–2005. Nat Hazards 42(1):125–148
5. EM-DAT (2006) The OFDA/CRED international disaster database. Universite Catholique de Louvain-Brussels- Belgium. http://www.emdat.be
6. Varazanashvili O, Tsereteli N, Amiranashvili A, Tsereteli E, Elizbarashvili E, Dolidze J, Qaldani L, Saluqvadze M, Adamia S, Arevadze N, Gventsadze A (2012) Vulnerability, hazards and multiple risk assessment for Georgia. Nat Hazards 64(3):2021–2056
7. Hancilar U, Tuzun C, Yenidogan C, Erdik M (2010) ELER software – a new tool for urban earthquake loss assessment. Nat Hazards Earth Syst Sci 10:2677–2696
8. Lagomarsino S, Giovinazzi S (2006) Macroseismic and mechanical models for the vulnerability and damage assessment of current buildings. Bull Earthq Eng 4:415–443
9. GEOSTAT (2007) Population. GDP and other indicators of national accounts. National Statistics Office of Georgia. http://geostat.ge/index.php?action=page&p_id=116&lang=eng
10. Balassanian S, Ashirov T, Chelidze T, Gassanov A, Kondorskaya N, Molchan G, Pustovitenko B, Trifonov V, Ulomov V, Giardini D, Erdik M, Ghafory-Ashtiany M, Grunthal G, Mayer-Rosa D, Schenk V, Stucchi M (1999) Seismic hazard assessment for the Caucasus test area. Annali di Geof 42(6):1139–1164
11. Bender B, Perkins DM (1987) Seisrisk III: a computer program for seismic hazard estimation. US Geol Surv Bull 1772:1–48
12. Smit P, Arzoumanian V, Javakhishvili Z, Arefiev S, Mayer-Rosa D, Balassanian S, Chelidze T (2000) The digital accelerograph network in the Caucasus. Bull, Open-File Reptvol 98, no 16, Earthquake hazard and seismic risk reduction. Kluwer Academic Publishers, Dordrecht, pp 109–118
13. Slejko D, Javakhishvil Z, Rebez A, Santulin M, Elashvili M, Bragato PL, Godoladze T, Garcia J (2008) Seismic hazard assessment for the Tbilisi test area (eastern Georgia). Boll Geof Teor Appl 49(1):37–58
14. Varazanashvili O, Tsereteli N (2010) Seismic situation's probability prediction in Great Caucasus during the period 2005–2025. J Seismol Earthq Eng 12(1):1–11
15. Tsereteli N, Varazanashvili O, Arabidze V, Mukhadze T, Arevadze N (2011) Seismic risk assessment for Tbilisi – new approaches. In: Avagyan A, Barry DL, Coldewey WG, Reimer DWG (eds) Stimulus for human and societal dynamics in the prevention of catastrophes, vol 80. IOS Press, Amsterdam, pp 109–130

16. Tsereteli N, Tanircan G, Safak E, Varazanashvil O, Chelidze T, Gvencadze A, Goguadze N (2012) Seismic hazard assessment for Southern Caucasus—Eastern Turkey energy corridors: the example of Georgia. In: Barry DL, Coldewey WG, Reimer DWG, Rudakov DV (eds) Correlation between human factors and the prevention of disasters, vol 94. IOS Press, Amsterdam, pp 96–111
17. Varazanashvili O, Tsereteli N, Tsereteli E (2011) Historical earthquakes in Georgia (up to 1900): source analysis and catalogue compilation. Pub Hause MVP, Tbilisi
18. Lee W, Larh S (1975) HYPO-71 (Revised), a computer program for determining hypocenter, magnitude and first motion patting of local earthquakes. US Geol Surv Open File, Report 75-31
19. Arefiev S, Mayer-Rosa D, Parini I, Pomanov A, Smit P (1991) The Rachi (Georgia, USSR) earthquake of 29 April 1991: strong-motion date of selected aftershocks 3 May 1991-30 June 1991. Publ Ser Swiss Seismol Serv 103:1–211
20. Arefiev S, Mayer-Rosa D, Parini I, Pomanov A, Smit P (1991) Spitak (Armenia, USSR) 1988 earthquake region: strong-motion data of selected earthquakes June 1990-April 1991. Publ Ser Swiss Seismol Serv 104:1–102
21. Jibladze E, Gurguliani I, Tsereteli N (2000) The prediction of earthquake peak ground acceleration in Caucasus. J Georgian Geophis Soc 5:68–76
22. NGA (2008) Special issue on the next generation attenuation project. Earthq Spectra 24(1):1–341
23. Cotton F, Scherbaum F, Bommer JJ, Bungum H (2006) Criteria for selecting and adjusting ground-motion models for specific target applications: applications to Central Europe and rock sites. J Seismol 10(2):137–156
24. EMME (2013) Earthquake model of Middle East: hazard, risk assessment, economics and mitigation. Brochure Prepared for GEM (Global Earthquake model), Istanbul
25. Scherbaum F, Cotton F, Smit P (2004) On the use of response spectral-reference data for the selection and ranking of ground-motion models for seismic-hazard analysis in regions of moderate seismicity: the case of rock motion. Bull Seismol Soc Am 94(6):2164–2185
26. Scherbaum F, Delavaud E, Riggelsen C (2009) Model selection in seismic hazard analysis: an information-theoretic perspective. Bull Seismol Soc Am 99(6):3234–3247
27. Kale O, Akkar S (2012) A new procedure for selecting and ranking ground-motion prediction equations (GMPEs): the Euclidean-Distance based Ranking (EDR) method. Bull Seismol Soc Am 103(2A):1069–1084
28. Akkar S, Sandıkkaya MA, Bommer JJ (2013) Empirical ground-motion models for point- and extended-source crustal earthquake scenarios in Europe and the Middle East. Bull Earthq Eng. doi:10.1007/s10518-013-9461-4
29. Chiou BSJ, Youngs RR (2008) An NGA model for the average horizontal component of peak ground motion and response spectra. Earthq Spectra 24(1):173–215
30. Akkar S, Cagnan Z (2010) A local ground-motion predictive model for Turkey and its comparison with other regional and global ground-motion models. Bull Seismol Soc Am 100(6):2978–2995
31. Zhao JX, Zhang J, Asano A, Ohno Y, Oouchi T, Takahashi T, Ogawa H, Irikura K, Thio HK, Somerville PG, Fukushima Y (2006) Attenuation relations of strong ground motion in Japan using site classification based on predominant period. Bull Seismol Soc Am 96(3):898–913
32. Mukhadze T, Arabidze V, Chanadiri J, Eremadze N, Korkia G (2008) Establishment of the damage grade and assessment of seismic risk of dwelling houses in Tbilisi. In: Proceedings of the first international conference on seismic safety of Caucasus region population, cities and settlements, September 8-11, 2008. Tbilisi, pp 153–156
33. Sobolev G (ed) (1997) Otsenkaseicmicheskoiopacnostiiriska. BSTS, Moskva
34. Varazanashvili O, Tsereteli N, Muchadze T (2008) Seismotectonic conditions and seismic risk for cities in Georgia. In: Apostol I, Coldewey WG, Barry DL, Reimer D (eds) Risk assessment as a basis for the forecast and prevention of catastrophes. IOS Press, Washington, DC, pp 26–35, http://www.lrc.fema.gov/starweb/lrcweb/servlet.starweb?path=lrcweb/STARLibraries1.web&search=R%3d182463

35. Dilley M, Chen RS, Deichmann U, Lerner-Lam AL, Arnold M, Agwe J, Buys P, Kjekstad O, Lyon B, Yetman G (2005) Natural disaster hotspots: a global risk analysis. The World Bank, Washington, DC
36. Tsereteli E (2007a) Mapping of mass-movement potential on the territory of Georgia: criteria of destabilization. In: Chelidze T (ed) Atlas of GIS-based maps of natural disaster hazards for the Southern Caucasus. Tbilisi, pp 13–15. http://www.coe.int/t/dg4/majorhazards/activites/DocumentsFrancesc/Etudes/2009/GEO_Atlas_SouthCaucasus.pdf
37. Tsereteli N (2007b) Principles of multi-risk calculation for natural hazards: the case of Georgia. In: Chelidze T (ed) Atlas of GIS-based maps of natural disaster hazards for the Southern Caucasus. Tbilisi, pp 26–29. http://www.coe.int/t/dg4/majorhazards/activites/DocumentsFrancesc/Etudes/2009/GEO_Atlas_SouthCaucasus.pdf
38. Chelidze T, Tsereteli N, Tsereteli E, Kaldani L, Dolidze J, Varazanashvili O, Svanadze D, Gvencadze A (2009) Multiple risk assessment for various natural hazards for Georgia. In: Apostol I, Barry DL, Coldewey WG, Reimer DWG (eds) Optimization of disaster forecasting and prevention measures in the context of human and social dynamics. IOS Press, Amsterdam, pp 11–32
39. Coburn A, Spence R (2002) Earthquake protection, 2nd edn. Wiley, Chichester
40. Samardjieva E, Badal J (2002) Estimation of the expected number of casualties caused by strong earthquakes. Bull Seismol Soc Am 92(6):2310–2322
41. Bramerini F, Di Pasquale G, Orsini A, Pugliese A, Romeo R, Sabetta F (1995) Rischiosismico del territorioitaliano. Proposta per unametodologia e risultatipreliminari. Rapportotecnico delServizioSismicoNazionale SSN/RT/95/01, Roma
42. OpenQuake, a computer program for seismic hazard estimation http://www.globalquakemodel.org/openquake/about/
43. EZFRISK, Software for Earthquake Ground Motion Estimation http://www.ez-frisk.com/
44. Petrovski J (2005) Tbilisi earthquake of April 25 2002. Materials of Tbilisi earthquake of April 25 2002. Prepared for UNDP. Tbilisi, pp 133–188
45. Gabrichidze G, Lomidze G, Mukhadze T, Odisharia A, Timchenko I (2004) April 2002 Epicenter earthquake in Tbilisi, Georgia. In: 13 the world conference on earthquake engineering, Vancouver, BC, Canada, 1–6 August 2004, Paper No. 10635

Chapter 21
Sensors Based on Chaotic Systems for Environmental Monitoring

Victor P. Cojocaru

Abstract This chapter reviews the literature on chaotic sensors and presents new results related to a titrimetry sensor for monitoring the conductivity of the water, with applications to environment pollution monitoring. A front-end device based on the chaotic sensor is also presented.

21.1 Introduction

Nowadays, environment pollution is one of the main threats to the well-being. The quality of the water is essential not only for the human population, but also for the entire supporting ecosystem. Water is susceptible to both natural and human pollutants [1, 2]. The control of the pollution degree for both the surface and the underground water networks is of particular interest for environment researchers, biologists and government officials responsible for water quality.

There are several researches reported in the literature related to chaotic dynamics in measurements, for example in electronic noses [3] and in other measurement applications [4]. Also, there are numerous chaotic circuits proposed in the literature, for various application fields, for example [5–7]. In this chapter, we describe after [4, 8–11] the concepts of some intelligent biomimetic sensors with nonlinear (chaotic) features, for measurements in the environment. Such a sensor for salts concentrations measurements was modeled and tested previously and was proposed in [8]. This chapter is a review of the results already reported and an extension

V.P. Cojocaru (✉)
D. Ghitu Institute of the Electronic Engineering and Nanotechnologies of the Academy of Sciences of Moldova, Academiei Street, Chisinau MD-2028, Republic of Moldova
e-mail: vcojocaru@nano.asm.md

H.-N. Teodorescu et al. (eds.), *Improving Disaster Resilience and Mitigation - IT Means and Tools*, NATO Science for Peace and Security Series C: Environmental Security,
DOI 10.1007/978-94-017-9136-6_21,

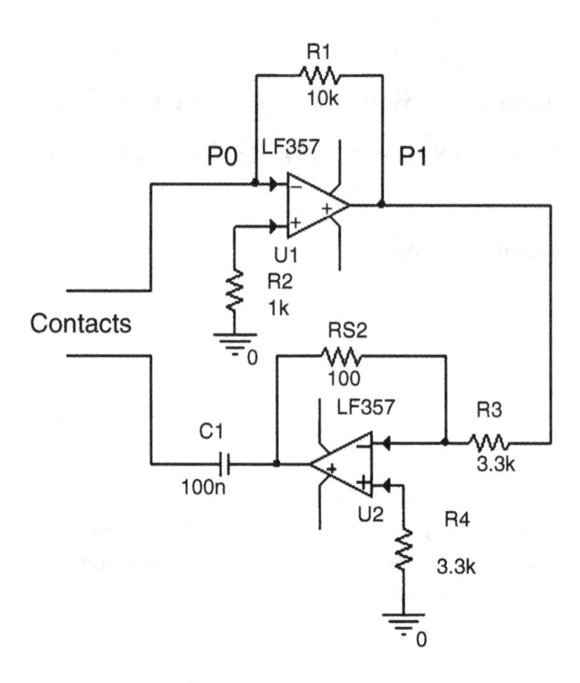

Fig. 21.1 Version of the chaotic circuit used in the sensor

of the researches presented in [8] and several other related papers, with additional technical and measuring details and new ideas regarding future applications.

The circuit in Fig. 21.1 and various variants of it were introduced and discussed in [12] and in the previous conference papers referenced in that one. Similar circuits were used in [13] and in [14], in relation to pseudo-random signal generators used in microsystem protection against side attacks.

The circuit uses the fractional-order electric behavior of the electrolytic capacitor and resistor composed by the electrolytic cell to produce a nonlinear dynamic behavior. The analysis of various fractional-order systems has recently emerged as a new chapter in chaotic system theory with interesting applications; see for example [15].

21.2 Concept of the Sensor

We present new results obtained with a simple sensor based on chaotic dynamics for the determination of components NaCl, $CuSO_4$, $Na_2S_2O_2$. The sensor is based on the change in dynamics produced in a nonlinear dynamic circuit, which has a circuit element composed of the measurement cell (measured saline solution together with the electrodes). In the literature, a large number of chaotic circuits have been proposed, for example [16, 17]; some of these circuits can be realized in CMOS technology [5], with low power consumption favoring portable or sensor network solutions as required in environment monitoring. The design of the sensor is based on a chaotic system using only two operational amplifiers in a positive feedback loop, with the sensing cell included in series in the loop, see Fig. 21.1. The circuit of the sensor is formed by two inverter amplifiers, which are connected through the H_2O solution and the respective electrodes, thus generating a non-linear feedback in the circuit.

The design is simple and easy to implement. The circuit uses two low cost components amplifiers, either type LF 157 N or 574UD3 (JFET input operational amplifiers, with low input voltage noise) powered at ±6 V. We preferred the noisier version, 574UD3, because its higher noise favors the starting of the chaotic oscillations in the circuit, moreover the noise contributes to maintaining the overall chaotic process. In the current implementation, the sensing element consists of two Fe/Zn electrodes immersed in a recipient (bath) containing the measured solution. The electrodes and the solution compose a measurement cell. The level of impurities influences the value of the resistance of the measuring cell, which is a part of the circuit, and therefore modifies the chaotic behavior of the sensor circuit. The investigated signal is captured at points P 0 and P 1 on the scheme in Fig. 21.1.

21.3 Monitoring Method for the Electrical Resistance and Results

The results are exemplified with pictures of the nonlinear dynamics when the measuring cell is immersed in tested solutions of three salts. The measured samples were various salts diluted in water, with various concentrations. The salts are NaCl, $CuSO_4$, $Na_2S_2O_2$. The dynamics differ from one salt concentration to another. It was found that the dynamics also differ for different salts at similar resistivities, possibly due to changes in electromotive force (chemical potentials), which also modifies the nonlinear dynamics. The various concentrations of the solution were obtained based on a solution of 0.5 g NaCl in 12 ml of H_2O, then by using from 0.2 to 1.8 ml of this solution for 1 l of H_2O, with steps of 0.2 ml. In this way, the precision of the solution concentration increment was insured.

In Figs. 21.2, 21.3, and 21.4 we represent the signals from the points P0 (X input oscilloscope) and P1 (Y input oscilloscope) as well as the XY phase trajectory for the solution of 12 ml of H_2O + 0.4 ml to 0.5 g NaCl for a liter of H_2O. We have the signals obtained under laboratory conditions at a temperature of 20 °C. The characteristic of the chaotic regime is illustrated by the phase diagram, which is the bi-dimensional representation of the trajectory of the system in the state space; the diagram is built in the plane using the voltages at points P0 and P1 of the circuit. The two signals, corresponding to P0 and P1 in the circuit, are significantly different, yet they are correlated. Their mutual dependency continuously varies, producing the diagram.

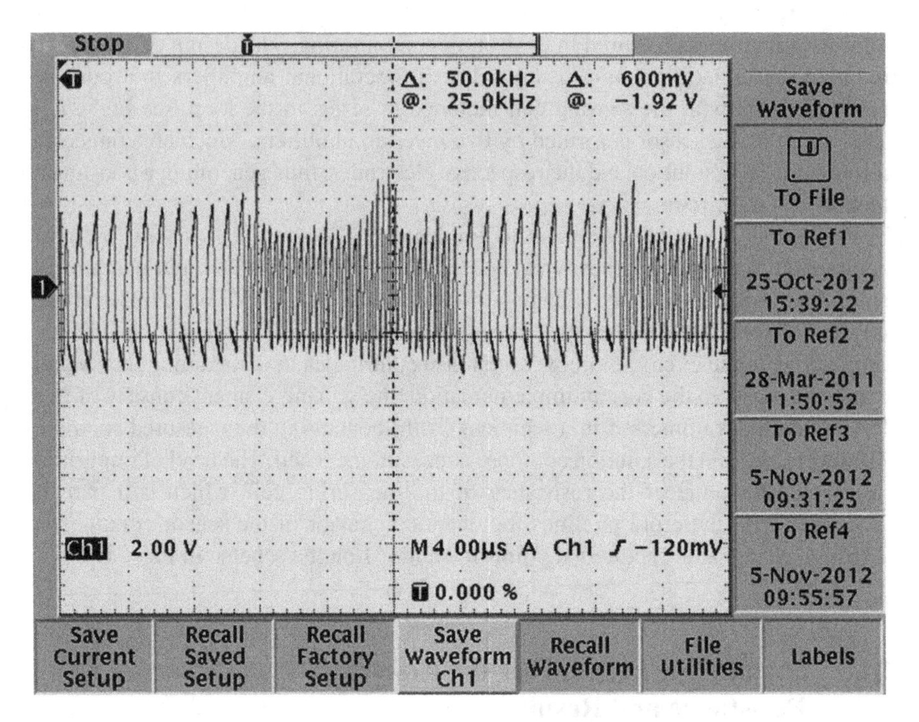

Fig. 21.2 Chaotic signal X(PO), 12 ml H_2O + 0.4 ml sol 0.5 g NaCl, at 20 °C

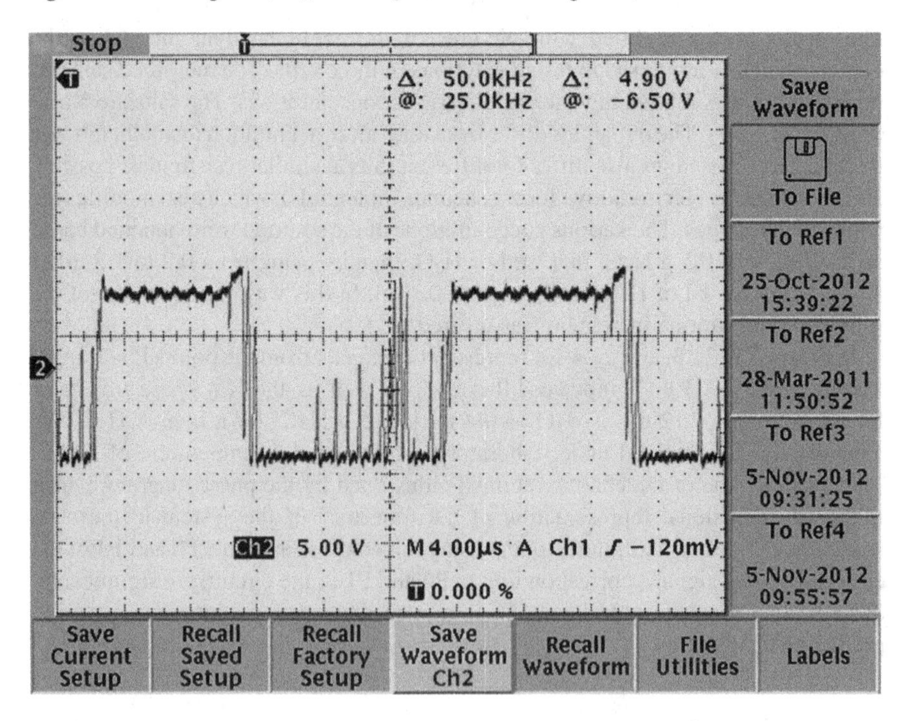

Fig. 21.3 Chaotic signal Y(P1), 12 ml H_2O + 0.4 ml sol 0.5 g NaCl, at 20 °C

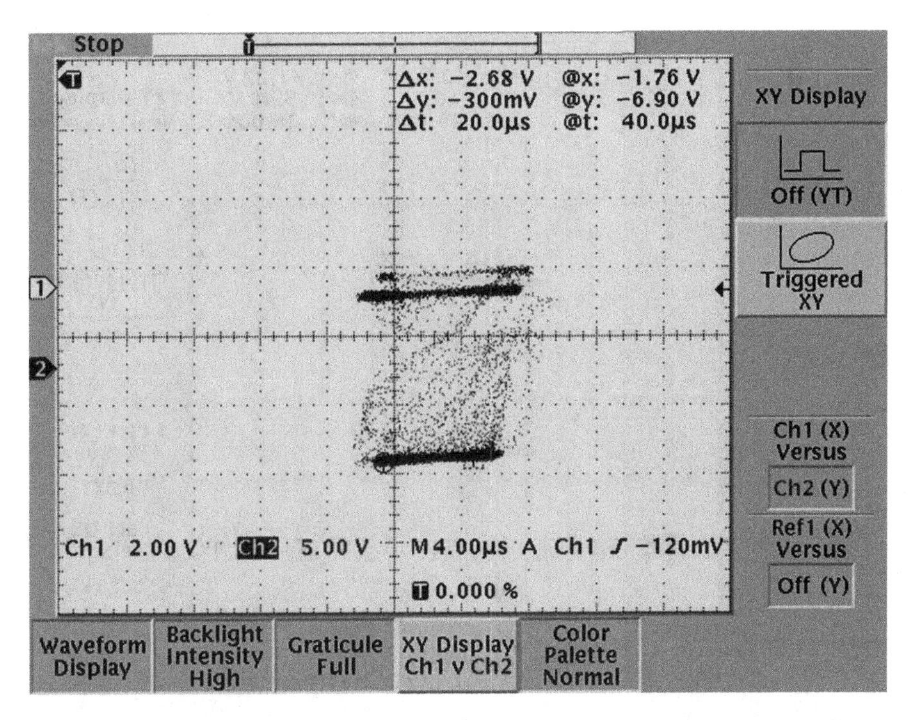

Fig. 21.4 Phase diagram, signal Y vs. signal X, 12 ml $H_2O + 0.4$ ml sol 0.5 g NaCl, at 20 °C

To investigate the influence of temperature on the chaotic process, the circuit was heated at constant temperatures; for 35 °C, the result is illustrated in the Fig. 21.5. We can state that the temperature did not alter significantly the signals and obviously not the phase portrait of the circuit.

Next, we exemplify the signals obtained for the solutions of $CuSO_4$, which is used in agriculture but acts as a polluter for phreatic waters. In Figs. 21.6, 21.7, and 21.8 we present the signals from the points P0 (X input oscilloscope) and P1 (Y input oscilloscope) as well the XY phase trajectory for the solution of 12 ml of H_2O +0.4 ml to 0.5 g $CuSO_4$ for a liter of H_2O. The pictures show the signals obtained under laboratory conditions at a temperature of 20 °C.

Notice in Figs. 21.6, 21.7, and 21.8 that the signals obtained from the sample of $CuSO_4$ differ from the ones obtained for the same concentration of NaCl, due to the different conductibility of the two solutions at equal concentrations. The next picture illustrates the signals obtained for the solution of the salt $Na_2S_2O_2$.

Figures 21.9, 21.10, and 21.11 present the signals from the points P0 (X input oscilloscope) and P1 (Y input oscilloscope) and XY phase trajectory for the solution of 12 ml of $H_2O + 0.4$ ml to 0.5 g $Na_2S_2O_2$ per liter of H_2O. The signals reported in all the pictures are also determined under laboratory conditions (at 20 °C). The obtained signals with this sample differ as well as from the signals with the tests with other two used salts.

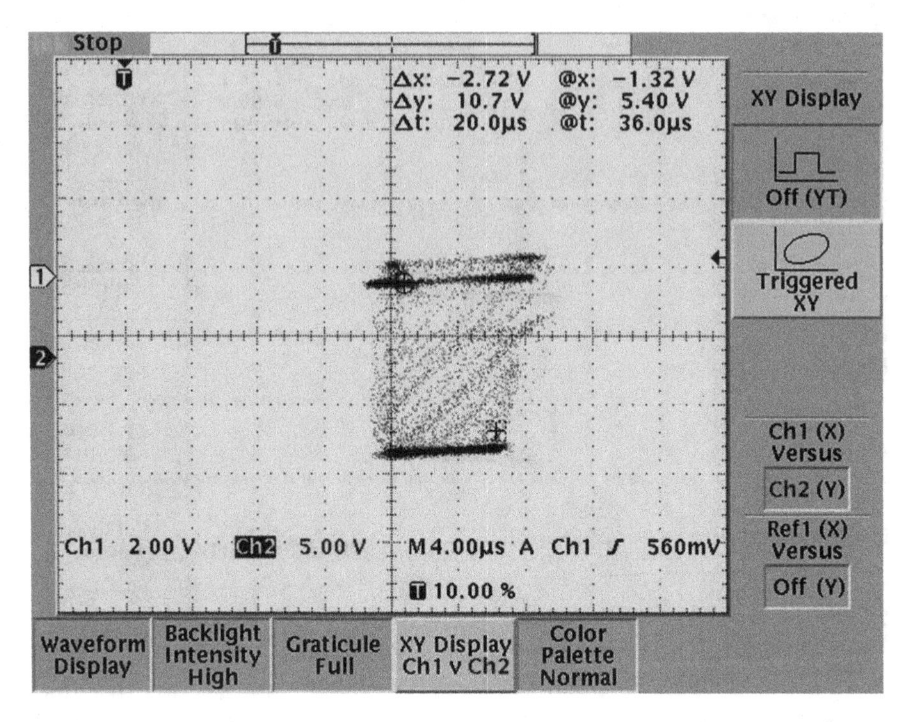

Fig. 21.5 Phase diagram, signal Y vs. signal X, 12 ml H_2O + 0.4 ml sol 0.5 g NaCl, at 35 °C

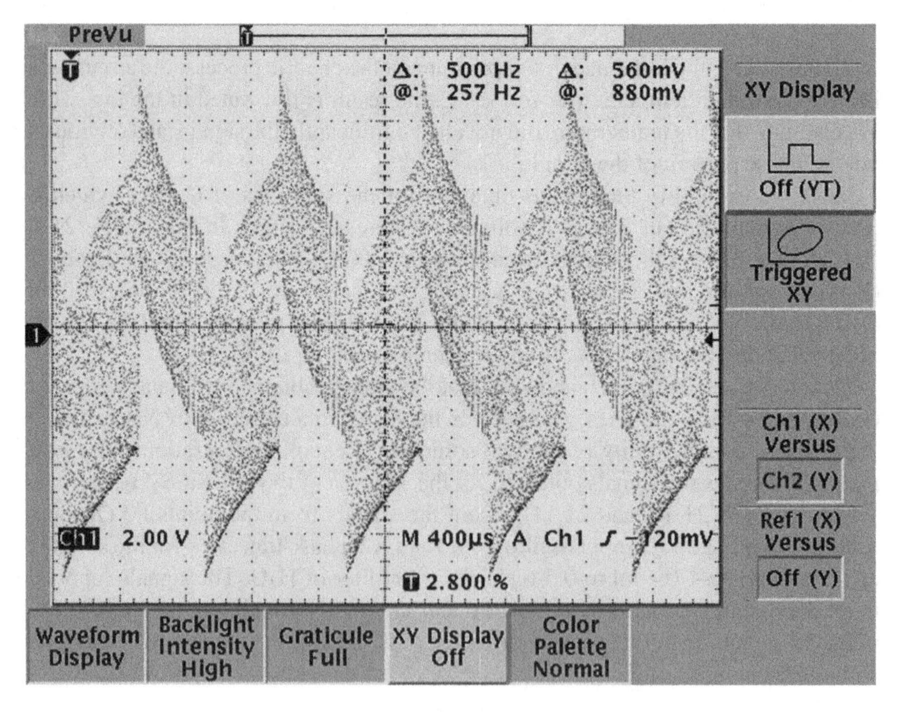

Fig. 21.6 Chaotic signal X, 12 ml H_2O + 0.4 ml sol 0.5 g $CuSO_4$, at 20 °C

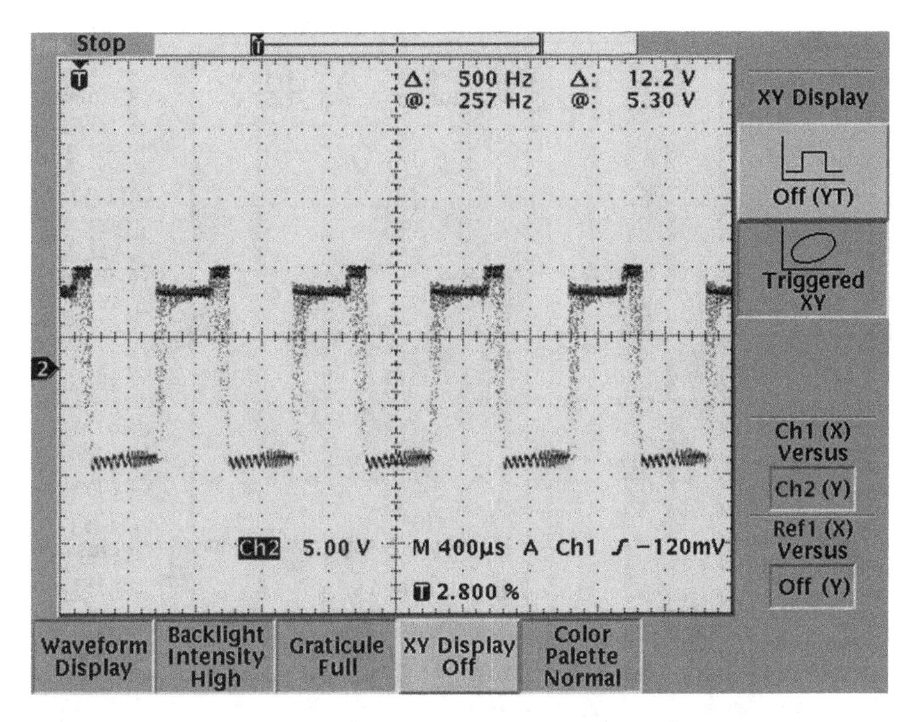

Fig. 21.7 Chaotic signal Y, 12 ml H_2O + 0.4 ml sol 0.5 g $CuSO_4$, at 20 °C

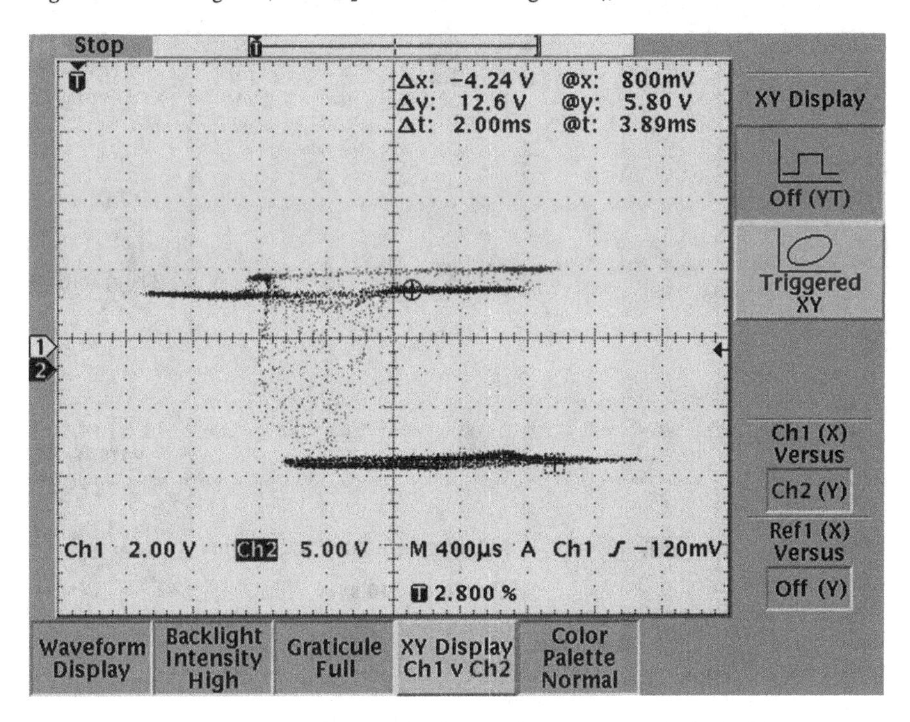

Fig. 21.8 Phase diagram, signal Y vs. signal Y, 12 ml H_2O + 0.4 ml sol 0.5 g $CuSO_4$

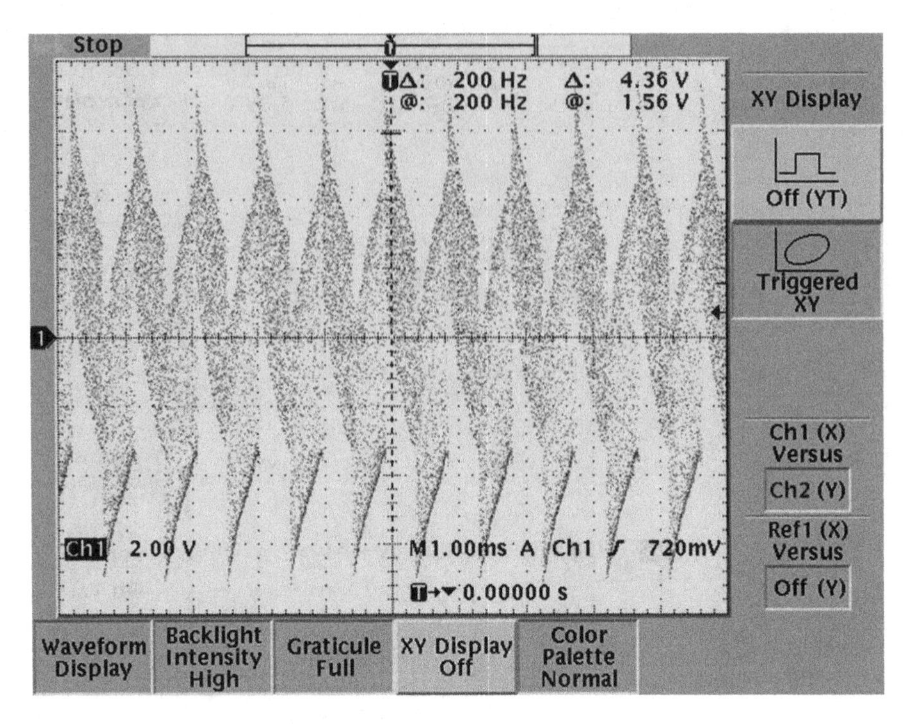

Fig. 21.9 Chaotic signal X, 12 ml $H_2O + 0.4$ ml sol 0.5 g $Na_2S_2O_2$

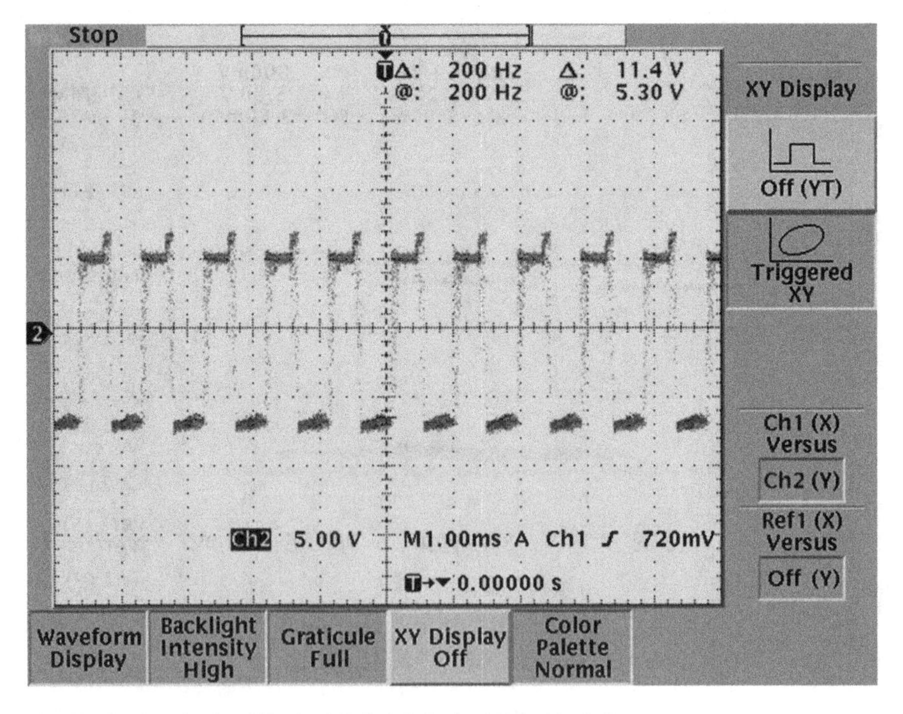

Fig. 21.10 Chaotic signal Y 12 ml $H_2O + 0.4$ ml sol 0.5 g $Na_2S_2O_2$

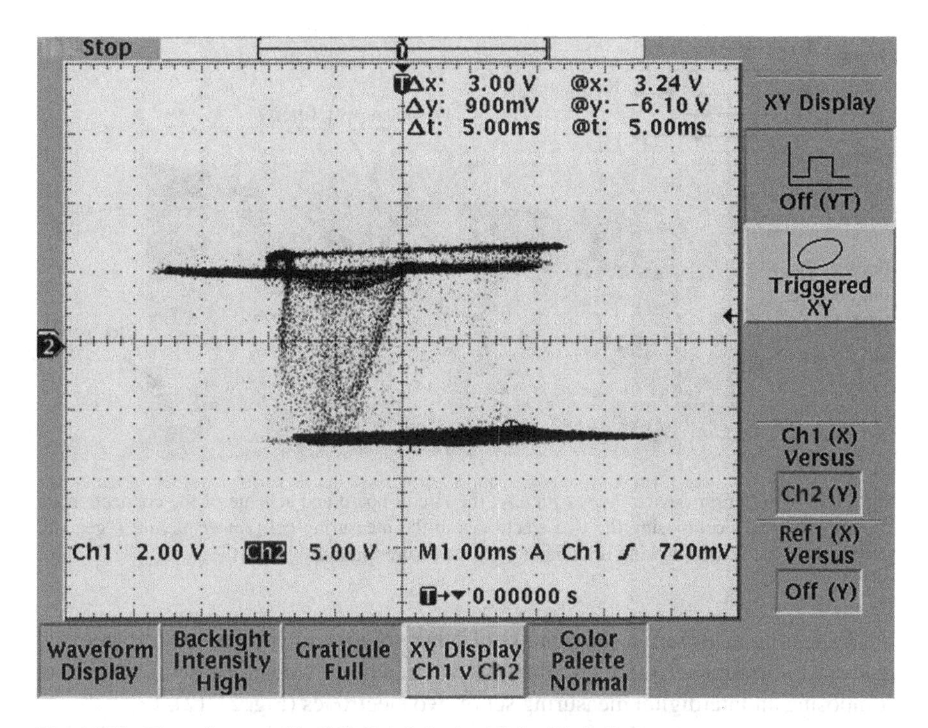

Fig. 21.11 Phase diagram, 12 ml $H_2O + 0.4$ ml sol 0.5 g $Na_2S_2O_2$

21.4 The Measuring Device

Figure 21.12 illustrates the sensor we built and used in the measurements reported in this paper and in [8]. It is made of accessible materials and has a compact design. The sensor head, consisting of the Au-plated electrodes must be completely submerged in the solution, to ensure a stable, reproducible measurement. The signal generated by the nonlinear circuit is sent for processing and decoding to a computer.

The output of the bio-mimetic sensor presented is not easy to decode by standard means [9]. The information it provides in the form of corroborated outputs, either as the attractor or as two sequences of samples, must be further process to connect the shape of the attractor to the resistance of the liquid. Only this correlation, to be through appropriate means not discussed in this article, allows the recovery of the information and completes the resistance monitoring procedure. Various methods for decoding the chaotic signal and for recovering the measured values were presented in ([10]; Teodorescu 1998, [9, 18–20]); suggestions on using fuzzy logic in relation to chaotic dynamics were made in [19].

We made several experiments in order to study the behavior of the chaotic circuit in various environments, for example, a solution of water and common soil was tested. This solution has not behaved as a nonlinear element and the oscillator was not triggered to generate chaotic signals. As future research, the behavior of the

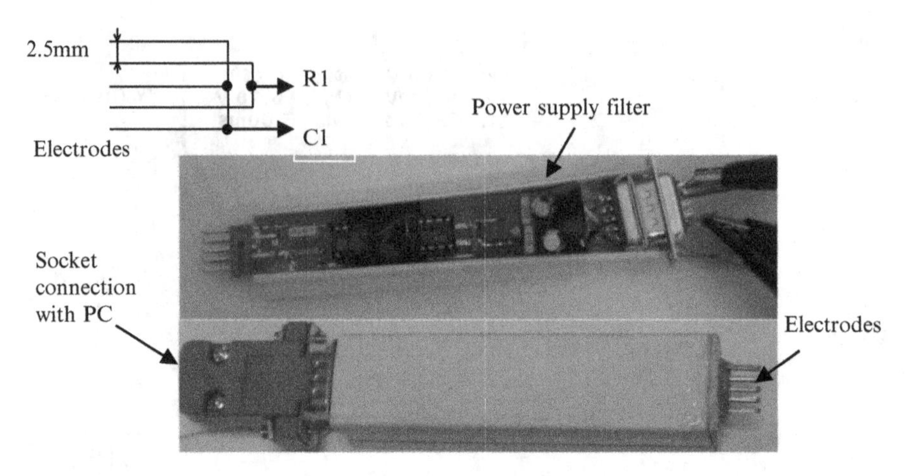

Fig. 21.12 The sensor device. *Upper picture*: the circuit board and scheme of the connection of the four electrodes composing the two electrodes in the measuring cell; *lower picture* – general view in the sealed case allowing immersion in the measured liquid

sensor will be analyzed in the context of other solutions with various instances of industrial impurities. Notice that the four electrodes are connected two by two and composing an interdigital measuring set of two electrodes (Fig. 21.12).

One may expect that the combination of conductivity cells, as reported in this paper, and of chemical sensors in a single nonlinear dynamic oscillator, when supplemented with an appropriate decoding system, could lead to a "taste" sensor with applications in efficient, detailed environment monitoring in regions where chemical plants may be a threat (private communication, HN Teodorescu, and [9]).

21.5 Conclusions

We reviewed a conductive titrimetry 'smart' sensor based on the nonlinear dynamics of a circuit that includes the measuring cell and brought new experimental evidence for its operation. The sensor has several biomimetic characteristics, including the internal nonlinear dynamics and the manner of coding the measured parameter as a specific dynamic regime; it is believed that sensorial neurons and neuronal networks perform similarly. Further research is needed to simplify the decoding of the chaotic signals and determine in a simpler manner the value of the measured parameter.

In the future, it might be possible to implant salinity sensors inside the human body for monitoring the salts in blood and in a stomach and possibly regulate the salt elimination.

Acknowledgments The methodology and the schemes of the experiment were conceived by Horia-Nicolai Teodorescu (see references); the detailed design, the implementation and the measurements were achieved by the author and HN Teodorescu, in a previous collaboration. This research was realized in the frame of the Group of Laboratories for Bio-Medical Engineering and Intelligent Systems, "Gheorghe Asachi" Technical University of Iasi during the inter-academy exchange visits between AS Moldova Republic and AR Romania. Some ideas in this chapter are based on Patent Number RO115316-B [11].

References

1. Ghadouani A, Coggins LX (2011) Science, technology and policy for water pollution control at the watershed scale: current issues and future challenges. Phys Chem Earth, Parts A/B/C 36(9–11):335–341
2. Earnhart D (2013) Water pollution from industrial sources, reference module in earth systems and environmental sciences. Encyclopedia Energy Nat Resour Environ Econ 3:114–120
3. Zhang L, Tian F, Liu F, Dang L, Peng X, Yin X (2013) Chaotic time series prediction of E-nose sensor drift in embedded phase space. Sens Actuators B 182:71–79
4. Teodorescu HN, Yamakawa T (1997) Applications of chaotic systems: an emerging field. Int J Intell Syst 12(4):251–253
5. Mandal S, Banerjee S (2004) Analysis and CMOS implementation of a chaos-based communication system. IEEE Trans Circ Syst 51(9):1708–1722
6. Mitsubori K, Saito T (1997) Dependent switched capacitor chaos generator and its synchronization. IEEE Trans Circ Syst-L Fund Theor Appl 44(12):1122–1128
7. Sundarapandian V, Pehlivan I (2012) Analysis, control, synchronization, and circuit design of a novel chaotic system. Math Comput Model 55(7–8):1904–1915
8. Teodorescu HNL, Cojocaru VP (2012) Biomimetic chaotic sensors for water salinity measurements and conductive titrimetry. In: Stoica A, Zarzhitsky D, Howells G et al (eds). Proceedings of the 3rd international conference on emerging security technologies (EST), IEEE, Lisbon, Portugal, 5–7 Sept 2012, pp 182–185
9. Teodorescu HN (2012) Characterization of nonlinear dynamic systems for engineering purposes – a partial review. Int J Gen Syst 41(8):805–825
10. Teodorescu HN, Stoica A, Mlynek D, Kandel A, Iov JC (2002) Nonlinear dynamics sensitivity analysis in networks and applications to sensing. In: Filip FG, Dumitrache I, Iliescu S (eds) Large scale systems: theory and applications 2001 (LSS'01) Book Series: IFAC Symposia Series, pp 333–338
11. Patent Number RO115316-B (1999) Inventor: Teodorescu H. Proximity gauge with a nonlinear oscillator. 1999-12-30
12. Teodorescu HN, Cojocaru V (2011) Complex signal generators based on capacitors and on piezoelectric loads. In: Skiadas C, Dimotikalis I, Skiadas C (eds) Chaos theory – modeling, simulator and application. World Scientific, Singapore, pp 423–430
13. Teodorescu HNL, Iftene E-F (2010) Analysis of the code masking efficiency of chaotic clocks in microcontroller applications. In: Gaiceanu M (ed) Proceedings of the 3rd international symposium on electrical and electronics engineering (ISEEE), IEEE 2010, Galati, Romania, 16–18 Sept 2010, pp 261–266
14. Teodorescu H-N, Iftene E-F (2014) Efficiency of a combined protection method against correlation power analysis side-attacks on microsystems. Int J Comput Comm Contr 9(1):79–84
15. Zhou P, Huang K (2013) A new 4-D non-equilibrium fractional-order chaotic system and its circuit implementation. Comm Nonlinear Sci Numer Simulat 19(6):2005–2011
16. Kobayashi Y, Nakano H, Saito T (2006) A simple chaotic circuit with impulsive switch depending on time and state. Nonlinear Dynam 44:73–79, Springer 2006
17. Çam U (2004) A new high performance realization of mixed-mode chaotic circuit using current-feedback operational amplifiers. Comput Electr Eng (Original Research Article) 30(4):281–290

18. Teodorescu H-NL, Hulea MG (2013) NNs recognize chaotic attractors. In: Dumitrache I, Florea AM, Pop F (eds) Proceedings of the 19th international conference on control systems and computer science (CSCS 2013), Bucharest, Romania, 29–31 May 2013, pp 52–57
19. Teodorescu HN, Kandel A, Schneider M (1999) Fuzzy modeling and dynamics. Fuzzy Set Syst 106(1):1–2
20. Teodorescu HN (2000) Method for measuring at least one parameter and device for carrying this method. European patent application, EP 0 981 038 A1, date of publication 23.02.2000 Bulletin 2000/08, date of filing 19.08.1998. Patent Application #988-10802.3, Aug. 1998/23.02.2000, EP0981038 (A1) Priority number(s): EP19980810802 19980819, patent: EP 981 038 A1/23.02.2000)

Chapter 22
Wearable, Assistive System for Monitoring People in Critical Environments

Livia Andra Amarandei and Marius Gheorghe Hăgan

Abstract People often operate in critical environments determined by disaster events as earthquakes, floods, or fires. The surveillance of the persons represents a difficult task for the rescue teams, while any information about the people in danger could be vital for the people life. A wearable device, that is incorporated in the clothes, was conceived and implemented as experimental model. This device is based on capacitive sensors that are able to detect the movements of the human body parts and to determine a so-called Global Activity Index, in order to inform about the physical status of the supervised person. It was suggested that the persons activating in a critical environment are wearing some special clothes incorporating a wearable device that acquires data from the capacitive sensors, these sensors being painted directly on the textile and giving information about the position of the human body related to the clothes. We describe in this chapter such a wearable system.

22.1 Prior Art

Wearable sensors and equipment for monitoring the movements of both rescuers and rescued in emergencies were hypothesized ([13], personal communication) to be useful in limiting injuries and fatalities. Various wearable devices have been proposed in the literature for monitoring the posture and movements of people. For example, a garment was presented in [11], which measures parameters of the seated posture, in order to prevent disorders that can arise from an incorrect posture

L.A. Amarandei
Technical University "Gheorghe Asachi", Iasi 437343, Romania

M.G. Hăgan (✉)
Institute of Computer Science of the Romanian Academy – Iasi Branch, Iasi, Romania
e-mail: marius.hagan@ramira.ro

H.-N. Teodorescu et al. (eds.), *Improving Disaster Resilience and Mitigation - IT Means and Tools*, NATO Science for Peace and Security Series C: Environmental Security,
DOI 10.1007/978-94-017-9136-6_22,

maintained for long time. There also exist a system that can detect the freezing of the gait of patients suffering from Parkinson's disease, in order to insure support for these patients [11].

Another kind of application is in sports domain, where activity recognition with budworm sensing is important for monitoring athletes in order to observe their physical condition [11]. As an example, there are available on the market heart rate belts, commonly used by sports enthusiasts. This kind of functionality can be obtained by inserting in the textiles conductive fibers. A simple shirt, easy to wear can be made and used for monitoring heart rate [11].

Smart sensors can be placed almost anywhere on the clothes: in the jacket (collar, sleeves, chest etc.), within the shoes, or inside other accessories. This system can be adaptive and it should provide information when needed, or deliver assistive information when are detected particular events [11].

A system that is able to recognize different movement patterns must accomplish some requirement as usability and power. So, the system must be small and easy to wear, it shouldn't be obtrusive or visible. The user should move free, and should not be affected by long-term usage of this system. The sensor must be chosen taking in account its power consumption. Higher amounts of energy will call for large capacity batteries that can affect the wearability of the system, by making it uncomfortable [11]. The miniaturized system can be interconnected wireless or whit conductive fibers inserted into textiles.

In order to improve the functionality of this system, there must be a clear distinction between different categories of activity, so that a correct decision over the type and number of sensors can be made. There are many criteria that can be used for classify the daily activities. As an example, a common criteria is the nature of the activity (periodic, static or sporadic activities) [11]. Accelerometric sensors were used in order to determine the movement distinctions between walking on a flat surface and climbing stairs. Using this sensor, the distinction between various movement patterns can be made [11].

Another implementation of these monitoring systems is based on capacitive sensors. Using capacitive sensors we can measure distance, pressure or conductivity, with various applications. Capacitance sensing can be used for heartbeat measuring, perspiration and respiration or human gesture recognition.

Lower limb locomotion was previously determined using mainly two ways: EMG and inertial sensors. Recently, an approach was to use the human body capacitance for determine the limb movements [16]. This system proposes the using of sensing bands, a signal-processing unit and a gait event detector. These bands can be placed on the lower limbs, directly on the skin. The signals are transmitted through wires to the sensing circuit, which is connected to a power circuit, alimented by a battery. The main idea was to use the human body as the dielectric and two metal films as the electrodes [16].

The usage of intelligent textiles became more and more common in medical application. One of these applications is textile electrodes. The advantage of these electrodes is that they are comfortable to wear, have no allergic reaction when interacting with the skin and can be used inside or outside clinics or hospitals. Their disadvantage is that the lack of contact increases impedance [10].

Capacitive sensors were also used for ECG measurements. The main advantage is that the direct contact with the skin can be avoided, so the risk of accidentally electroshocks is at minimum. All the necessary electrodes can be placed on a belt, forming a unitary network. The belt contacts the skin [4].

Smart sensors are also used in non-medical areas, with the precise scope of monitoring people and their vital parameters. For example, it is implemented and tested a system made on textile that incorporates sensors, wires and electrodes in order to monitor respiration, ECG, EMG, body posture and movements of the rescue team members. In a dangerous situation, like fires, it is important to know any information about the physical status of the persons that are involved in such an environment [5].

Capacitive sensors were also used to create wearable systems capable of measuring temperature and alert the persons that were them about the danger that are exposed to, depending on the heat measured and the humidity of that specific environment. It also can give information about the level of physical activity, the intensity of that activity and the influence over the body temperature [5].

Various researchers consider that the manufacturing the clothes should be a separate process, not interfering with sensor manufacturing, meaning that the sensors should be made separately and then applied on the garment. Basically, there will result some patches that can be placed and remove from the cloths. That gives us the liberty to move the sensor wherever we need [3]. The sensor position depends on what we want to record. It can be placed on the collar or on the sleeves. The sensor placed as patches on the sleeves can distinct between chewing, swallowing and talking [3].

Capacitive sensors can be made based on multifunctional carbon nanotube that has an excellent durability. It is flexible and robust, and it has various potential applications in smart electronics that are human friendly and can be made wearable. It can be integrated in the cloths or placed directed on the skin, making it suitable for applications in rehabilitation and health monitoring [2].

There is a large range of applications where capacitive sensors are suitable. Even in robotics industry, these smart sensors are used to emulate tactile sensing on humanoid robots. These new generations of robots are used in unstructured environments, in close interactions with humans. To create these sensors, a deformable 3D dielectric was used, covered by a conductive and a protection layer [9].

An application of textile sensors is a wearable system that can be used for self-monitoring, noninvasive, of biomedical, biochemical or physical parameters. It is an efficient system because it gives information that can be used to get an early diagnosis of several diseases (in particular physical rehabilitation, respiratory or cardiovascular) and can give support for elderly or disabled people [1].

The communication between this smart sensors and computer is an important aspect in wearable system development. All the communication needs to be wireless, in order to increase the independency of the supervised person. The supervisor is able to send triggers to activate some processes. Also, this communication between the sensor network and remote computer needs to be secured. Physiological signal (ECG, EEG, limb movement) are monitored with intelligent sensors and their state need to be supervised, in order to prevent damages that can occur in the system [8].

In neuro-rehabilitation, the movement detection represents the main interest for researches. Innovative technology is used to evaluate the motor functions, including trunk motions. Sensorised garment can be used in home-rehabilitation settings that can classify the motor tasks, inform the monitored person and store motor performance for further remote control by therapist [15].

The next parts of the chapter presents the concept of network of intelligent clothes, the use of capacitive sensors painted on clothes, and wearable equipment for monitoring the movements. The last parts of the chapter present results and conclusions.

22.2 Network of Intelligent Clothes

In a network of intelligent clothes, each person is identified based on ID, the information is sent to a concentrator, called RFC, this concentrator is able to collect data from short distances, a range of about 400 m, and to send its remote using a GSM network to a supervising center. The main part of the system consists of accelometric and capacitive sensors that are integrated in the clothes. Each of these sensors has a unique Id that is recognized by the RFC during the RF communication (Fig. 22.1).

The network of Supervising Device with Intelligent clothes has the role to supervise the person, and collect the physical parameters. These parameters consist of movements of the hands, trunk or neck, his position and the Global Activity Index [13]. The monitoring would be performed all the time the person is in a critical environment.

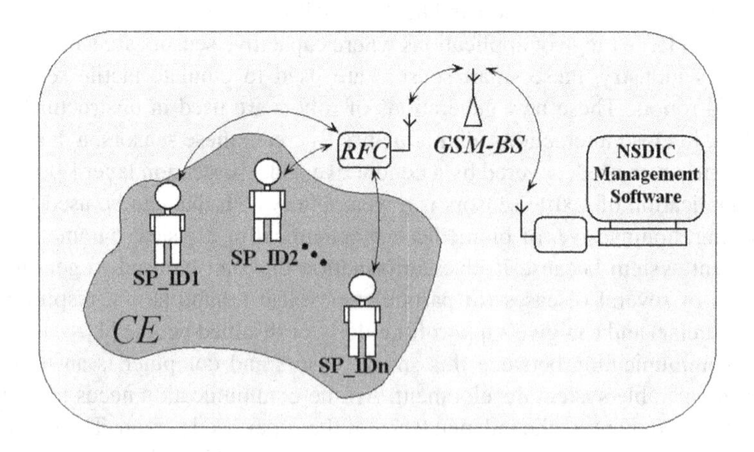

Fig. 22.1 A supervised group of people who activate in a critical environment

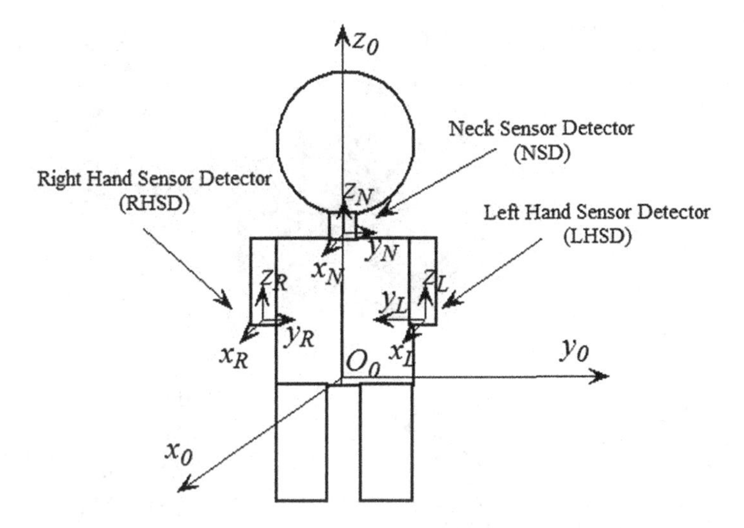

Fig. 22.2 Sensor network positioning on the clothes

The sensors are positioned on the collar and on the sleeves of the clothes, being able to detect the roll-pitch-yaw rotations of the neck and hands. Neck Sensor Detector consists of an accelorometric sensor, which determines the neck and trunk movements, and a set of capacitive sensors dedicated to determine the neck movements. The accelerometric sensor is placed on the collar, in the cervical zone, closely of the capacitive sensors. The Right hand sensor and left hand sensor are capacitive sensors and are placed on the cuffs in order to determine the movements of the hands ([13, 14]) (Fig. 22.2).

All these signals help to determinate the Global Activity Index [13] that inform about the global physical activity of the supervised person.

22.3 Capacitive and Accelerometric Data Acquisition

The signals generated by the sensors are acquired by the microcontroller using an I2C interface [6]. The inference block determine the particular movements of the body and calculate the GIA. The communication block consists of wire communication modules (I2C, SPI, UART) and RF communication module (868,3 or 915 MHz).

The capacitive components of the sensors are converted to digital numbers using a capacitive-to-number convertor (CDC), AD7747, which has a capacitive input. The circuit is mounted on a small circular PCB (diameter 20 mm, thickness 0.5 mm) being glued near the pads of the capacitive sensor in order to avoid an influence of the long connection wires. An I2C set of functions was defined for each circuit, because AD7747 has a single, unchangeable I2C address. The maximum data-rate is about 50SPS.

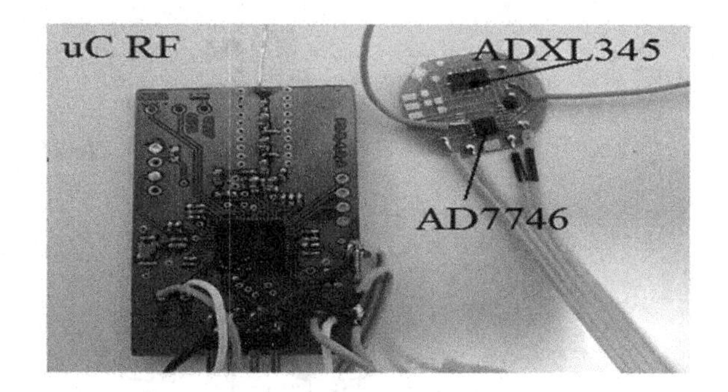

Fig. 22.3 Front-end circuit with an accelerometer – *left*. The system including wireless transmitter – *right*

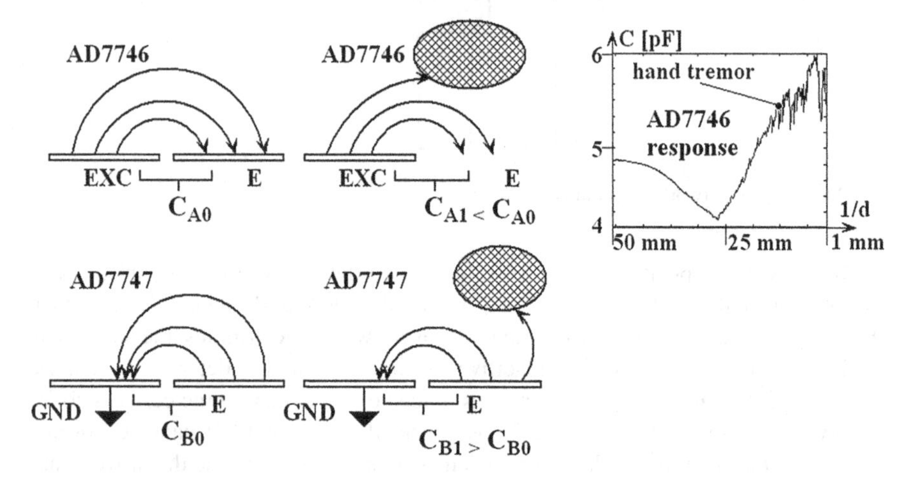

Fig. 22.4 AD7746 versus AD7747 during the hand close to the sensor

ADXL is a 3D accelerometer having implemented a fall detecting algorithm; it is responsible by trunk movements and position detection. This configuration enables us to determine the main changes of the human body position (Fig. 22.3).

The AD7147 circuit is an integrated CDC with on-chip environmental calibration. The CDC has 13 inputs channeled through a switch matrix to a 16-bit, 250 kHz sigma-delta ($\sum - \Delta$) converter. They need only 2.6–3.3 V supply voltage, with an operating current of 1 mA in full power mode and 21.5 μA in low power mode, being quite suitable with power consumption less than 3.3 mW in the worst case. The module uCRF is a microcontroller that incorporates the processing unit and the RF module in the same chip.

We should remark some issues related to the capacitive-to-number convertors. The circuit AD7746 uses an active electrode as reference, while AD7747 considers the ground as the reference. A decreasing of the capacitance value occurred when the hand (or other object) approaches to the sensor, in case of A7746 circuit. This effect is due to the loss of the electric field generated by the EXC electrode (Fig. 22.4 top), as only part of the electrical field lines end to the E electrode. If the distance will decrease (less than 25 mm, in this example) the capacitance

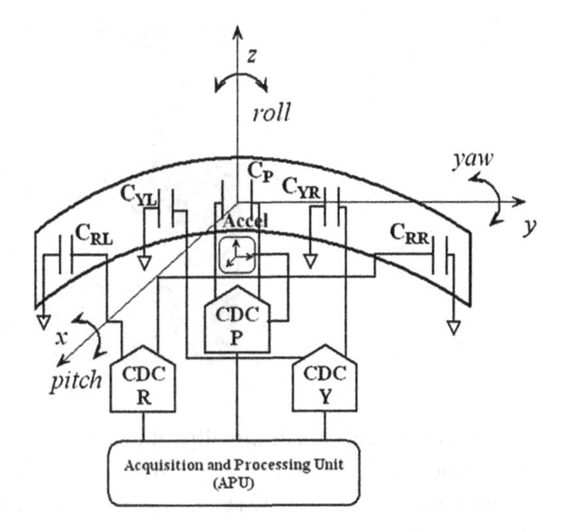

Fig. 22.5 Capacitive sensors mounted on the collar – block scheme

value starts to increase because part of the potential line from the object will be directed to the E electrode. In case of AD7747 circuit, approaching an object to the sensor, the loss of the electric field will produce an increasing of the capacitance value (Fig. 22.4, bottom). Further technical issues related to the use of capacitive-to-number convertors are dealt with in [7].

22.4 Painted Capacitive Sensors Configured on the Collar of a Shirt

The inter-digital configuration for the capacitive sensor painted directly on the textile is proper to detect the position between the body parts and the clothes [13]. A capacitive sensor was manually made, using silver and nickel conductive paints. The silver painted traces have a resistance of about 1 Ω, while the nickel resistances of the traces is about 20 Ω. Because Ni traces are less flexible than those based on silver, the silver paint was preferred to implement the experimental models of the capacitive sensors.

The "roll pitch yaw" rotations of the neck are detected by a network of capacitive sensors mounted on the shirt collar, following [13]. The network is depicted in Fig. 22.5 and its realization in Fig. 22.6.

22.5 Results

The movement of the neck in the coronal plane, respectively yaw rotation, is detected by the capacitive sensor mounted on the collar that generate a sinusoidal signal according with a left-right movement. During this movement, the

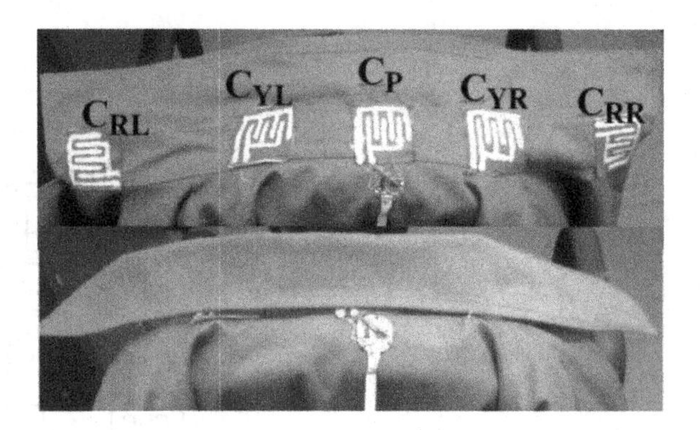

Fig. 22.6 Capacitive sensors mounted on the collar – experimental model implemented

accelerometric sensor is minimally influenced. The trunk movement during walk has an influence on both sensors, capacitive and accelerometric. The nonlinear filtering methods presented in [14, 12] can be applied in order to improve the quality of small accelerometric signals.

The movements determine a variation of the accelerometric signals on the Oz axis and Ox axis, moreover a movement of the neck related to the capacitive sensor is detected. During gait, the body behaves as an inverted pendulum. The center of the gravity (COG) was determined using a pedometric platform [7], but has not taken into account the movements of the articulation of the pelvis. A correlation between trunk movement (related to the COG) and the COG trajectory need to be done.

The sensors network configuration presented in the Figs. 22.5 and 22.6 generated the signals shown in the Figs. 22.7 and 22.8. The sensors' sensitivity is good (we obtained large amplitudes of the signals from the capacitive sensors for small amplitudes of the neck movements. After repeated wear of the sensors due to the movements, the painted surfaces were interrupted and the sensors functionality was damaged. A capacitive sensor using an embroidered electrode made of metallic wire was also tested.

22.6 Conclusions and Future Results

The capacitive painted sensors used in order to detect the movement of the human body parts is a feasible method for the laboratory experiment only (at least for paints based on silver and nickel) due the rigidity of the painted layers which leads to the interrupts of the electrical routes. A better solution could be to use electrodes based on the embroidered electrodes, which are more reliable. The method to determine the roll-pitch-yaw rotations of the neck is feasible and the obtained signals could be the basis for an algorithm in order to calculate the global activity index (GIA), its value being the parameter which give information about the conditions of a person which is supervised.

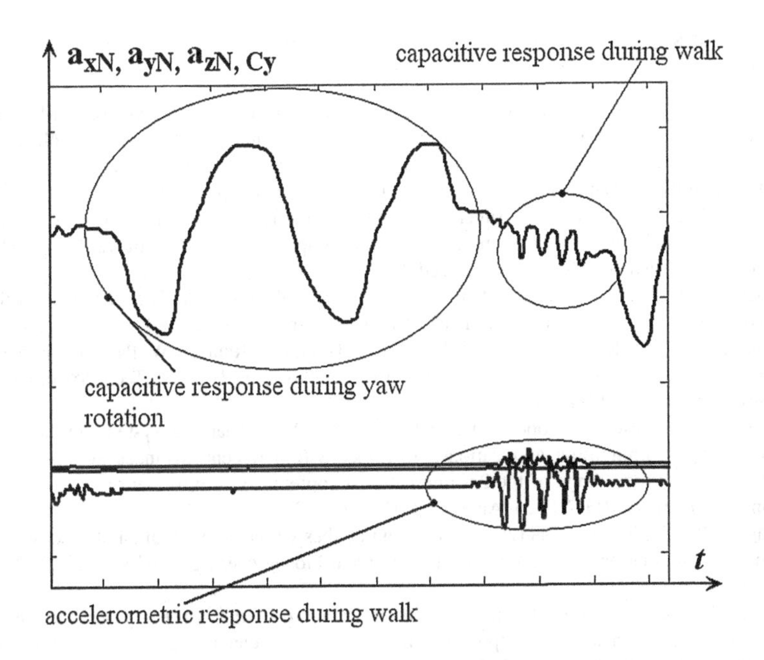

Fig. 22.7 Capacitive and accelerometric signals during neck and trunk movements

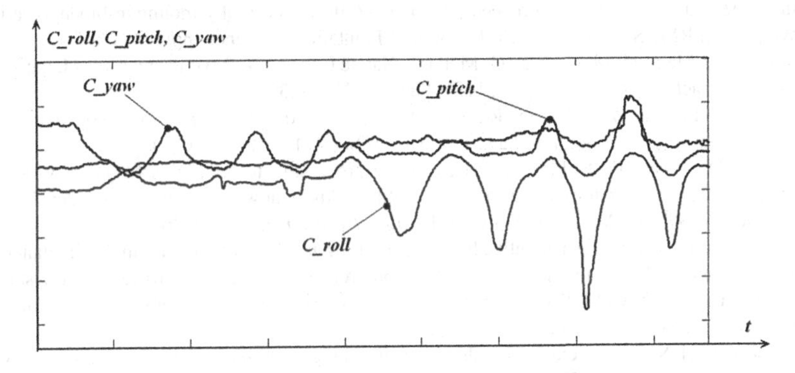

Fig. 22.8 The signals generated by the capacitive sensors during roll, pitch, and yaw movements of the neck

The next steps of the research will include the implementation of the inference block for determining in real time the particular movements of the neck and hands and to calculate the GIA index. We will test a set of subjects and will analyze their behaviors based on roll-pitch-yaw rotations during daily activity. We intend to simulate some specific surveillance operations. Also, we will incorporate in the clothes devices for alerting the people about dangers in a critical environment.

References

1. Andreoni G, Fanelli A, Witkowska I, Perego P, Fusca M, Mazzola M, Signorini MG (2013) Sensor validation for wearable monitoring system in ambulatory monitoring: application to textile electrodes. In: Pervasive computing technologies for healthcare (Perva-siveHealth), 2013 7th international conference on, 5–8 May 2013, Venice, pp 169–175
2. Cai L, Song L et al (2013), Super-stretchable, transparent carbon nanotube-based capacitive strain sensors for human motion detection. http://www.nature.com/srep/2013/131025/srep03048/full/srep03048.html. Accessed 21 Nov 2013
3. Cheng J, Lukowicz P, Henze N, Schmid A, Amft O, Salvatore GA, Tröster G (2013) Smart textiles: from niche to mainstream. IEEE Pervasive Comput 12(3):81–84
4. Despang HG, Weder A, Pietzsch M, Lindner O, Heinig A, Rentsch W, Paul A, Fischer W-J (2013) A modular system for capacitive ECG-acquisition. Biomed Tech 58(Suppl 1):2. doi:10.1515/bmt-2013-4218
5. Florea G, Dobrescu R, Popescu D, Dobrescu M (2013) Wearable system for heat stress monitoring in firefighting applications. In: Lopez S (ed) Recent advances in computer and net-working. Proceedings of the 2nt international conference on information technology and computer networks (ITCN'13), Antalya, 8–10 Oct 2013
6. Hăgan MG, Teodorescu H-N (2013) Intelligent clothes with a network of painted sensors. The 4th IEEE international conference on E-health and bioengineering – EHB 2013, 21–23 Nov 2013, Iaşi
7. Hagan M, Teodorescu H-N, Sirbu A (2010) Data processing for posturography and gait analysis. 3rd international symposium on electrical and electronics engineering (ISEEE), Galati, 16–18 Sept 2010, pp 267–272
8. Jovanov E, Raskovic D, Price J, Chapman J, Moore A, Krishnamurthy A (2001) Patient monitoring using personal area networks of wireless intelligent sensors. Biomedical Sciences Instrumentation, vol. 38. In: Proceedings of the 38th annual rocky mountain bioengineering symposium RMBS 2001, April 2001, Copper Mountain Conference, pp 373–378
9. Maiolino P, Maggiali M, Cannata G, Metta G, Natale L (2013) A flexible and robust large scale capacitive tactile system for robots. IEEE Sensor J 13(10):3910–3917
10. Marcias R, Fernandez M, Bragos R (2013) Textile electrode characterization: dependencies in the skin-clothing-electrode interface. J Phys Conf Ser 434, conf. 1, pp 1–4
11. Roggen D, Tröster G, Bulling A (2013) Signal processing technologies for activity- aware smart textiles. In: Kirstein T (ed) Multidisciplinary know-how for smart-textiles developers. Elsevier, Woodhead Publishing Series in Textiles, April 2013, pp 329–365
12. Teodorescu H-N (2006) An adaptive heuristic filter for acceleration measurements in planetary atmospheres. AHS 2006: first NASA/ESA conference on adaptive hardware and systems, proceedings (1st NASA/ESA conference on adaptive hardware and systems, Istanbul, Turkey, 15–18 June 2006, pp 101–108
13. Teodorescu H-N (2013) Textile-, conductive paint-based wearable devices for physical activity monitoring. The 4th IEEE international conference on E-health and bioengineering – EHB 2013, Iaşi, 21–23 Nov 2013
14. Teodorescu H-N, Hagan M (2007) High accuracy acceleration measuring modules with improve signal processing capabilities. IDAACS 2007: Proceedings of the 4th IEEE workshop on intelligent data acquisition and advanced computing systems, Dortmund, Germany, 06–08 Sept 2007, pp 29–34
15. Tormene P et al (2012) Estimation of human trunk movements by wearable strain sensors and improvement of sensor's placement on intelligent biomedical clothes. http://www.biomedical-engineering-online.com/content/11/1/95. Accessed 15 Nov 2013
16. Zheng E, Chen B, Wei K, Wang Q (2013) Lower limb wearable capacitive sensing and its applications to recognizing human gaits. Sensors 13(10):13334–13355